Unmanned Aircraft Systems Innovation at the Naval Research Laboratory

Unmanned Aircraft Systems Innovation at the Naval Research Laboratory

Jay Gundlach
Richard J. Foch

LIBRARY OF FLIGHT

Ned Allen, Editor-in-Chief
Lockheed Martin Corporation
Bethesda, Maryland

Published by
American Institute of Aeronautics and Astronautics, Inc.
1801 Alexander Bell Drive, Reston, VA, 20191-4344

Figure credits:

Aurora Flight Sciences: 9.24, 9.31, 11.13.

Sean Bain: 2.1, 2.24, 2.27, 2.50, 3.12, 3.35, 3.36, 7.16, 11.5, 11.7.

CIA Museum: 6.13.

Richard J. Foch: 2.2, 2.6, 2.9, 2.13, 2.16, 2.23, 2.25, 2.26, 2.28, 2.30, 2.32, 2.47, 2.49, 2.51, 2.53, 3.2, 3.6, 3.11, 3.13, 3.19, 3.27, 3.30, 3.34, 3.37, 3.39, 4.1, 4.10, 4.12, 4.15, 4.17, 4.22, 4.27, 4.29, 4.31, 4.33, 4.35, 4.37, 4.41, 4.44, 4.46, 4.53, 5.3, 5.6, 5.14, 5.21, 5.24, 5.26, 5.29, 5.34, 5.39, 5.41, 6.1, 6.6, 6.7, 6.11, 6.14, 6.17, 7.2, 7.5, 7.7, 7.11, 7.13, 7.15, 7.17, 7.22, 7.27, 7.29, 7.31, 8.5, 8.10, 8.13, 8.25, 8.30, 8.33, 8.45, 8.52, 9.7, 9.9, 9.11, 9.14, 10.20, 11.11, B.1, B.2.

Jay Gundlach: 6.29.

NAVSEA-AUTEC: Chapter 10 press release photograph.

Theiss Aviation: 7.24.

U.S. Marine Corps: 5.33 (photo by Lance Cpl. Brian A. Jaques).

Unattributed figures are held by the Naval Research Laboratory.

American Institute of Aeronautics and Astronautics, Inc., Reston, Virginia

1 2 3 4 5

Library of Congress Cataloging-in-Publication Data

Gundlach, Jay, 1975-

Unmanned aircraft systems innovation at the Naval Research Laboratory / Jay Gundlach, Richard J. Foch.

pages cm. – (Library of flight series)

Includes bibliographical references and index.

ISBN 978-1-62410-259-2 (print)

1. Drone aircraft–Research–United States–History. 2. Naval Research Laboratory (U.S.)–Research–History. I. Foch, Richard J. II. Title.

UG1242.D7G864 2014

623.74′690973–dc23

2014005281

ISBN 978-1-62410-259-2

To the memory of
Richard J. Foch
1957–2014

A Note from the Editor-in-Chief

The other skunkworks: Kelly Johnson and Ben Rich led the famed Lockheed Skunk Works for decades in the middle of the 20th century. Unmanned aircraft systems (UAS) technology was known only to a few for most of the 20th century, but its epicenter was the Vehicle Research Section (VRS) at the Naval Research Laboratory (NRL). Here, many of the most creative, productive, and fruitful thinking about UAS technology, as well as design theory and practice for small aerial robots (mini-, micro-, and nano-unmanned air vehicles), occurred. VRS was staffed by some of the most imaginative, able, and eventually most experienced aeronautical engineers of the late 20th century—for every full-scale aircraft project in those years, the VRS did at least 10. NRL management recognized early that small, inexpensive unmanned aircraft operating in concert with other systems and in swarms could and would become uniquely capable military systems as the solid-state and quantum electronics revolution accelerated. VRS made developments in low Reynolds number (LRN) aerodynamics, stability and control, data links for small flight vehicles and their scaled-down structures, and materials and manufacturing techniques, including the associated systems engineering processes. These developments set the stage for, and then spawned, a whole new class of useful systems with gross takeoff weights less than a few 10s of kilograms, including many less than 1 kilogram. These were essentially militarized model airplanes.

The visionary engineers staffing NRL's numerous VRS research and development programs were inspired and led for most of the decades before his retirement by the legendary imagination of Rick Foch, VRS engineer, director, and then senior scientist from the late 1970s until his retirement in 2013. Foch had a small budget to support a few talented staff, a handful of committed contractors, and collaborators in many universities. His ideas and infectious enthusiasm for aerial robotics and his Skunk Works–like program management style helped shape the designs, technologies, and programs for UASs at all the military services, the Defense Advanced Research Projects Agency (DARPA), and many—perhaps most—UAS programs both in the

United States and abroad. Those years of Foch's breakthrough after breakthrough—especially in the smallest weight classes—have proven pivotal in the history of aeronautical engineering. Many critical technical solutions for UAS systems just now coming into prominence have roots reaching deeply into the VRS and the brilliant insights of Foch and his colleagues. *Unmanned Aircraft Systems Innovation at the Naval Research Laboratory* by Jay Gundlach and Richard J. Foch is the previously untold and generally unknown story of the VRS, its people, its projects, and its ideas. The quiet anonymity of the VRS is rare in our age of disclosure, especially since it was one of the United States' most important centers of aeronautical innovation and national security solutions—our other skunkworks. And thus, this story earns an especially prominent position in AIAA's Library of Flight series.

The Library of Flight is part of the growing portfolio of information services from AIAA. It extends our offerings with the best from a growing variety of topics in aerospace from policy, to history, to law, to management, and beyond. The Library of Flight documents the crucial role of aerospace in enabling, facilitating, and accelerating our technological society and its global commerce, communication, and defense.

Ned Allen
Bethesda, Maryland

Contents

Preface xiv
Acknowledgments xvii

Chapter 1 Introduction 1
Unmanned Aircraft Research at the Naval Research Laboratory (NRL) *1*
VRS Program History *3*
NRL Technology Advancements *5*

Chapter 2 Ship-Launched Decoys and Targets 9
Active Electronic Decoy (AED), 1975–1978 .... *9*
Fairchild Spinner AED, 1975 .... *10*
Fairchild Sailwing AED, 1975–1976 .... *12*
DSI AED, 1977–1978 .... *15*
Teledyne Brown AED, 1977–1978 .... *17*
E-Systems AED, 1977–1978 .... *19*
Long Duration Expendable Decoy (LODED), 1978–1982 .... *21*
Interim Active Electronic Decoy (IAED), 1980–1982 .... *30*
Shear Wing RF Decoy, c. 1981 .... *33*
Ship-Launched Electronic Decoy (SLED), 1984–1985 .... *34*
Radar Expendable Active Decoy (READY), c. 1984 .... *36*
Flying Radar Target Advanced Technology Demonstrator (FLYRT ATD), 1991–1993 .... *38*
Transonic Missile Engagement Target (TMET), 1994–1995 .... *55*
Submarine-launched Tactical Airborne Reconnaissance Expendable (STARE), 1995 .... *57*
Vertically Launched Imaging Infrared Decoy Technology (VLIIRDT), 2000 .... *60*

Chapter 3 Air-Launched Fixed Wing Aircraft 63
Airborne Active Expendable Decoy–Forward Fired (AAED-FF), 1986–1990 .... *63*
Pendulus, 1987 .... *66*
NRL Raptor, 1992 .... *71*
Expendable Multi-mission Marine Aircraft (EMMA), 1999 .... *76*
Extender, 1999–2000 .... *79*

Flight-Insertable Detection Expendable for Reconnaissance (FINDER), 1999–2010 85
Air Launched Integrated Countermeasure, Expendable (ALICE), 2001–2005 91
Duster, 2004 94
Air Deployed Long Endurance Reconnaissance (ADLER), 2005 96
Insertion of Cargo Expendable (ICE), 2005 98
Disposable Air Vehicle Experiment (DAVE), 2006 100

CHAPTER 4 RESEARCH VEHICLES 103
Shearwing, 1980 104
Low Altitude/Airspeed Unmanned Research Aircraft (LAURA), 1986–1990 106
LAURA Three Surface 113
LAURA Joined Wing 116
LAURA Tandem Wing 120
LAURA Variable Span 125
PLICATE Wing ASW UAS, 1987 130
Semi-Rigid Delta Wing, 1987 131
Flying Wing Research Testbed (FLYRT), 1987–1988 132
Shadow Displaced Tail (DT), 1988–1991 135
Semi-Autonomous Research Aircraft (SARA), 1990–1994 140
CG Shift Demonstrators, 1992–1995 143
First Skew Wing CG Shift, c. 1992 145
Second Skew Wing CG Shift, c. 1992 146
High Aspect Ratio (HAR) CG Shift, c. 1992 148
CG Shift DT, 1992 150
CG Shift Bird, 1993–1995 151
Ramwing, 1993 155
Venture, 1994–1995 158
Sail-A-Plane, 2005–2006 160

CHAPTER 5 GROUND-LAUNCHED FIXED WING MISSION AIRCRAFT 167
Self-Navigating Drone Expendable/Recoverable (SENDER), 1994–1998 167
SENDER 2, 1998–1999 174
Crawler Delivery Aircraft, c. 1995 177
Swallow, 1996–1998 178
Searcher, 1997 185
Ghost, 1997–1998 186
Para Eyes, 1997–1998 190
Moth, 1999 193
Nexus, 1999–2000 196
Dragon Eye, 2000–2002 201
Sea Airborne Lead Line (Sea ALL), 2001–2002 207
Golden Eagle, 2002 210
Sea Quell, 2003 211

Spotlight, 2005 .... *213*
Counter-UAS UAS, 2006 .... *215*
Sea Wind, 2008 .... *216*
Trade Wind, 2010–2011 .... *220*
Flying Radar Reflector UAS, 2010 .... *225*

CHAPTER 6 MICRO AIR VEHICLES 227
Flying Printed Circuit EW Aircraft, 1982 .... *229*
Micro Tactical Expendable (MITE), 1995–2002 .... *230*
Pink Eye, 1996–1997 .... *234*
Silicon Bug, 1996 .... *236*
Winged Autonomous Silent Platform (WASP), c. 1997 .... *237*
Dragonfly, 1997 .... *239*
Micro Air Vehicle Demonstrators, 1997 .... *240*
Cube of Sky, 1997 .... *241*
Nexus Surrogate MAV, 1999 .... *242*
Semi-Autonomous Mobile Platform Expendable (SAMPLE), 2002–2004 .... *244*
Close-In Covert Autonomous Disposable Aircraft (CICADA), 2004–2011 .... *251*

CHAPTER 7 UTILITY AIRCRAFT 261
Arsea, 1981–1982 .... *261*
Drone Aircraft Radio Controlled (DARC), 1982–1984 .... *264*
Baseline Aircraft Radio Controlled (BARC), 1985–1990 .... *267*
NDM I, II, and III, 1991–Present *270*
Test Radar Calibration Experiment (TRACE), 1993 .... *275*
Telemaster, c. 1993–Present *277*
Dakota, 1995–1998 .... *280*
Dakota 2, 1998–2002 .... *285*
Unmanned Titan Tornado, 1995–1996 .... *288*
Science Instrumentation Environmental Remote Research Aircraft (SIERRA), 2002–Present *291*
Arcturus T-15, T-16, and T-20, c. 2009–Present *293*
BAI Tern, c. 2001 .... *295*
Thorpe P10B, c. 2005 .... *296*
Magpie, c. 2008 .... *297*
UASUSA Tempest, 2011 .... *298*

CHAPTER 8 VERTICAL TAKEOFF AND LANDING (VTOL) AND ROTORCRAFT 299
Ship's Tethered Autogyro Platform (STAPL), 1975–1980 .... *299*
Libelle, 1986 .... *304*
Commander's Overhead Vehicle for Elevated Reconnaissance (COVER), 1986–1989 .... *309*
Small Ships Compatible Decoy (SSCD) Autogyro, 1992–1993 .... *313*
Rotary Wing Decoy (ROWDY) 1, 1995 .... *322*
Eager, 1995–1997 .... *323*

High Point, 1999–2000 .... *328*
Dragon Warrior/Vantage, 2001–2004 .... *330*
Scientific Payload Insertion Device Electric Rotor (SPIDER), 2004–2005 .... *336*
Threat Radar Active Countermeasure Expendable (TRACE), 2008 .... *339*
Expendable Radar Active Countermeasure Electric Rotorcraft (ERACER), 2008 .... *340*
Rotary Wing Decoy (ROWDY) 2, 2010 .... *341*
Stop Rotor, 2010–Present *342*

CHAPTER 9 PLANETARY FLIGHT 349
Kitty Hawk, 1997–1998 .... *350*
NASA 723, 1999 .... *356*
Kitty Hawk 2, 2000–2001 .... *358*
Mars Astronaut Companion (MAC)-1, 2003–2004 .... *360*
Matador, 2004–2006 .... *362*
Titan Flyer (AVIATR), 2011 .... *368*
Venus Flyers, 2010–2011 .... *378*
Mars Helicopter, 2012 .... *378*

CHAPTER 10 FUEL CELL AIRCRAFT 381
Spider Lion, 2005 .... *381*
Experimental Fuel Cell (XFC), 2006–Present *384*
Ion Tiger, 2008–2013 .... *397*
Jake Spirit, 2011 .... *405*

CHAPTER 11 MANNED AIRCRAFT 409
Fly Back Ejection Seat, 1991 .... *409*
Veeness, 1992 .... *410*
Rapid Descent Vehicle, 1993 .... *411*
Man-Operated Light Expendable (MOLE), 1993 .... *413*
Submersible Manned Air-Launched Expendable (SMAX), 1993 .... *416*
Personal Extraction Aircraft (PEA), Mid-1990s *418*
Ion Dasch, 2011–Present *418*

CHAPTER 12 SUBSYSTEM TECHNOLOGY 427
Laser Tethered Decoy System, 1992–1993 .... *427*
Energy Harvesting by Power Line Recharging, 2004–2005 .... *429*
Advanced Turbine-Driven Generator, 2004–Present *432*
Vision-Based Recovery System, 2006–2008 .... *439*
Autonomous Locator of Updrafts (ALOFT), 2007–2009 .... *442*
Low Cost Micro Autopilot (LCMA), 2008–2012 .... *446*

CHAPTER 13 CONCLUSION 449
Lessons Learned and Guiding Principles *449*

Appendix A Engineering Background 455
Methods for Quick Initial Estimates of Performance and Size *457*

Appendix B Airfoil Technology 463
Ribless Wing Construction Technique, 1977–Present *463*

Afterword 465
Jay Gundlach, President, Gundlach Aerospace LLC, December 2013 .... *465*
Richard J. "Rick" Foch, Senior Scientist for Expendable Vehicles, Naval Research Laboratory, Washington, DC, May 2013 .... *466*

Subject Index 471

Aircraft Index 483

Supplemental Materials 487

PREFACE

This book is an encyclopedia of lessons learned and aircraft development war stories. More than 90 unmanned aircraft are detailed, covering nearly four decades of development (c. 1975 to 2012) in the Naval Research Laboratory (NRL) Vehicle Research Section (VRS). The systems discussed cover a variety of missions, levels of autonomy, configurations, propulsion systems, structural materials, and other technologies. These vehicles were designed to fly in the atmospheres of Earth, Mars, and Saturn's moon Titan.

The book is written from engineering and aircraft design perspectives. However, it is also intended to be accessible for the nontechnical aviation enthusiast. Except for a brief engineering background discussion in Appendix A, no mathematical equations were necessary to tell this story. The subject matter should be of interest not just to the unmanned aircraft designer, but also to those involved with other aspects of aviation.

The projects in NRL's Vehicle Research Section have been sponsored primarily by the U.S. Navy, with many of the fundamental research projects developed specifically under the sponsorship of NRL's parent organization, the Office of Naval Research (ONR). In addition, numerous other government organizations have been strong sponsors. These include the U.S. Marine Corps (USMC), U.S. Army, Defense Advanced Research Projects Agency (DARPA), Defense Threat Reduction Agency (DTRA), Rapid Reaction Technology Office, Office of the Secretary of Defense, NASA, and others. While sponsor needs and requirements are important, this book does not delve into details of sponsor relationships, missions, or payloads.

NRL's unmanned aircraft have served a variety of purposes. Many were research aircraft intended to advance specific technologies. Others were mission aircraft or mission aircraft demonstrators. Missions included chemical/biological/nuclear agent detection, reconnaissance, surveillance, antisubmarine warfare, ship electronic warfare protection, signals intelligence, scientific missions, and sensor emplacement. Many utility aircraft were developed and operated for systems integration support and payload testing.

Unmanned aircraft systems (UASs) are systems. These systems typically incorporate numerous elements beyond the unmanned aircraft, including payloads, communication systems, ground control stations, and support equipment. In this book, the authors have elected to concentrate on the unmanned aircraft portion of the system. The VRS has made its largest contribution to flight vehicles while leveraging state-of-the-art system elements.

The majority of unmanned systems developed by the VRS are unique. It could be argued that no other organization—whether in industry or government—has come close to this level of productivity for unmanned aircraft development.

This book covers not only unmanned aircraft that achieved flight, but also many notable conceptual designs. In many cases, these conceptual designs influenced later programs. In other cases, the designs had innovations that may have been ahead of their times. For every program that went forward, there were approximately 10 system concepts that never left the drawing board. The authors do not attempt to provide comprehensive coverage of all the conceptual designs developed in the VRS.

The book chapters are arranged by system category. Many of the unmanned aircraft fall within multiple categories, so the most applicable is chosen. For example, the XFC is a fuel cell aircraft and a ground- or ship-launched fixed wing aircraft. Since it is most relevant to the fuel cell story, it is placed in that chapter. Within each chapter, the vehicles are described in chronological order.

The VRS has enjoyed a large degree of staff continuity, which has contributed in no small measure to the group's productivity. Many of the group's senior engineers have been employed by NRL since the 1980s. This continuity is important because the valuable experience of the early programs is retained and applied to new efforts. The VRS mission has remained constant, and the technical challenges and enjoyable workplace environment have helped maintain the personnel continuity.

Thomas Edison, who was instrumental in establishing NRL in the early 20th century, once said, "Genius is 1% inspiration and 99% perspiration." The UASs covered in this book represent the realization of many dreams and aspirations that only became reality thanks to the "inspiration and perspiration" of many dedicated, highly talented engineers and technicians. This book honors and showcases the accomplishments of this tireless and gifted team, who made these innovative aircraft and their technologies possible.

NRL Vehicle Research Section (VRS) Personnel

Present-Day Staff Members

Sean Bain*
Mike Baur
Chris Bovais
Steve Carruthers
Pete Chaplin
Al Cross
Christine Dailey
Peggy Davidson
Bob Eber
Dan Edwards
Kim Goins
Matt Hazard
Dave Indyke
Aaron Kahn
Matt Kelly
Joe MacKrell
Rudy Monteleone
Greg Page*
Eric Peddicord
Drew Rogers
Mike Schuette*
Jeremiah Smith
John Southwick
Steve Tayman
Mike Vick
John Watson*
Trent Young

Former VRS Staff Members

Tim Aberg*
Kevin Ailinger
Bobby Baker*
Mike Bender
Drew Bishop*
Kenny Boothe*
Joe Brannon*
Ed Clabaugh
Vic Curtis*
Ray Cyr
Jill Dahlburg
Brian Donnelly
Bill Dunsavage*
Dick Eklund*
Carlos Figueroa
Chris Fisher*
Rick Foch
Tim Funk*
Ron Gilman
Lee Goldberg*
Angelo Guarino*
Jay Gundlach*
Mark Haupt
Don Hayes
Jim Heidemous*
Rich Hoch
Dave Kanazawa*
Jim Kellogg
Frank Klemm
Stan Kozloski
Jack Meiling*
Hugo Mellace
Leslie Monforton
Paul Oppenheimer
Chet Ozamina
Nandy Pizarro*
Naveed Quraishi
Dave Rannigan*
Don Srull*
Carol Sullivan
Hugh Thacker
Eric Tich
Robert Vogt
Andi Walden
Barry Walden
Jim Ward*

Plus many dozens of summer interns and temporary hires.

(*) Engineering Support Contractor

Acknowledgments

This book would not have been possible without the support of many Vehicle Research Section and other NRL staff. The authors would like to express special thanks to Dr. Frank Klemm, Dr. John Montgomery, Dr. Warren Schultz, Al Cross, Joe MacKrell, Dr. Mike Vick, Chris Bovais, Peggy Davidson, Kim Goins, and Steve Tayman. Sean Bain made important contributions to this book by generating photo-realistic CAD models of several aircraft. Claire Peachey greatly improved the quality of the book on behalf of NRL in her extensive editing prior to submission to AIAA. We would like to thank AIAA for their high-quality and accelerated publication. The authors are especially grateful to our families for providing encouragement, support, and significant time to write this book.

Chapter 1

Introduction

Unmanned Aircraft Research at the Naval Research Laboratory (NRL)

The concept of high-performance, autonomous, small unmanned aircraft systems (UASs) began at NRL in the mid-1960s in direct response to emerging electronic warfare (EW) technologies that enabled significant miniaturization of radar systems. Obvious benefits of the NRL-developed miniature radar technologies included smaller and lighter airborne radars for Navy aircraft and seekers for guided missiles. A few far-sighted NRL scientists and engineers realized that these technologies also supported the development of a new class of countermeasures for EW.

Radar chaff had been used as a "passive" EW countermeasure since World War II. The chaff (called Window) consisted of thousands of metal foil strips thrown overboard by Allied bombers to confuse German radar defense systems. The foil strips were cut to a length corresponding to the wavelengths of the threat radar so each was a resonant dipole antenna and the resultant radar reflection was so huge it masked the locations of the individual bombers. The new technology envisioned by NRL would exploit the miniaturization of radar components to enable "active" countermeasures—that is, compact radar jammers and small electronic packages that created signals that mimicked a radar reflection of a ship or aircraft, thus creating a realistic false target to confuse threat radars. At NRL, the Tactical Electronic Warfare (TEW) Division was formed in 1971 to develop these new technologies, and it was apparent that once miniature active countermeasure electronic systems were available, an efficient method of deploying them from the defending ship or aircraft would be needed.

To develop both the platforms and their compact active countermeasures payloads, the Offboard Countermeasures (OCM) group was formed within the TEW Division in 1971. The OCM group was first led by Dr. John A. Montgomery, a visionary and aspiring physicist. He established the notional requirements and long-term development goals for small, high-performance military platforms and their electronic payloads for Navy missions. He envisioned that the best way to rapidly deploy EW payloads from ships or aircraft was to package them within small air vehicles that would be stored in a compact volume, be launched by pushing a button, automatically become efficient flying machines, autonomously perform their EW mission, and then be expended.

The requirements for these offboard systems, which were called Radio Frequency (RF) decoys or Active Electronic Decoys (AEDs), were very ambitious for the early 1970s. The only autonomous unmanned flying machines at that time were guided missiles, which were highly sophisticated, expensive, and not very reliable. RF decoys would need to be autonomous, highly reliable to afford good protection for the defending platform, able to fly well-defined mission profiles under a broad range of environmental conditions, and sufficiently low cost to be expendable following their mission. Montgomery realized that highly innovative, out-of-the-box solutions would be needed, but not having an aerospace engineering staff, he turned to industry for the development of this challenging new type of unmanned air vehicle. Unfortunately, the notional requirements, especially endurance and reliability, could not be met by industry at the time. This clearly indicated the need for significant, high-risk/high-payoff technology development, which government laboratories such as NRL are well suited to perform.

Rising to the challenge, Montgomery began building a staff of aerospace engineer researchers in the TEW Division. In 1975, he hired Dr. Frank Klemm, a graduate of Georgia Institute of Technology, and assigned him to head a new Vehicle Research Section (VRS). Klemm identified the requisite technologies for AEDs. These included low Reynolds number (LRN) aerodynamics to enhance airfoil efficiency and improve flight performance; molded airframe structures for rapid, low-cost production, especially of precise aerodynamic shapes; automatic deployment and fully autonomous flight and navigation to enable rapid response and to eliminate human error; efficient propulsion for maximizing mission duration with near-instant starting for rapid and reliable deployment; compact and electrically efficient EW payloads that were optimized for packaging into small air vehicles and that could withstand anticipated launch loads and environmental conditions; and expendability, which both eliminated the need for post-mission recovery and stressed low production cost.

Klemm and a small staff of engineering support contractors developed a set of AED notional requirements. They then worked with industry to develop three AED demonstrator systems that were successfully flown but did not incorporate all the new technology advancements identified for practical RF decoy systems. To develop the requisite technologies, additional NRL staff was needed, as well as a source of funds to pay staff salaries. Fortunately, the Office of Naval Research (ONR), the primary sponsoring organization of NRL projects, recognized the potential of RF decoys for Fleet defense and provided financial support for technology development. In 1979, the first engineer Klemm hired was Mr. Richard Foch. The only government engineer in the VRS besides Klemm, Foch was assigned a broad range of engineering and testing tasks of increasing challenge. Soon, he was becoming overloaded with tasks, so Klemm instructed him to recruit a few more engineers. Foch began finding and hiring many talented engineers and engineering technicians to work at NRL, and the VRS steadily grew in size and engineering capability.

In 1985, Montgomery was promoted to head the TEW Division, Klemm was promoted to head the OCM group, and Foch became head of the VRS. As the experience and success of the VRS grew, the group was able to obtain financial sponsorship that enabled it to continue to grow and flourish, peaking in size in the 1990s with more than 40 government and engineering support contractor employees. In 2005, Foch became the Navy's Senior Scientist for Expendable Vehicles. Mr. Al Cross now heads the VRS, guiding a staff of more than two dozen skilled employees. Also, as this book is being written, Klemm heads the TEW Division and Montgomery leads NRL as its Director of Research.

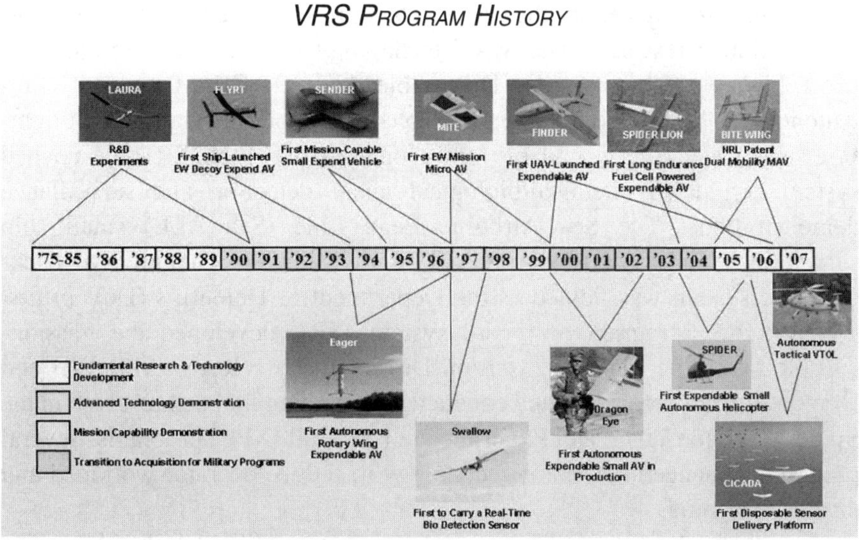

Fig. 1.1 VRS technology development cycle.

From 1975 through 1989, the VRS conducted fundamental research and technology development (Fig. 1.1). The early mission of interest was offboard countermeasures for ships to support EW. This spawned numerous

EW decoy programs such as AED, Interim Active Electronic Decoy (IAED), Ship's Tethered Autogyro Platform (STAPL), and Long Duration Expendable Decoy (LODED). Also during this time, the VRS conducted research on enabling technologies such as electric propulsion, composite structures, autonomy, in-flight reconfiguration, and low Reynolds number (LRN) aerodynamics. Important research programs during this period included Semi-Autonomous Research Aircraft (SARA), Low Altitude/Airspeed Unmanned Research Aircraft (LAURA), Libelle, and Small Ships Compatible Decoy (SSCD).

After many of the enabling technologies matured, the VRS shifted to advanced technology demonstrations in the early to middle 1990s. Two key programs were the Flying Radar Target (FLYRT) and Eager ship protection decoys. These two programs were substantially larger in scope and complexity than earlier programs.

In the middle to late 1990s, NRL explored other classes of mission aircraft. The Self-Navigating Drone Expendable/Recoverable (SENDER) and Swallow programs demonstrated the feasibility of chemical and biological agent detection. The VRS conducted leading research in Micro Air Vehicle (MAV) technology through the Micro Tactical Expendable (MITE) program.

The first decade of the 21st century saw refinement of the technologies, operational employment of the systems, higher-performance vehicle designs, and continued research. The MAV technology was enhanced through the Close-In Covert Autonomous Disposable Aircraft (CICADA) and Semi-Autonomous Mobile Platform Expendable (SAMPLE) programs. The Flight-Insertable Detection Expendable for Reconnaissance (FINDER) air-launched system conducted chemical/biological agent detection and surveillance demonstrations. The Sea Airborne Lead Line (Sea ALL) small ship surveillance aircraft saw operational service and the Dragon Eye back-packable system was fielded as the Department of Defense's (DoD's) first "Small UAS" program of record system. NRL developed the mission-capable Scientific Payload Insertion Device Electric Rotor (SPIDER) and Dragon Warrior rotorcraft, and conducted pioneering fuel cell research in the Spider Lion, Ion Tiger, and Experimental Fuel Cell (XFC) programs. Several programs enhanced autonomy technology to reduce operator workload and enable swarming.

Since the early 1990s, NRL has explored planetary flight with NASA. The main objective has been flight on Mars. Numerous Mars flyer concepts have been developed, with increasing robustness and reduced risk. The most recent program was a Titan Flyer that would use nuclear propulsion.

As of 2012, the NRL VRS had conducted more than 60 programs and flight-tested more than 300 unmanned aircraft. Approximately $175M of

Navy technology investment and over $70M of non-Navy sponsorship have been provided for these efforts.

NRL Technology Advancements

The vehicles developed by the VRS placed great demands on new technologies. Most of the designs required large strides in emerging technology development. The system requirements typically included autonomous operation, high performance (range and endurance), rapid deployment, high reliability, operations in hostile environments, affordable expendability or reusability, and organic operations with the host platform. These requirements drove key advancements in autonomy, structures, configuration design, propulsion, aerodynamics, and payloads. Technologies developed at NRL are in use today in many military and commercial unmanned systems. Many system developers may not know the origins of the technologies that are commonplace today.

Autonomy

NRL began with analog and fluidic autopilots in the 1970s and early 1980s. The initial missions were to fly autonomous flight paths for ship protection decoys without a pilot in the loop. GPS was not available then, so the pseudo-inertial sensors had to have sufficiently low drift rates to enable a preprogrammed trajectory. Typical autopilot control included wing leveling and rate-damped heading hold. Through the Flying Radar Target Advanced Technology Demonstrator (FLYRT ATD) program, NRL developed a digital autopilot with low-drift-rate, three-axis, fiber-optic laser rate gyros. This was followed by the pioneering Light Weight Autopilot (LWAP) on the SENDER program, which was a compact GPS-based autopilot. NRL further refined miniaturization through the various MAV efforts. The Eager program used real-time kinematic (RTK) relative GPS to enable precision flight relative to a moving platform. Later, the VRS worked closely with industry to perfect modern miniature autopilots for rotorcraft and autonomous flight behaviors. Most early demonstrators used radio control (RC) due to the prohibitive cost and complexity of available autopilots, but today the systems nearly all conduct autonomous operations very early in the programs. NRL also developed image processing technology to enable vision-based autonomous flight in GPS-denied environments. Many commercially available autopilots have a direct technology heritage in the pioneering NRL research.

STRUCTURES

NRL made advances in composite structures technology for unmanned aircraft. Contemporary efforts were aimed at homebuilt aircraft and advanced manned military aircraft. Early NRL unmanned aircraft used fiberglass and foam core construction. Later designs utilized Kevlar and carbon fiber materials. The construction started with hand-layup techniques. Today most NRL systems utilize computer numerically controlled (CNC) machined female molds with hollow structures. With the small size of the unmanned aircraft, even advanced aerospace composites can be applied at relatively low cost.

CONFIGURATION DESIGN

The VRS conducted research on a wide range of configurations for LRN vehicles. The LAURA program compared four LRN configurations against a single requirement. The MITE and SAMPLE programs developed several MAV configurations, often with synergistic aerodynamic and propulsion technologies. Foch advanced the displaced tail configuration in numerous efforts. Many of the systems had to reconfigure in flight with high reliability. The AED, FLYRT ATD, Extender, FINDER, XFC, and Mars flyer programs established mechanisms and configurations that could dynamically deploy, leveraging aerodynamic, mechanical, and inertial forces. The systems had to deploy from containers such as decoy launch systems, sonobuoy tubes, backpacks, atmospheric entry vehicles, or vertical launch system canisters.

PROPULSION

NRL conducted pioneering research in propulsion systems for small unmanned aircraft. Propulsion technologies included turbines, electric motors, batteries, fuel cells, and bio-inspired systems.

The VRS developed small-diameter turbojet engines in the early 1990s for high-speed targets. A later project developed a high-efficiency recuperated turbogenerator for hybrid electric propulsion.

NRL developed and integrated high-performance electric motors. Early motors were brushed samarium cobalt rare earth magnet motors with efficiencies ranging from 70% to 80%. Early electric vehicle programs solved vexing electromagnetic interference (EMI) problems. NRL invented neodymium-iron-boron magnets to reduce the cost of motors, along with potting and sealing techniques to protect them against salt water corrosion. NRL transitioned metal oxide semiconductor field-effect transistor (MOSFET) high-power, high-speed digital switching technology from its radar systems to compact, high-efficiency motor controllers. This breakthrough enabled brushless motors that converted DC current to three-phase

AC current. So the Tesla motor could be applied to unmanned aircraft with 90% to 95% efficiency. Today, motor technology has been advanced to the point at which limited additional gains are possible, so NRL has shifted research to energy storage.

NRL's electric unmanned aircraft have pushed battery technology since the 1980s. Early designs used secondary (rechargeable) nickel cadmium (NiCd) batteries, which enabled flight durations of 3 to 20 minutes. The SARA program experimented with primary (nonrechargeable) high specific energy silver zinc (AgZn) and lithium sulfur dioxide ($LiSO_2$) batteries, which enabled a record-setting 2-hour and 45-minute flight in 1992. NRL was an early adopter of rechargeable lithium ion and lithium polymer (LiPo) batteries in designs starting around 2000.

The VRS has pioneered unmanned aircraft fuel cell propulsion development over the past decade. The research has advanced hydrogen-fueled proton exchange membrane (PEM) fuel cell efficiency and weight reduction, fuel cell balance of plant design, and hydrogen storage. Key programs include Spider Lion, Ion Tiger, XFC, and Ion Dasch.

Aerodynamics

The VRS unmanned aircraft programs advanced LRN aerodynamic design. NRL experimented with advanced LRN airfoils that operated at Reynolds numbers between 60,000 and 500,000. Through the Airborne Active Expendable Decoy-Forward Fired (AAED-FF) and Mars flyer programs, NRL developed configurations that operated simultaneously in the LRN and transonic regimes. The various MAV programs explored LRN vehicle design and propulsion interactions. The in-flight reconfiguring aircraft such as the Variable Span LAURA and Air Launched Integrated Countermeasure, Expendable (ALICE) had airfoils that could operate effectively when nested or deployed. LRN airfoil technology was also applied to custom propellers to improve efficiency.

Payloads

Various groups at NRL led the development of miniaturized payloads for small unmanned aircraft applications. The TEW Division developed miniaturized electronic payloads that could spoof or jam threat radars. NRL developed miniaturized chemical and biological agent detection systems for *in situ* detection and sampling of airborne agents.

Chapter 2

Ship-Launched Decoys and Targets

By the mid-1970s, NRL's Warfare (TEW) Division had developed prototype miniature electronic warfare (EW) payloads that used traveling wave tube (TWT) amplifiers to produce radar signals similar to the reflection from a large Navy ship. NRL realized that placing such a payload on a small, ship-launched unmanned aircraft could yield a potent ship protection system. Such a system could produce a false target to spoof an incoming anti-ship missile. The missile would then be led away from the ship and harmlessly toward the unmanned aircraft.

The conceptual vision was to have a decoy that could be rapidly launched from a ship at the push of a button. However, there were numerous technological barriers in the late 1970s. GPS did not yet exist. Electric propulsion was not yet practical, and only externally started reciprocating engines were available. Unfolding airframe technologies were not mature, with the exception of spring-loaded fins on contemporary missiles. No contemporary unmanned aircraft were optimized to fit within small containers, integrate onto ships, fly autonomously, or be affordably expendable. NRL focused on developing and integrating these needed technologies to create functional vehicle systems. This chapter details NRL's ship-launched Radio Frequency (RF) decoy development from about 1975 to 2000.

Active Electronic Decoy (AED), 1975–1978

The goal of the Active Electronic Decoy (AED) program was to develop an autonomous, expendable aircraft that could rapidly lift EW radar repeater payloads off a ship to decoy threat missiles. NRL developed a set of requirements for an AED. It was to be a boost glider that was rocket launched from the defending ship. The launch angle was 60 degrees from level and oriented away from the ship. The initial acceleration was 120 *g* for 15 milliseconds, followed by 8-*g* acceleration for 2 seconds. The wings unfolded at apogee, and the gliding commenced. The gliding flight lasted about 90 seconds, followed by a nonrecoverable ditching. The system had to be operable to sea state 5 conditions.

NRL did not have an aerospace engineering staff in the TEW Division at the beginning of the program, so four different industry teams were tasked to develop their own designs to the NRL requirements. The teams were Fairchild Industries in Maryland, Developmental Sciences International (DSI) in California, Teledyne Brown in Alabama, and E-Systems Melpar Division in Virginia. Each resulting design was unique, with innovative features, and all but one were successfully flown.

An aircraft ejection seat rocket motor for the booster was provided to each industry contractor as government-furnished equipment (GFE). Each aircraft carried a 40-pound EW payload, also provided by NRL.

The system had to be autonomous. However, for the winged vehicles, a very simple control scheme was devised. The aircraft were set to be trimmed in pitch so that only roll control via a wing leveler was required, to keep the aircraft from rolling over in a spiral dive. The aircraft only had to fly away from the ship in a straight line, not in a complicated flight path.

FAIRCHILD SPINNER AED, 1975

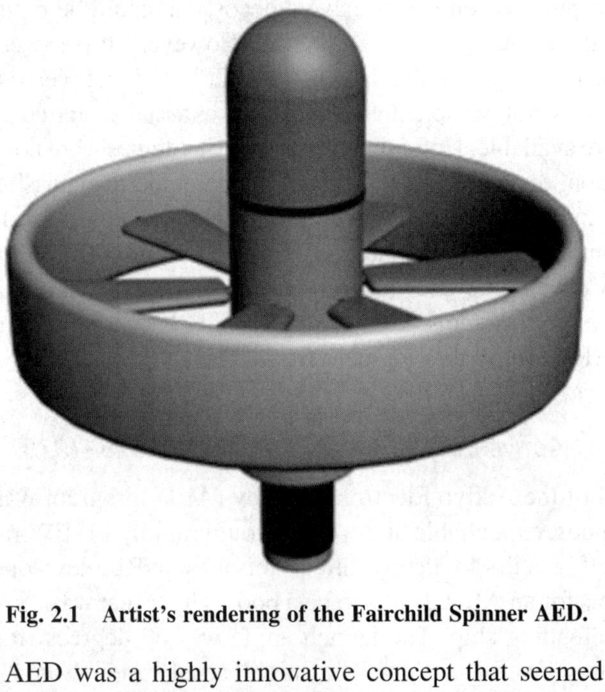

Fig. 2.1 Artist's rendering of the Fairchild Spinner AED.

The first AED was a highly innovative concept that seemed especially attractive for an expendable unmanned vehicle. Called Spinner (Fig. 2.1), it was inspired by a simple toy known as a flying disk. The flying portion of the toy consists of a central shaft with three or more rotor blades that are surrounded by a safety ring. This simple flying machine is gyroscopically stabilized by the rotation of the ring as the entire vehicle rotates. To launch the flying disk toy, it is set into a keyed socket that rotates inside a handheld launcher. The operator then quickly pulls a string wound around the socket inside the launcher, causing the vehicle to spin at a high enough rpm for the rotors to generate more lift than its weight. The vehicle ascends skyward, gyroscopically maintaining the attitude of the launcher. Finally, as the rpm winds down, the vehicle descends to the ground.

The Spinner AED was designed to use the same basic physics as the toy. Specifically, the launcher would spin the airframe to an rpm sufficient for the rotors to lift the airframe, and the structural ring around the rotor blades would store sufficient angular momentum to enable the Spinner AED to fly for approximately 1 minute. To control the moment of launch, the rotor blade pitch was collective controlled, with the spin-up being performed at the minimum drag angle to facilitate acceleration to the flight rpm. Upon reaching the flight rpm, the blades were then commanded to the maximum lift setting to enable the highest initial rate of climb. All propulsive power in the airframe was stored angular momentum, and the only control system was that which collectively set the two blade pitch angles. It was an amazingly simple, expendable, autonomous, vertical takeoff (VTO) aircraft, and it was to be NRL's first AED. This was certainly a very "out-of-the-box" solution to rapidly deploying an EW payload away from a ship. A wankel rotary engine was selected as the launcher power plant to spin-up the Spinner. The design and fabrication contract was awarded to Fairchild, and flight testing was much anticipated. Figure 2.2 shows a two-view drawing, and Table 2.1 lists the technical characteristics for the Spinner AED.

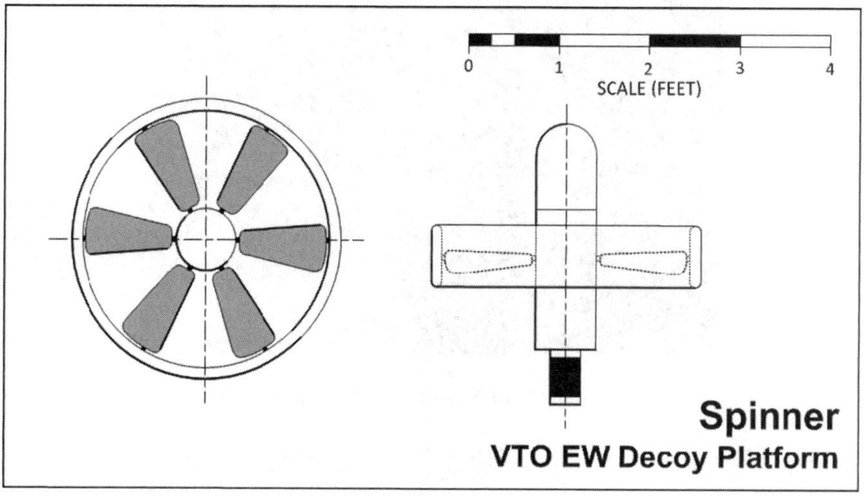

Fig. 2.2 Spinner AED two-view drawing.

Table 2.1 Spinner AED Technical Characteristics

Parameter	Value
Diameter	3.0 ft
Disk area	7.1 ft^2
Length	3.0 ft
Takeoff weight	Unknown
Payload weight	40.0 lb
Endurance	1.0 min
Launch method	Rotary engine in launcher spins vehicle to launch rpm
Recovery method	Expendable
Propulsion type	Stored angular momentum rotor

Unfortunately, the prototype was not sufficiently robust to handle the high structural loads caused by the required high rpm and gyroscopic loads. Upon nearing lift-off speed, the Spinner disintegrated in the launcher. Meager financial resources prevented further development to address the design flaws. Perhaps with today's composite structures, a modern version of the Spinner AED might prove successful. However, the 1970s requirement of 1 minute flight endurance is no longer sufficient for Navy EW missions, so a modern version would be used for a different type of mission.

FAIRCHILD SAILWING AED, 1975–1976

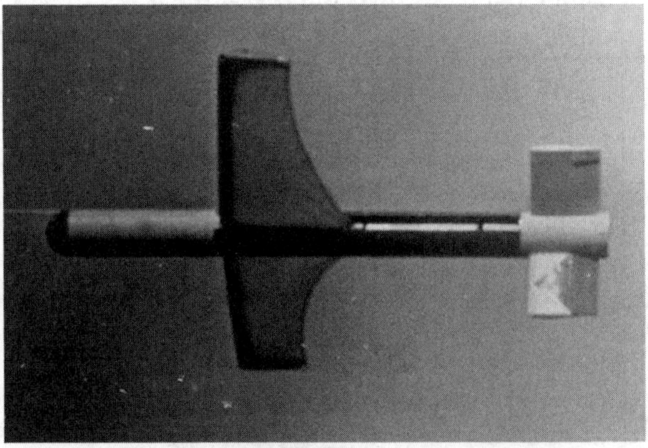

Fig. 2.3 Fairchild Sailwing AED in gliding flight.

The Fairchild Sailwing AED (Fig. 2.3) employed the Princeton Sailwing technology: a rigid D-tube formed the leading edge of each wing, a catenary cable formed the trailing edge, and flexible upper and lower surfaces

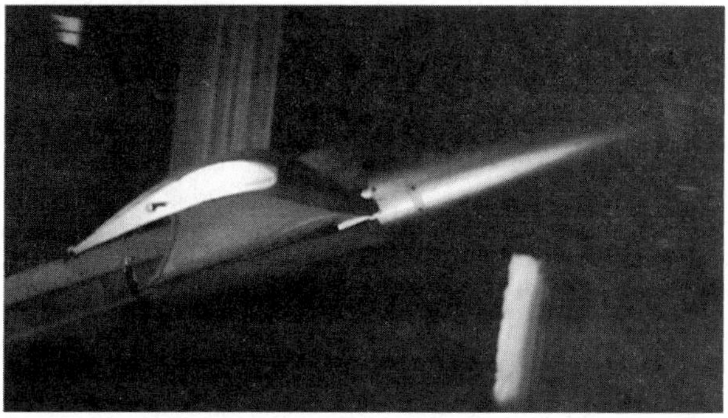

Fig. 2.4 Side view of the Fairchild Sailwing AED wing twist distribution.

extended from the D-tube to the trailing edge. Roll was controlled by adjusting the tension of the trailing edge cable. There was a root rib along the fuselage and a wing tip rib. When stowed, the right and left side leading edge D-tubes each folded along the side of the fuselage, and the tip rib and wing skin fabric fit within the cavity. The fuselage had a constant circular cross section. The tail had a Y-tail configuration.

The Fairchild Sailwing AED was launched and flown several times but was not fully successful. Wind tunnel tests revealed some unusual behavior during wing deployment. The aircraft tended to have an uncommanded roll-bias. The trailing edge wire tension could not be controlled precisely enough, and even slight asymmetric tension generated asymmetric wing twist and a large corresponding rolling moment (Fig. 2.4). This resulted in inconsistent flight performance (Fig. 2.5). The wing also had effective wash-in: The wing tips had a slightly higher lift coefficient than the root, which can cause wing

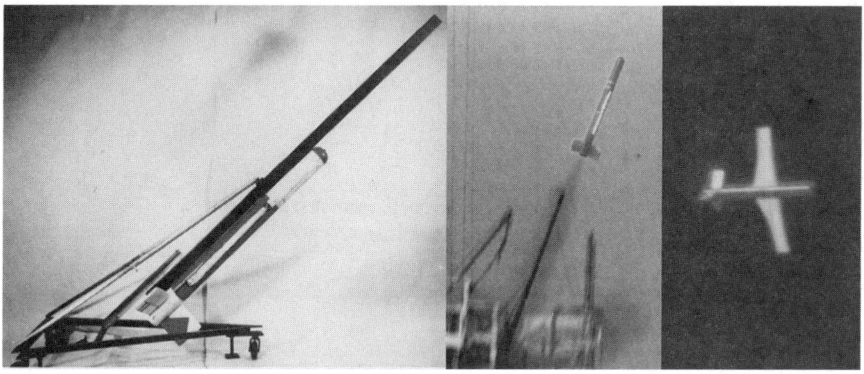

Fig. 2.5 Fairchild Sailwing AED launch sequence.

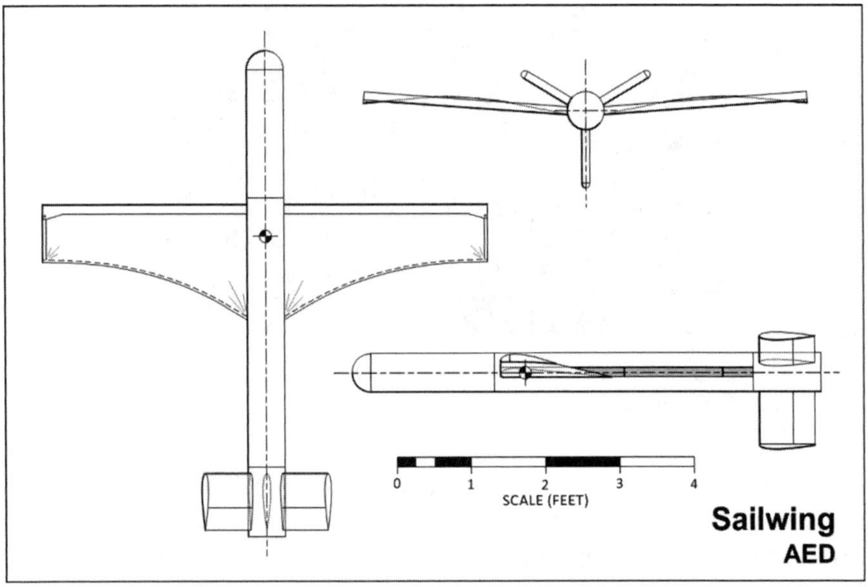

Fig. 2.6 Fairchild Sailwing AED three-view drawing.

tip stall. Foch discovered the aerodynamic wash-in effect 15 years after the program concluded, too late to make any changes that might have made the The Fairchild Sailwing AED three-view drawing is shown in Fig. 2.6, and the technical characteristics are listed in Table 2.2.

Table 2.2 Fairchild Sailwing AED Technical Characteristics

Parameter	Value
Wing span	6 ft
Wing area	6 ft^2
Length	7.5 ft
Takeoff weight	110.0 lb
Payload weight	40.0 lb
Cruise speed	77 KTAS
Endurance	1.6 min
Launch method	Solid propellant rocket launch
Recovery method	Expendable
Propulsion type	Rocket glider

DSI AED, 1977–1978

Fig. 2.7 DSI AED launch.

The DSI AED design had a horizontal tail, an inverted vertical tail, and a single-piece scissoring wing that ran the length of the body when stowed. The wing pivoted out during launch (Figs. 2.7 and 2.8). The tails had pitch control surfaces that were set up for trim according to a trim schedule during launch. Key trim schedule events included initial launch and expending the rocket

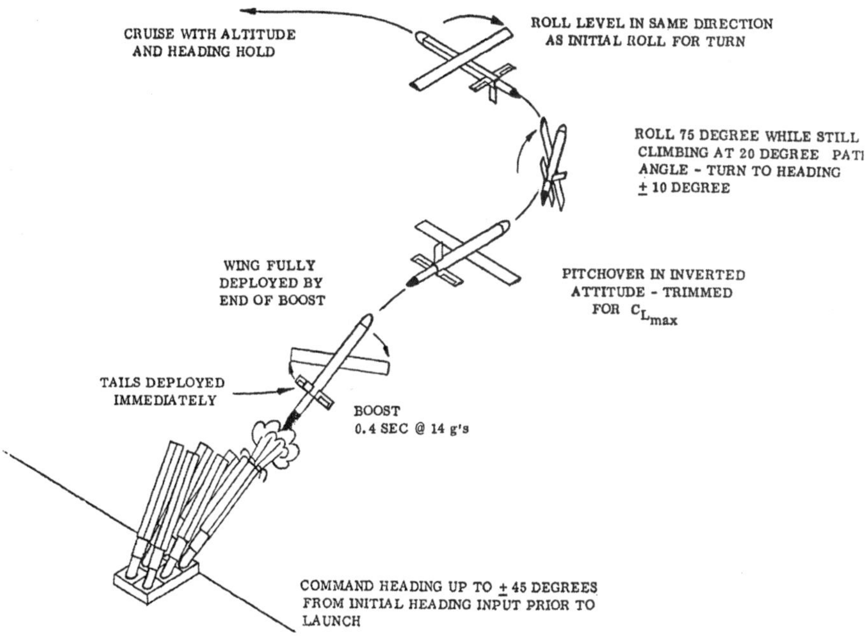

Fig. 2.8 DSI AED launch sequence.

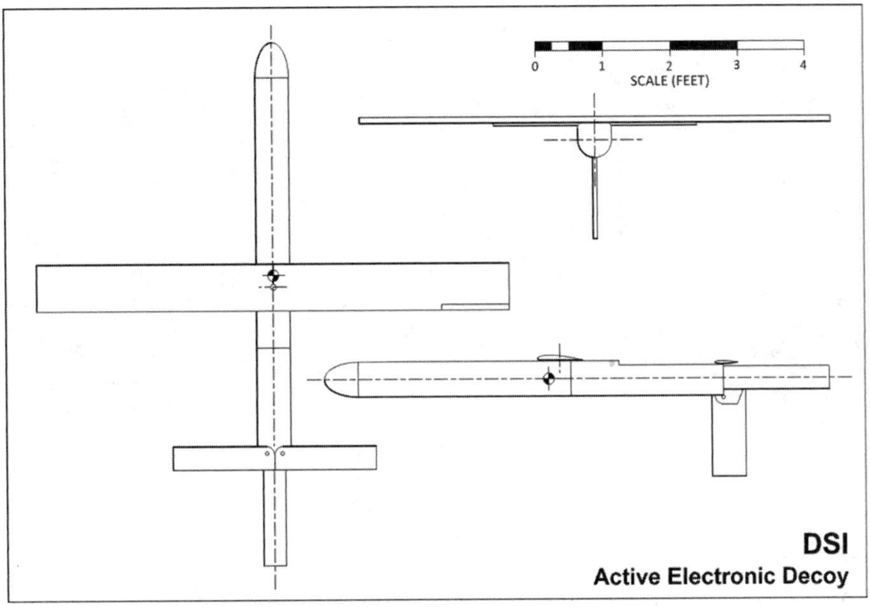

Fig. 2.9 DSI AED three-view drawing.

motor. Roll control was accomplished with a single aileron on one side of the wing. Although downward aileron deflection is more effective than upward deflection, the control system worked effectively. The system was successfully demonstrated. Figure 2.9 shows a three-view drawing of the DSI AED, and Table 2.3 lists its technical characteristics.

Table 2.3 DSI AED Technical Characteristics

Parameter	Value
Wing span	7.0 ft
Wing area	4.67 ft^2
Length	7.5 ft (including booster)
Takeoff weight	110.0 lb
Payload weight	40.0 lb
Cruise speed	90 KTAS
Endurance	1.6 min
Launch method	Solid propellant rocket launch
Recovery method	Expendable
Propulsion type	Rocket glider

Teledyne Brown AED, 1977–1978

Fig. 2.10 Teledyne Brown pleated wing/tail design.

Teledyne Brown developed a patented design with pleated wings and tail (Figs. 2.10 and 2.11). After launch, two wings swung out from the bottom of the fuselage. When stowed, each wing chord occupied half the fuselage width. The chord of the wing increased when deployed due to the pleated airfoil design. The leading edge had an open C-tube box. The remainder of the chord nested inside the C-tube leading edge like multiple slats or shingles. Teledyne Brown named this design the "Bird Wing" AED (Figs. 2.12 and 2.13, Table 2.4). This aircraft flew successfully.

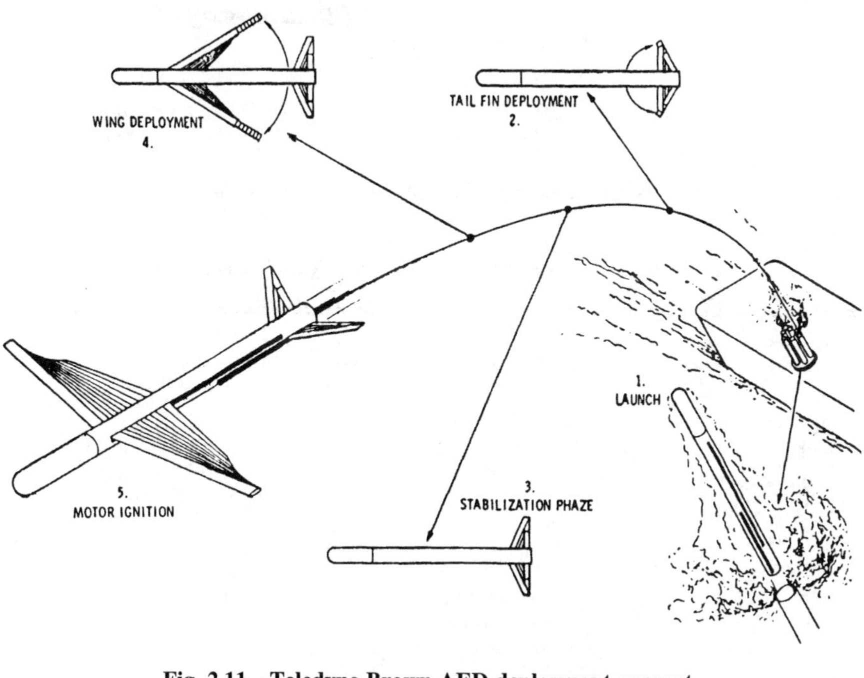

Fig. 2.11 Teledyne Brown AED deployment concept.

Fig. 2.12 Wind tunnel testing the Teledyne Brown AED.

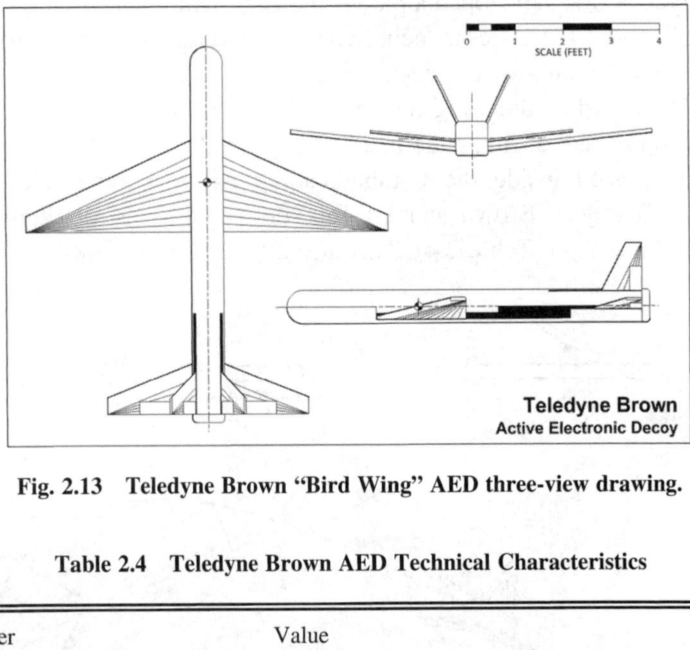

Fig. 2.13 Teledyne Brown "Bird Wing" AED three-view drawing.

Table 2.4 Teledyne Brown AED Technical Characteristics

Parameter	Value
Wing span	7.5 ft
Wing area	9.37 ft^2
Length	7.5 ft
Takeoff weight	120.0 lb
Payload weight	40.0 lb
Cruise speed	86 KTAS
Endurance	1.6 min
Launch method	Solid propellant rocket launch
Recovery method	Expendable
Propulsion type	Rocket glider

E-Systems AED, 1977–1978

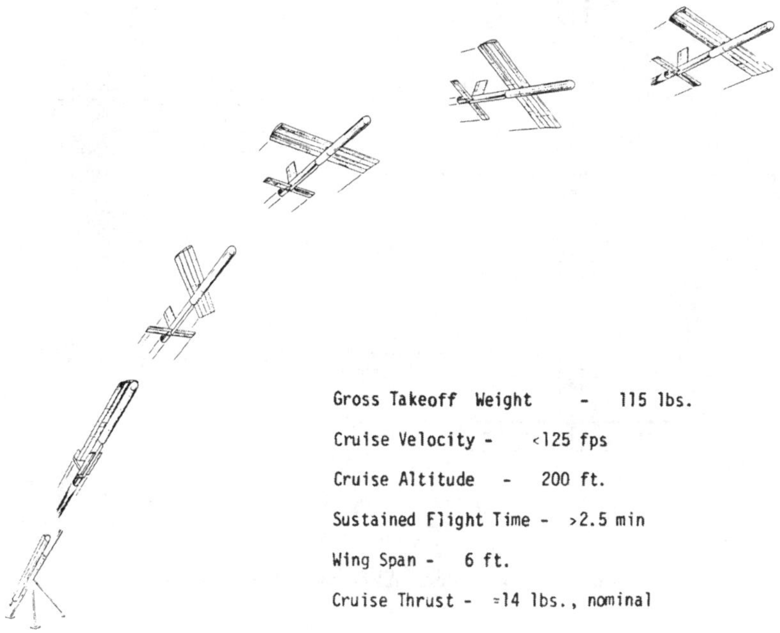

Fig. 2.14 E-Systems AED launch sequence.

The E-Systems AED design had a conventionally shaped airfoil that was flexible so that it wrapped around the fuselage when stowed. When deployed, the wing chord was approximately three times the diameter of the circular fuselage cross section (Fig. 2.14). This greater wing area decreased the wing loading and reduced the descent rate (Fig. 2.15). The leading edge section

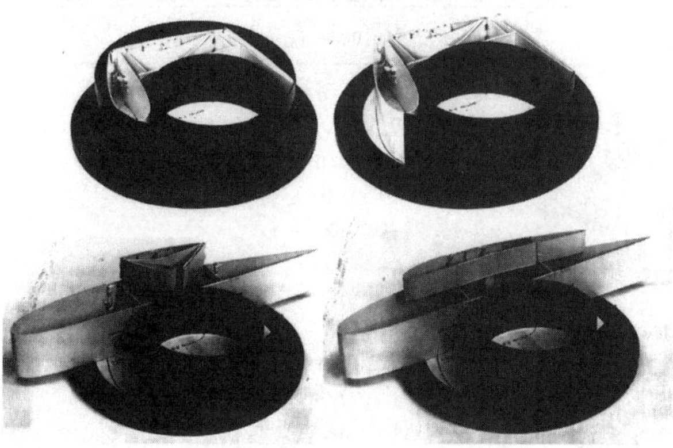

Fig. 2.15 E-Systems flexible airfoil concept model.

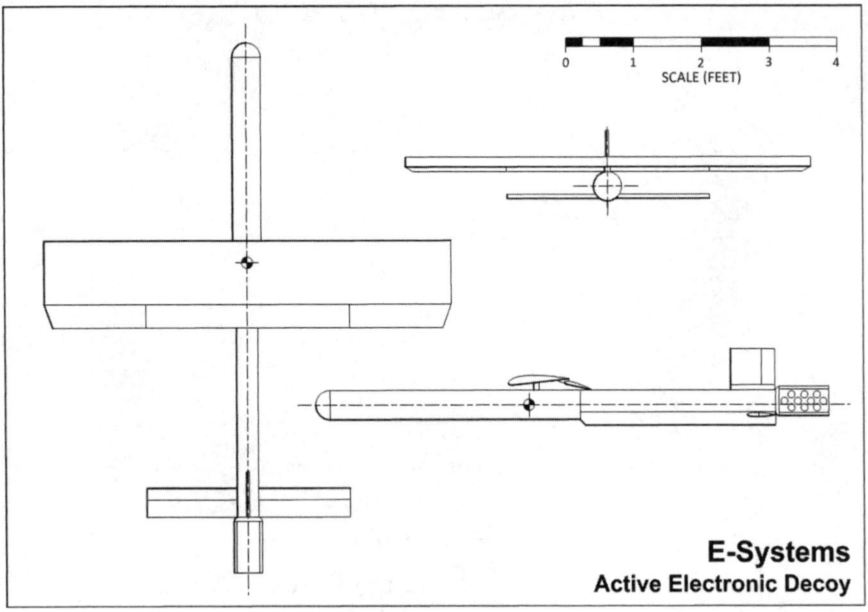

Fig. 2.16 E-Systems AED three-view drawing.

curved around the fuselage and sprung up when deployed. The trailing edge section was a slotted flap that was hinged to fit next to the fuselage side when stowed. The single top-mounted wing rotated for deployment. Figure 2.16 shows a three-view drawing, and Table 2.5 lists the technical characteristics. The E-Systems AED was successfully demonstrated.

Table 2.5 E-Systems AED Technical Characteristics

Parameter	Value
Wing span	6.0 ft
Wing area	7.5 ft^2
Length	7.6 ft (including booster)
Takeoff weight	115.0 lb
Payload weight	40.0 lb
Cruise speed	67 KTAS
Endurance	2.5 min
Max L/D (lift/drag)	8.2
Launch method	Solid propellant rocket launch
Recovery method	Expendable
Propulsion type	Rocket glider

LONG DURATION EXPENDABLE DECOY (LODED), 1978–1982

Fig. 2.17 LODED 56.25% scale prototype on launcher.

Long Duration Expendable Decoy (LODED) represents the starting point for NRL's ship-deployed, long-endurance unmanned aircraft systems (UAS) technology (Fig. 2.17). The design endurance for LODED was 9 hours. This program made numerous important advances, although a retrospective evaluation of the project reveals how much additional work was required to fulfill the vision of a practical, high-performance system. The reader is encouraged to compare this system with the XFC UAS (Chapter 10). Three decades of research separating these programs yielded numerous technology advances.

NRL first approached industry to develop a custom aircraft to meet the mission need of a long-endurance EW decoy. The contractor responses yielded prohibitively expensive solutions, and the development timelines were excessive. Also, industry believed that there was an insufficient market to warrant company investment or dedication of resources. With no viable industry path, NRL decided to develop this capability in-house. Such an approach was initially met with considerable skepticism by industry, because government laboratories typically did not take on the role of system developer and were not known for being lower cost than industry. Many of the projects described in this book initially sought industry bids, but the proposed costs were considered excessive, or companies did not bid due to the perception of high risk. NRL often conducted projects internally to reduce costs and improve schedule.

LODED requirements, constraints, and key system elements were defined. A significant design driver was the need to handle the system aboard a ship. Sailors needed to carry the system between decks. The storage container that included the air vehicle and launcher had to be of a size and weight that allowed it to be portable by only two sailors. The box would be brought up to the deck, secured to the deck, and opened. The rectangular box would unfold

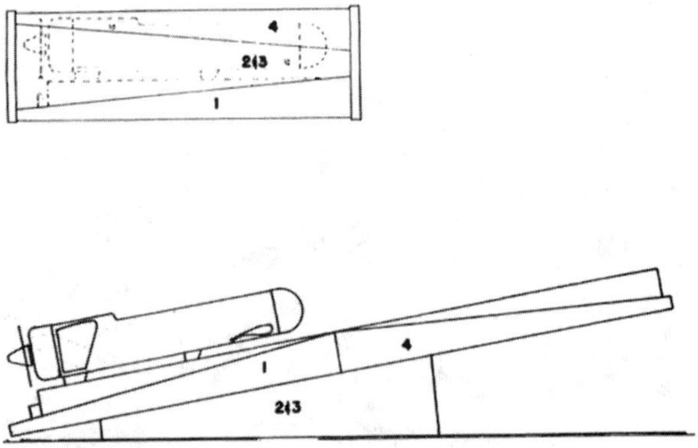

Fig. 2.18 Mission LODED launcher concept.

to form an integral launch rail, which would be almost twice the length of the box (Fig. 2.18). The aircraft would be assembled.

The LODED air vehicle had a tandem wing configuration. While new at the time, this configuration is becoming more prevalent today for tube-launched UASs because of its compatibility with compact packaging (Fig. 2.19). Notable modern examples of tandem wing designs include the AeroVironment Inc. Switchblade and the Advanced Ceramics Research (ACR) Coyote. One motivation for designing tandem wings for LODED was to have four

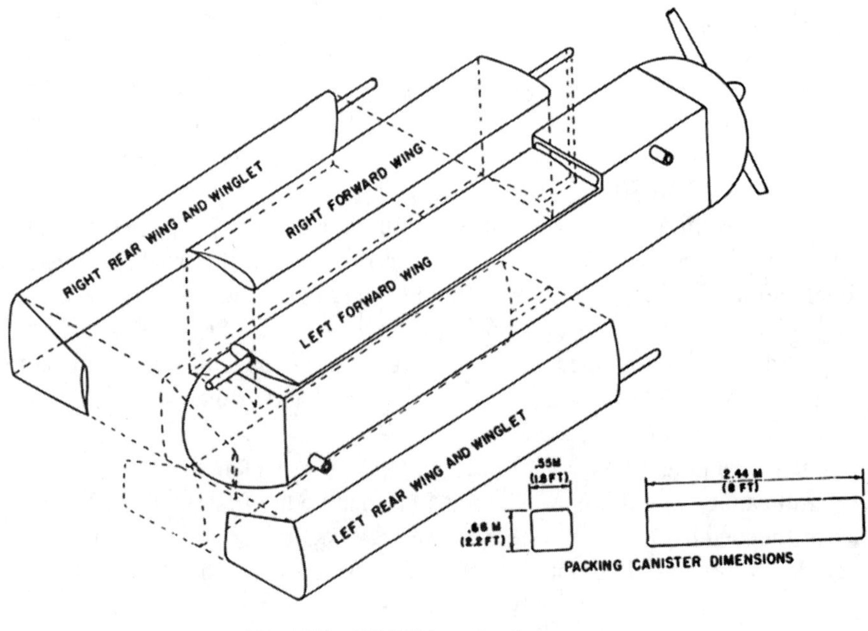

Fig. 2.19 LODED packaging concept.

Fig. 2.20 LODED original upward winglet orientation.

identical wing panels that could be easily attached to the aircraft body during preflight assembly. The wings were sized both for flight performance and for packaging. The front wing was located below the rear wing so that the wake from the front wing would not interfere with the rear wing.

Two winglets were fitted onto the wing tips of the rear wings to provide directional stability. Winglets were a new invention by aeronautical engineer Richard Whitcomb of NASA Langley Research Center at the time of the LODED program, and he was consulted on their application to this vehicle. The winglets extended upward from the rear wing tips on the first LODED designs, but lateral-directional handling challenges later drove a modification to downward facing winglets, as discussed in more depth on the following pages.

The winglets served three functions on the LODED vehicle (Fig. 2.20). First, they reduced the induced drag on the rear wing. The aft wing operated at a lower lift coefficient than the forward wing, so the induced drag reduction alone would not have been a satisfactory trade-off for the weight penalties and additional complexity of adding winglets if not for the second and third functions. Second, the winglets provided directional stability much like a traditional vertical stabilizer. The aft location provided a sufficient moment arm for yaw stability. Third, winglets increased the lift curve slope of the rear wing to improve the longitudinal dynamic stability. In other words, a change in angle of attack increased the change in lift coefficient on the aft wing faster than on the front wing. This provided a nose-down stabilizing moment when the angle of attack increased. NRL obtained a U.S. patent for the combined use of winglets for directional stability and performance.

LODED used an unconventional control system (Fig. 2.21). The front and aft wings had adjustable incidence angles. The front wings adjusted incidence in the same direction on each side to generate a pitching moment. This full wing pitch control design can be called a "wingevator." The aft

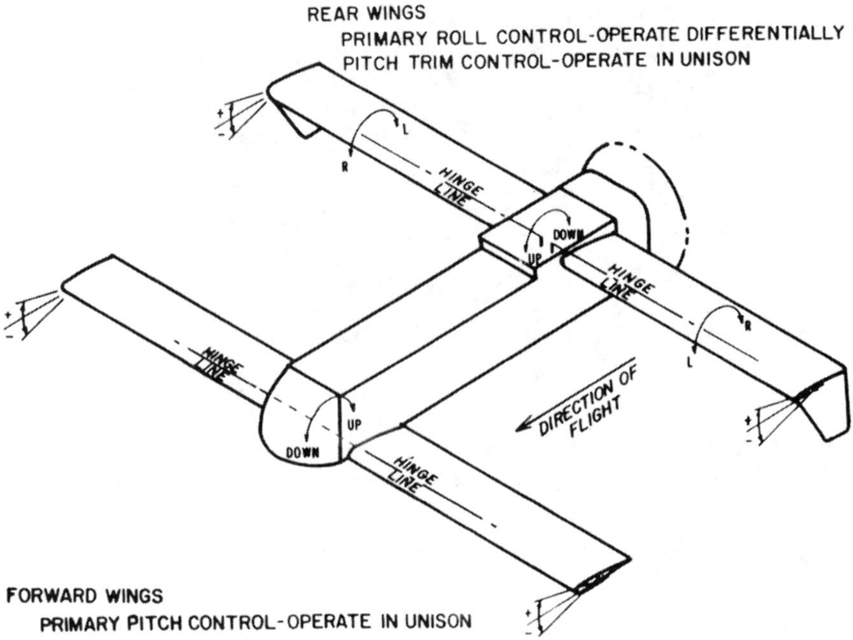

Fig. 2.21 LODED control system.

wings adjusted incidence angles in opposite directions on each side to generate a rolling moment. This full wing roll control design is commonly known as a "wingeron." There were no yaw effectors. All wings used a highly cambered Wortmann FX 63-137 airfoil without any control surfaces. The torque loads on the actuators were reduced by placing the pivot at approximately 30% of the chord length from the leading edge rather than at the quarter chord position. Note that the wings were untapered and unswept, so the mean aerodynamic chord was equal to the wing chord. The airfoil lifted about the quarter chord, so the aft pivot position provided a nose-up pitching moment from the lift that counteracted the nose-down pitching moment associated with the wing camber. This arrangement is intended to minimize torsional moment at the cruise lift coefficient (C_L).

The actuators were mechanically connected to sockets that rotated around ball bearings. A control horn stuck out of the socket, and the actuator push rod moved the control horn. The wing rods extending from the wing panels plugged into the socket in a tube-within-a-tube arrangement. The outer tube was moved by the actuator. The inner tube from the wing locked in with a pin and key. This attachment system was quick to install, but difficult to remove. The wing disassembly from the fuselage required unscrewing several fasteners, which was a time-consuming process. This limitation was deemed acceptable for an expendable vehicle.

The actuators were custom designed by NRL using industry components. Commercially available model aircraft actuators were not suitable due to the large torque loads of the rotating wings. The NRL actuators were of a screw jack configuration, in which a tooth gear on the motor drove a worm gear. This was an irreversible control system, in which the actuators could move the control surfaces, but the control surfaces could not drive the actuator. Any pre-loads on the actuator interface were transferred to the actuator structure rather than countered by motor torque. Therefore, the actuator did not require large power inputs to hold a load. Screw jack actuators are rarely used on small unmanned aircraft today; rotary actuators are generally employed.

The scale models and the full-sized LODED aircraft used homebuilt aircraft composite construction techniques that were popular at the time. In particular, the Rutan VariEze fabrication influenced the design. The fuselage was built as a sandwich with fiberglass plies over sheet foam. The fuselage had a nominal rectangular cross section. The wings had hot-wired foam core with fiberglass plies on the outer surface. No female molds were used. Higher-strength carbon fiber composite materials were not yet available. Each LODED wing had an aspect ratio of 10, which was high by contemporary unmanned aircraft standards.

Propulsion was a challenge for the LODED program. NRL ultimately selected a two-cylinder Herbrandson two-stroke engine for the mission aircraft. This power plant was originally developed for the new ultralight aircraft market, and it used a gas and oil mixture. No suitable heavy fuel engines were available. The 56.25% scale LODED models used a 0.40- or 0.45-cubic-inch two-cycle model airplane engine. The engine was cooled via a ram-air inlet and an annular exhaust around the spinner. The inlet was located ahead of the rear wing in line with the leading edge, which was a stagnation region with peak pressure. The fuel tank was located near the center of gravity (CG) in the fuselage, which was far from the pusher engine. The carburetor did not provide sufficient fuel draw, so a fuel feed pump was employed. It pumped fuel into the engine header tank. The flow was controlled with an NRL-developed header tank floating valve to maintain a constant fuel level.

Problems with electromagnetic interference and radio frequency interference (RFI) were corrected during development. The radio control system was an early-generation digital system with limited noise rejection capability. The servos were located far from the receiver with long leads. These picked up RF signals much like an antenna, which introduced signal noise. RF chokes consisting of magnets were placed on the leads. Repositioning servo mounting locations by trial and error also seemed to

Fig. 2.22 Views of LODED on the launcher used during flight tests.

remedy the problem. The system required significant tuning and adjustment to obtain adequate electromagnetic interference (EMI) and RFI performance, but the reliability was still not satisfactory for an operational system.

Onboard electrical power was provided by an engine-driven alternator. A lead acid battery on the power bus provided energy storage and also served as an electrical noise filter from the alternating current of the alternator. The battery could power onboard systems in the event of an alternator failure.

NRL built a 3/4-scale wind tunnel model and tested it at the University of Maryland Glenn L. Martin wind tunnel to verify aerodynamic performance, determine the stability derivatives, and verify the location of the neutral stability points.

The flight test program began with a 56.25% scale dynamically scaled model. The goal of the model was to evaluate the wing mechanisms, controllability, and CG. The reason for this scale was that the winglet root airfoil chord of the full-scale aircraft was 56.25% of the wing chord, and both used the same airfoil. Therefore the full-scale mock-up rib templates could be reused. The hand-cut aluminum templates took about two days to fabricate, so reuse allowed a significant labor savings. Fabrication was a small-scale operation: As a young engineer, Foch personally built several LODED aircraft at a rate of one aircraft every two weeks.

LODED was designed to be rail launched, but the early flight tests used a conventional tricycle landing gear arrangement. After the initial airworthiness and reliability of the 56.25% scale model were established, the program transitioned to the rail launcher system (Fig. 2.22). The LODED integral box launcher system was not fully developed, so an Army surplus Aquila pneumatic rail launcher was used for the flight test program as an interim capability. The Lockheed Aquila was a contemporary battlefield reconnaissance UAS program. A custom shuttle adapter that interfaced to the existing landing gear struts was developed for LODED. The wheels could be removed and replaced by solder pins. The pins had sufficient strength to hold the aircraft to the shuttle for engine start and full-throttle thrust loads. The shuttle would hit a snubber at the end of the launch stroke, which generated sufficient shock loads to shear the solder pins. The aircraft would then fly free of the launcher. This arrangement was used without any problems.

The scale model used lead ballast in the nose to simulate the payload. This ballast accounted for 25% of the launch weight. The nose payload location required this weight for all flights to maintain the CG position. Therefore, it was not possible to fly at lower weight and then progressively build up to higher weights. The aircraft was always weight critical, which made it difficult to fly during the early test program. Tandem wings require a CG closer to the front wing, so the heavy nose payload made balancing the aircraft straightforward. LODED carried two times the load on the front wing as on the rear wing. The fuel was located in the fuselage such that the CG did not change as the fuel depleted.

There were a large number of early unsuccessful flights on the 56.25% scale model. The aircraft would take off, start the initial climb-out, and then become laterally-directionally unstable due to Dutch roll divergence. (Dutch roll is described in further detail in Appendix A.)

In an attempt to understand this problem, NRL developed a six-degree-of-freedom (6-DOF) flight simulator to help engineers evaluate handling characteristics and facilitate autopilot development. NRL hired Mal Abzug to help develop the Fortran-based flight simulator. Abzug was the inventor of the Leading Systems Inc. Amber UAS, a predecessor to the General Atomics Predator UAS. The simulator did not show any problems with the LODED flight characteristics and no Dutch roll divergence. The team felt that the earlier crashes must have been caused by a stall or wind gust rather than any inherent problems with the design. A new subscale demonstrator was built and flown in calm conditions. Approximately 5 seconds after takeoff, the aircraft experienced Dutch roll divergence and crashed. After the computer model was scrutinized, it was discovered that there was an erroneous input for a product of inertia term. It was off by a factor of 32.2, which was caused by a conversion error related to the acceleration of gravity. Later simulation runs with the appropriate inputs predicted the Dutch roll problems. A lesson for the team was that it is risky to believe software outputs over observed flight behavior.

The winglets were put on the aft wings primarily to provide directional stability with yaw angle (stabilizing Cn, β). However, the upward orientation of the winglets provided too much negative Cl, β, or dihedral effect. The enhanced dihedral led to the Dutch roll divergence problem. The design was changed so that the winglets pointed downward from the wing tip. Richard Whitcomb had not considered this downward winglet orientation for manned aircraft because he was worried about ground strike during takeoff and landing. However, there were no aerodynamic reasons not to use this orientation on this unmanned aircraft that did not use runways in operations. The new configuration was tested on the 3/4-scale wind tunnel model, and the Dutch roll problems appeared to be solved. In practice, the new winglet configuration slightly increased the negative (nose-down) pitching moment on the wing, which increased the load on the wing actuator. With the downward winglets, the rear wing wingerons produced proverse yaw with differential deflection for roll.

After the winglet configuration change, the LODED flight test program continued. The aircraft was able to successfully take off and it was controllable in flight. However, actuator failures developed as the aircraft accumulated flight time and the weight increased due to design improvements. The higher weight required the aircraft to fly faster. This increased the dynamic pressure on the wings and therefore the wing torsion loads also increased. The loads on the actuators became excessive, leading to the failures. The new winglet configuration also contributed to the actuator failures. To solve the problem, NRL ganged two actuators to each wing panel instead of a single actuator. This was done in such a way that each actuator provided half of the wing incidence change. A complete failure of a single actuator would result in the loss of half the wing incidence deflection. The compromised aircraft would still be functional and able to continue flight and land.

The radio controlled (RC) tests used a custom fail-safe system that caused the aircraft to crash in the event of a lost communications link. Fail-safe systems are a form of flight termination system that is common for small unmanned aircraft flight test programs. It should be noted that inadvertent activation of flight termination systems has led to numerous crashes for various unmanned system test programs. The custom NRL fail-safe system activated a flight termination control mode when the link was lost. The system was intended to restore controlled flight if the link was reestablished. However, the system malfunctioned and locked the pilot out during an intermittent signal loss. This resulted in the loss of one LODED vehicle. The fail-safe system was fixed for later flights.

NRL instrumented the 56.25% scale model. The NRL custom flight test boom was a two-vane system that measured sideslip and angle of attack. A pitot probe was at the tip of the boom, and there were static ports on both sides of the fuselage.

A ground receiver unit received the commanded control deflection commands. Actuators driven by the receiver drove servos, which were connected to a needle and protractor. The movement of these servos was recorded on video. The ground-based servos were not under the same loads as the flight servos. The actual aircraft control surface deflection was not measured, only the commanded deflection. Later, miniature analog potentiometers were placed on the servos and the outputs were plotted on a strip chart recorder.

The two LODED test pilots were Maynard Hill and Rick Foch. Hill was a metallurgist and aircraft designer employed by the Johns Hopkins University Applied Physics Laboratory. He was a pioneering figure in small unmanned aircraft development, and is perhaps best known for development of the X-Drone and numerous model aviation endurance and range records, including the first transatlantic crossing by an RC model aircraft. Hill taught Foch how to fly large unmanned aircraft, which have flying characteristics far different than smaller RC model aircraft.

The final flight of the 56.25% scale model took place in 1982. NRL was ready to continue development of the full-scale aircraft, but the sponsor lost interest in the design. The LODED program was originally planned to take two years, but the effort had already taken almost four years. LODED sponsors felt that more development work needed to be done to demonstrate a reliable production prototype, and elected to stop supporting the project. The mission LODED three-view drawing is shown in Fig. 2.23, and characteristics of the subscale and mission LODED vehicles are listed in Table 2.6.

The LODED program made contributions to several technologies, including

- Modular vehicle concept that can fit in a box
- Plug-in wing mechanisms
- Lower winglets for directional stability and without Dutch roll divergence

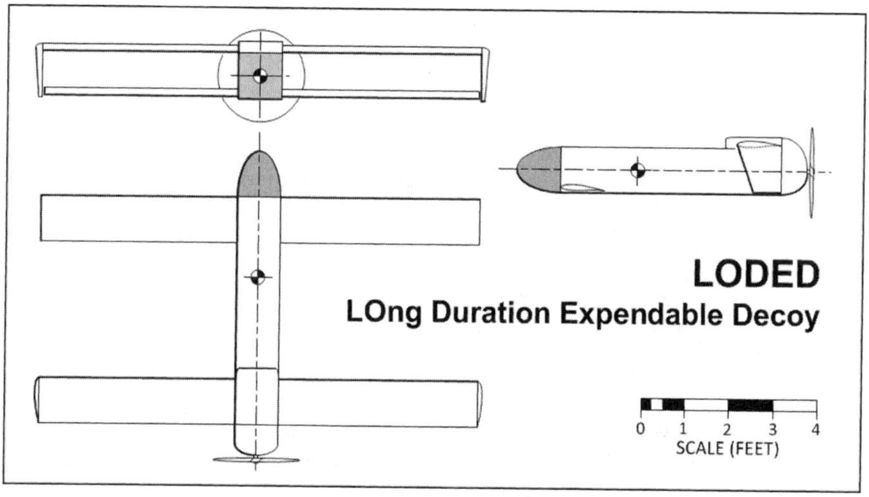

Fig. 2.23 Mission LODED three-view drawing.

Table 2.6 LODED Technical Characteristics

Parameter	56.25% Scale Value	Mission LODED Value
Wing span	5.66 ft	10 ft
Wing area	6.4 ft^2	20 ft^2
Length	4.08 ft	7.1 ft
Fuselage width	6.75 in.	12 in.
Takeoff weight	25.0 lb	140.0 lb
Empty weight	13.0 lb	70.0 lb
Payload weight	6.0 lb (nose ballast)	40.0 lb
Fuel weight	0.75 lb	30.0 lb
Loiter speed	33 KTAS	38–42 KTAS
Dash speed	64 KTAS	118 KTAS
Endurance	<0.25 hr	9 hr
Maximum L/D	13.5	14.0
Airfoils	Wortmann FX 63-137 on both wings and winglets	Wortmann FX 63-137 on both wings and winglets
Launch method	Conventional wheeled takeoff for early tests. Pneumatic rail launcher for later tests.	Pneumatic rail launcher
Recovery method	Conventional wheeled landing for early tests. Expendable for later tests.	Expendable
Propulsion type	Two-cycle reciprocating engine	Two-cycle reciprocating engine
Power	1 hp	20 hp

- First use of Whitcomb winglets to enhance aerodynamics and directional stability
- Successful catapult launcher integration

Interim Active Electronic Decoy (IAED), 1980–1982

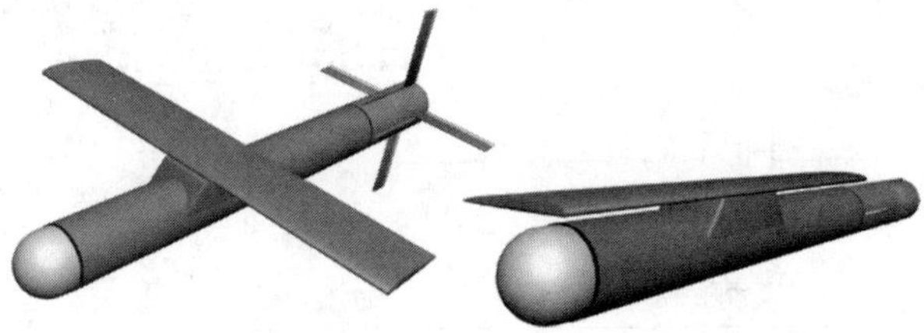

Fig. 2.24 IAED renderings.

The Navy was interested in a more advanced ship-launched EW decoy after the AED program. The Interim Active Electronic Decoy (IAED) program was intended to provide an initial capability while the technology was being advanced. The IAED aircraft still utilized a boost-glide approach (Fig. 2.24), as in the AED program, though a final, production system would use longer-endurance propulsion.

The IAED program proceeded as two separate phases. The first phase, conducted in 1980, consisted of the development of two NRL designs for IAED. The goal for the NRL designs was to establish a baseline to help assess industry designs for the second phase.

The two NRL IAED designs (IAED and IAED-2) had a single-piece pivot wing with a short span, and four tail surfaces that folded out of the fuselage. The wing was a relatively simple design, to improve reliability, although this simplicity came at the expense of endurance, due to the wing's limited span and area. IAED-2 had unique off-center wing pivot geometry to place the wing in the proper position with respect to the CG. This clever feature, invented by Dr. Klemm, allowed the wing to shift aft when it unfolded, all while rotating about a single, simple pivot.

The IAED fuselage had a constant 5.12-inch-diameter circular cross section. IAED-2 had an elliptical cross section enclosing the 5.12-inch-diameter cylindrical volume, so it could enclose a larger payload than IAED. The remaining volume alongside the payload was to be filled with custom-made, kidney-shaped lithium thermal batteries. This type of battery is explosively activated to melt its electrolyte and provide very high power density. Today, lithium thermal batteries are considered too dangerous for Navy shipboard use. The NRL IAED and IAED-2 were designed to have a flight duration of 1 min, giving a no-wind range of approximately 1 nautical mile.

Three vendors participated in the second phase of the IAED program from 1980 to 1982. E-Systems carried forward their earlier flexible-wing AED aircraft, and DSI offered the single-piece scissor wing concept from their AED also. Little documentation remains for the E-Systems and DSI designs.

The third vendor, Boeing, provided an innovative, injection-molded polyester the thermoplastics would airframe. Thermoplastics were injected into molds as hot liquid, and when cooled to take the final geometry, the thermoplastics would not take a new shape if reheated. The wings were folded forward on top of the fuselage when stowed. The wings pivoted back such that they had no sweep when deployed. The fuselage had a square-cross-section main fuselage, which transitioned to a smaller boom of square cross section. Multiple tails folded along the aft fuselage near the sustainer rocket motor. The avionics were located aft of the wing pivot. The battery was located ahead of the wing, and the payload was located at the nose.

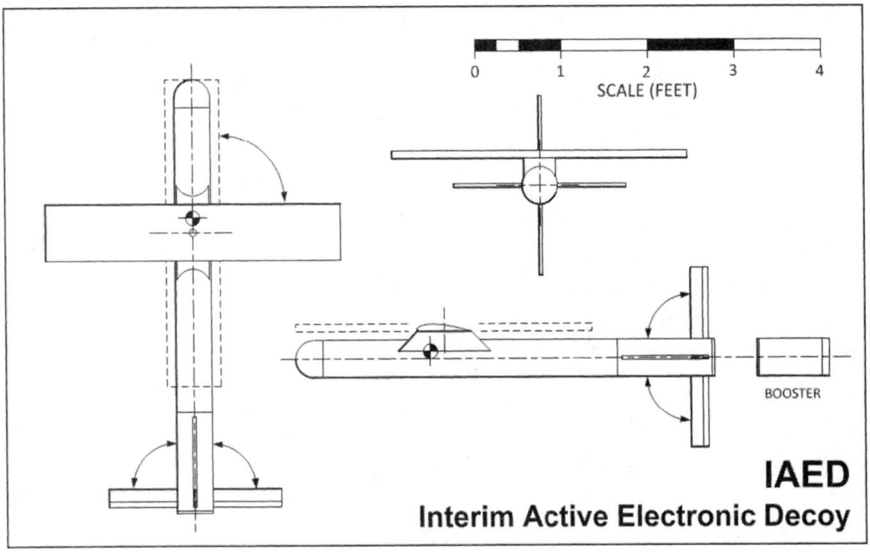

Fig. 2.25 NRL IAED three-view drawing.

The models underwent extensive calculations, flight simulations, and wind tunnel testing. Also, full-size engineering mock-ups were built. However, none of the IAED aircraft flew under this program. IAED (Fig. 2.25), IAED-2 (Fig. 2.26, Table 2.7), and the industry IAEDs were superseded by the SLED and READY programs (discussed later in this chapter).

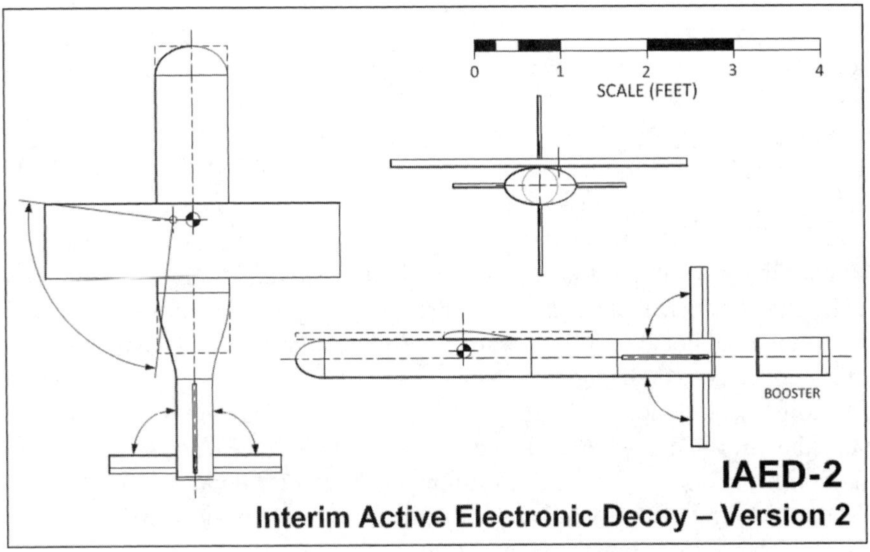

Fig. 2.26 NRL IAED-2 three-view drawing.

Table 2.7 NRL IAED and IAED-2 Technical Characteristics

Parameter	IAED Value	IAED-2 Value
Wing span	3.41 ft	3.42 ft
Wing area	2.14 ft^2	2.84 ft^2
Length	5.67 ft with booster	5.67 ft with booster
Launch weight	35.0 lb	45.0 lb
Empty weight	10.0 lb	10.0 lb
Payload weight	15.0 lb	25.0 lb
Glide weight	25.0 lb	35.0 lb
Cruise speed	60 KTAS	70 KTAS
Endurance	1.0 min	1.0 min
Range	1 nmi	1.2 nmi
Launch method	Rocket launch	Rocket launch
Recovery method	Expendable	Expendable
Propulsion type	Solid fuel propellant	Solid fuel propellant

SHEAR WING RF DECOY, C. 1981

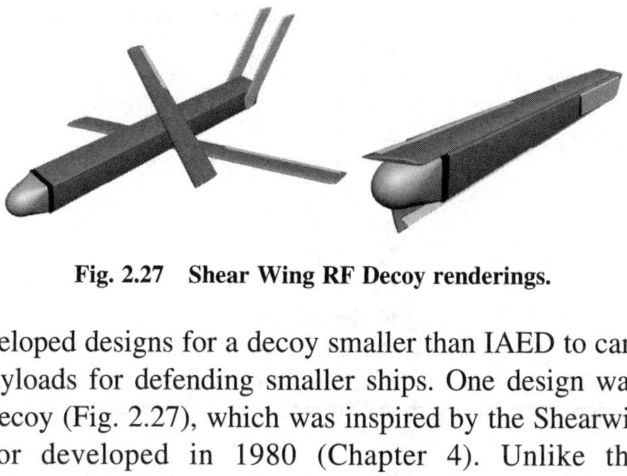

Fig. 2.27 Shear Wing RF Decoy renderings.

NRL developed designs for a decoy smaller than IAED to carry reduced-size EW payloads for defending smaller ships. One design was the Shear Wing RF Decoy (Fig. 2.27), which was inspired by the Shearwing research demonstrator developed in 1980 (Chapter 4). Unlike the research demonstrator, the decoy had two vertical tails on the fuselage to provide stability during launch when the wings were closed. The Shear Wing RF Decoy would be rocket launched from an MK 36 decoy launcher. The design was not pursued because it was thought to be too radical at the time. A convenient rationale was the crash of the Shearwing demonstrator, even though the crash was caused by pilot error and not a design flaw. Figure 2.28 shows a three-view drawing of the Shear Wing RF decoy, and its technical characteristics are listed in Table 2.8.

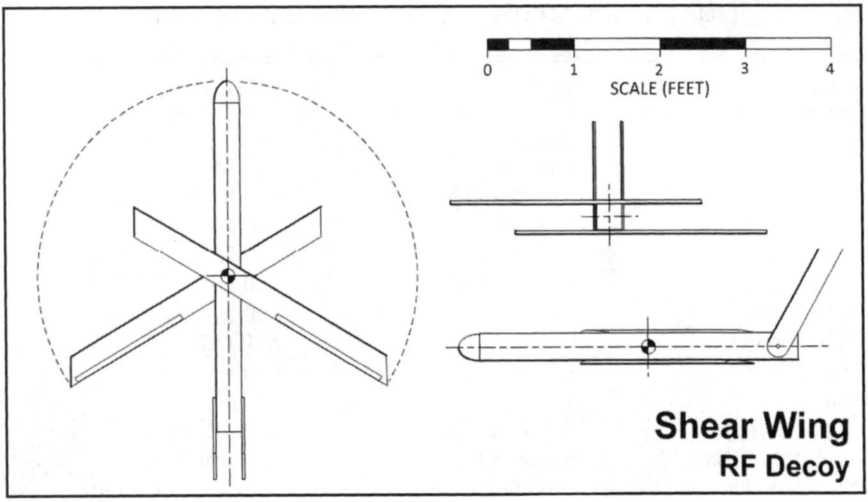

Fig. 2.28 Shear Wing RF Decoy three-view drawing.

Table 2.8 Shear Wing RF Decoy Technical Characteristics

Parameter	Value
Wing span	3.75 ft
Wing area	2.3 ft^2
Length	4.83 ft (including booster)
Takeoff weight	28.25 lb
Payload weight	8.6 lb
Cruise speed	56 KTAS
Endurance	1 min
Launch method	Rocket launch
Recovery method	Expendable
Propulsion type	Solid fuel propellant

SHIP-LAUNCHED ELECTRONIC DECOY (SLED), 1984–1985

Fig. 2.29 Engineering mock-up of SLED.

Ship-Launched Electronic Decoy (SLED) was intended to be a more advanced version of IAED. A key improvement was the development of a sustainer rocket motor for longer-endurance flight (Fig. 2.29). The sustainer motor could operate for several minutes, which enabled more sophisticated maneuvers and longer-duration ship protection. Improved autopilots were to be developed under this program to support the flight maneuvers.

The booster and sustainer motor were built as a single unit. The booster motor burned from the core to the casing, and it had a star pattern to increase the surface area for the burn. This provided high thrust for a short duration. Once the booster phase burned out, a sustainer motor took over. The sustainer motor had a recast grain. This motor had an end burn, in which the burn occurs along the entire diameter and evenly burns until the fuel is expended. A key challenge in the sustainer motor development was to prevent erosion along the sides to ensure that the cross section burned evenly. The motors were tested at the Naval Ordnance Station (now the Naval Surface Warfare Center), Indian Head, Maryland. Motor development took place from 1982 to 1984. This dual-operation rocket motor was a critical technology that enabled the SLED program.

NRL conducted conceptual design studies of a new platform, using the Boeing IAED configuration as a baseline for analysis, and conducted 6-DOF simulations of the design concept.

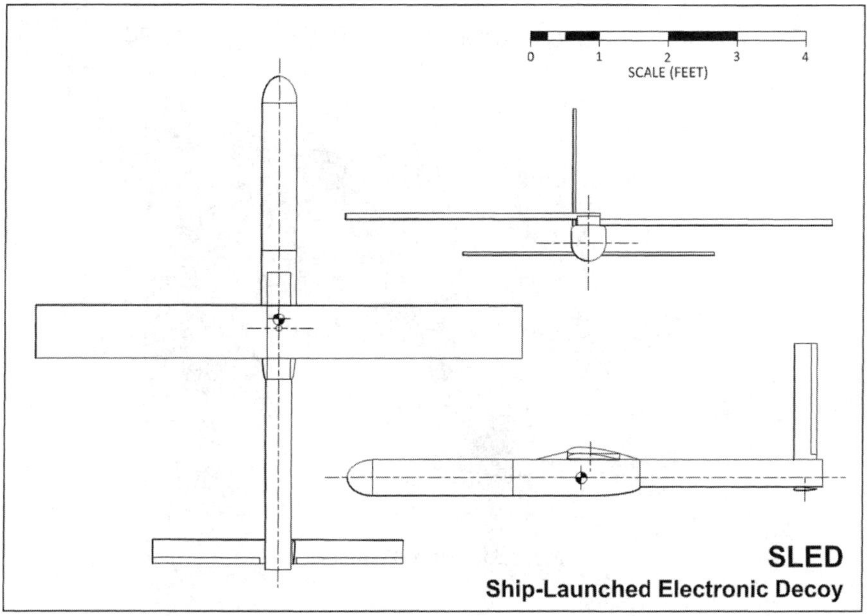

Fig. 2.30 SLED three-view drawing.

Table 2.9 SLED Technical Characteristics

Parameter	Value
Wing span	7.0 ft
Wing area	5.25 ft^2
Length	6.9 ft
Launch weight	55.0 lb
Empty weight	45.0 lb
Payload weight	20.0 lb
Cruise speed	50 KTAS
Endurance	3 min
Max L/D	14.0
Launch method	Solid propellant rocket launch
Recovery method	Expendable
Propulsion type	Rocket glider

NRL planned to contract with industry to develop SLED when funding became available, but sponsorship was not forthcoming and the program ended in 1985. In its place, development of a joint U.S.-Australian concept, the NULKA hovering rocket decoy system, was initiated. A SLED three-view drawing is shown in Fig. 2.30, and its technical characteristics are listed in Table 2.9.

Radar Expendable Active Decoy (READY), c. 1984

Fig. 2.31 READY model.

The ROCN (Taiwan) Navy approached NRL about co-developing a boost glide decoy similar to AED/SLED. The system would carry a small EW payload based on the AAED miniature TWT payload (Chapter 3). ROCN had to defend only smaller ships, so the payload could be smaller than that of AED/SLED. The plan was that NRL would design the system (Fig. 2.31) for the Radar Expendable Active Decoy (READY) and Taiwan would manufacture and operate it.

The original assumption was that the system had to be simple to manufacture, with tooling kept to a minimum. However, it was discovered that the Taiwanese customer had sophisticated aerospace manufacturing capabilities, as seen on license production of the Northrop F-5E Tiger fighter.

The READY fuselage had a 5.125-inch circular cross section near the front, and a square cross section aft of the wing. Scallop-shape fairings helped the transition between the two different cross-section geometries. The fairings were inelegant, but they served the purpose. The two fuselage sections used standard-size aluminum extrusions for simplicity.

The tail section had two vertical stabilizers and two horizontal stabilizers. These tail sections folded forward along the square fuselage edges when stowed. The folding geometry resulted in slightly asymmetric tail geometry. This asymmetry was viewed as risky at the time, although later NRL asymmetric designs would prove the arrangement to be workable.

The wings folded forward on top of the fuselage when stowed, stacked on top of each other to share the same pivot. The wings pivoted back during launch. A pop-up fairing reduced drag at the leading edge of the wing-fuselage

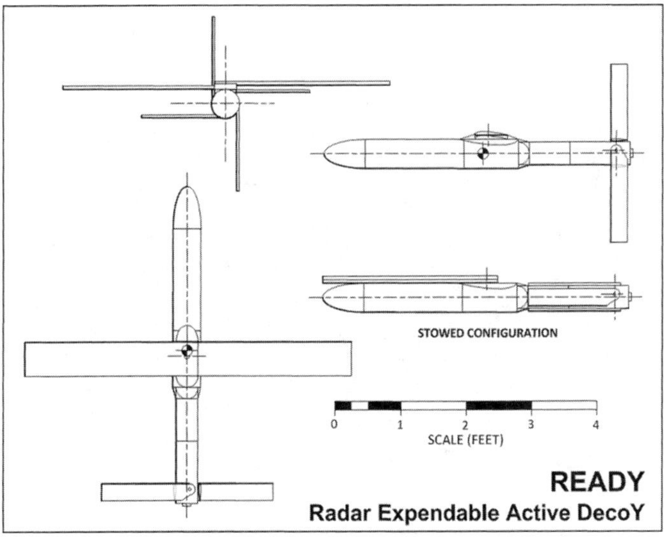

Fig. 2.32 READY three-view drawing.

Table 2.10 READY Technical Characteristics

Parameter	Value
Wing span	5 ft
Wing area	2.5 ft^2
Length	5.58 ft (including booster)
	4.63 ft (without booster)
Launch weight	29.0 lb
Empty weight	25.0 lb
Payload weight	8.0 lb
Cruise speed	50 KTAS
Endurance	1 min
Max L/D	12.0
Launch method	Solid propellant rocket launch
Recovery method	Expendable
Propulsion type	Rocket glider

interface. The system was to be launched from an MK 36 launcher. The booster was jettisoned after launch. Figure 2.32 shows a three-view drawing of READY, and Table 2.10 lists its technical characteristics.

NRL wrote a detailed proposal, but the program was not funded. The concept may have provided sufficient technology to allow similar indigenous designs to be developed. NRL built an aluminum desk model, which was the only hardware directly associated with the program. The design experience helped influence later programs such as FLYRT ATD, discussed next.

Flying Radar Target Advanced Technology Demonstrator (FLYRT ATD), 1991–1993

Fig. 2.33 FLYRT ATD in the anechoic chamber to measure antenna performance.

The purpose of the Flying Radar Target Advanced Technology Demonstrator (FLYRT ATD) program was to demonstrate the feasibility of an electronic decoy to defend ships against advanced RF-guided anti-ship missiles. The Naval Sea Systems Command (NAVSEA) sponsored the ATD to conduct a flight demonstration of the all-up system in a controlled, at-sea environment with simulated threat radars. FLYRT carried an advanced EW payload that would create a credible, but false, radar target. FLYRT was an expendable countermeasure, so it was not recovered at the end of the flight. It was estimated that this EW decoy system minus the electronic payload could have been produced for $50,000 per unit in 1991 dollars.

FLYRT (Fig. 2.33) was launched with a solid fuel rocket motor from an unmodified MK 36 Decoy Launching System (DLS), a standard launcher on the decks of all Navy ships. FLYRT fit in a 9-inch by 9-inch by 78-inch package, the maximum allowable dimensions that enable the firing of all rounds from the MK 36 DLS. The rocket motor was modified from an MK 137 chaff round mortar to reduce the shock loads due to its ignition occurring inside the closed breach barrel. The slower burn rate motor reduced the initial shock spike to 10,000 *g* followed by a 1000-*g* push until clear of the barrel. The shock was absorbed locally by the structure, so the impact on the design was not significant. The rocket booster generated an average axial acceleration of 10 *g* after clearing the launcher. The airframe had sufficient strength for the launch loads when folded and latched.

NRL designed three initial configurations. All underwent extensive wind tunnel testing. The best two configurations were selected for further evaluation. The final configuration was selected based on performance and deployment characteristics.

Multiple wind tunnel models were tested. A half-scale model was used to evaluate the aerodynamic characteristics during launch. The angle of attack and sideslip sweeps went up to 68 degrees to support the aerodynamic database. A full-scale aft fuselage model helped characterize the aerodynamics of tail deployment and the structural integrity. Deployment was tested at the 46 feet per second launcher exit velocity, and the deployed configuration was tested at the 80 feet per second flight speed. The propulsion modules were tested in a full-scale propulsion test model that consisted of a cowling and a shortened aft fuselage. The propulsion performance was evaluated with instrumentation measuring propeller rpm, thrust, torque, airspeed, motor temperature, motor voltage, and current.

The three configurations considered for the FLYRT ATD were a pylon-mounted high wing, a low wing, and a negative stagger multi wing.

The pylon-mounted high wing configuration folded its two wing panels along the side of the fuselage while the wing center section, which was attached to the pylon struts, rotated aft to sit atop the fuselage (Fig. 2.34).

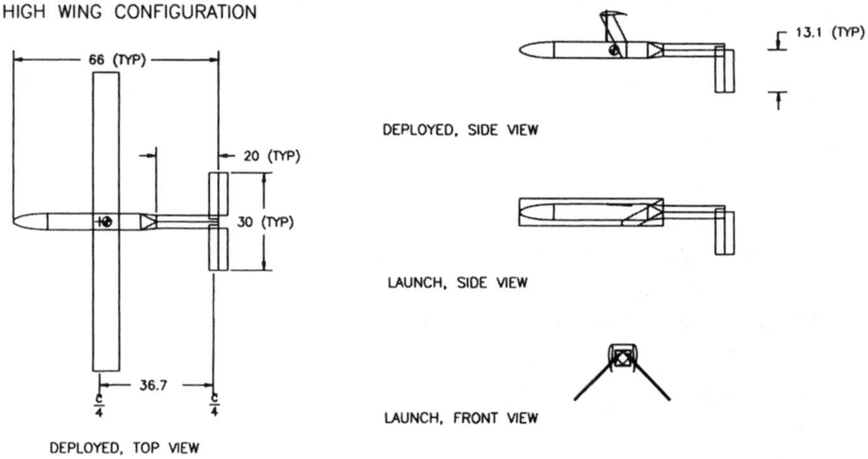

Fig. 2.34 High wing FLYRT ATD configuration.

This configuration was the first to be wind tunnel tested and had acceptable flying characteristics but had the lowest lift-to-drag ratio (L/D) due to the drag of the pylon struts. Tests also determined that while the inverted V-tail was good for the stability and control characteristics in flight, the control was inadequate during the ballistic launch phase. The V-tail could provide only mixed yaw/roll and pitch control, but fully independent three-axis control was needed. Consequently, although all three configurations were initially tested with the inverted V-tail, an X-tail layout was selected and provided the required control response for all flight modes.

The low wing configuration was shown to have the best L/D, as expected, but required a skewing center wing panel with two unfolding wing tips, which was more complex to sequence and deploy than the high wing or multi wing designs (Fig. 2.35). Its higher L/D was viewed favorably since it would give more latitude for antenna integration, which would feature high-drag cylindrical elements.

The negative stagger multi wing configuration enabled simple wing deployment by the single rotation of four wing panels (Fig. 2.36). It also had the largest wing area of the three configurations. This provided superior endurance but a lower L/D, which reduced range. As the system trade-offs were performed, issues arose with coordinating the wing panel opening. Each right/left pair had to open symmetrically, and because wing sweep was needed for trim, the forward pair of wings had to rotate 80 degrees simultaneously with the rear wings, which would be rotating 100 degrees. Mechanical linkages to coordinate the wing panel unfolding were relatively simple but also added a slight weight penalty.

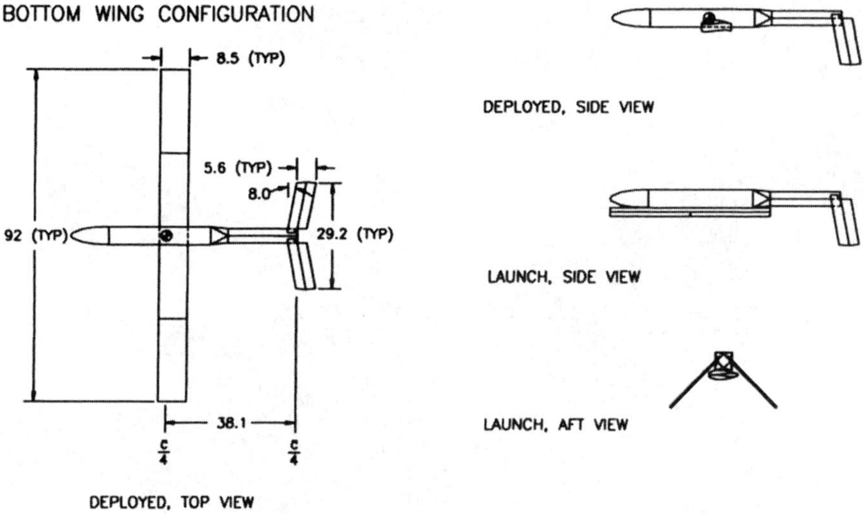

Fig. 2.35 Low wing FLYRT ATD configuration.

The low wing and multi wing configurations were superior to the high wing configuration. Each could be developed into a successful UAS. The low wing configuration had fewer components (Fig. 2.37), the highest L/D, and a simpler, albeit higher-risk, wing deployment. It was chosen for the FLYRT ATD.

FLYRT ATD was a relatively conventional low wing tractor monoplane. Its wing had three segments with dihedral in the outboard panels. The wing was mounted below the fuselage. The motor and propeller were mounted in the front of the fuselage. The tail had an X configuration.

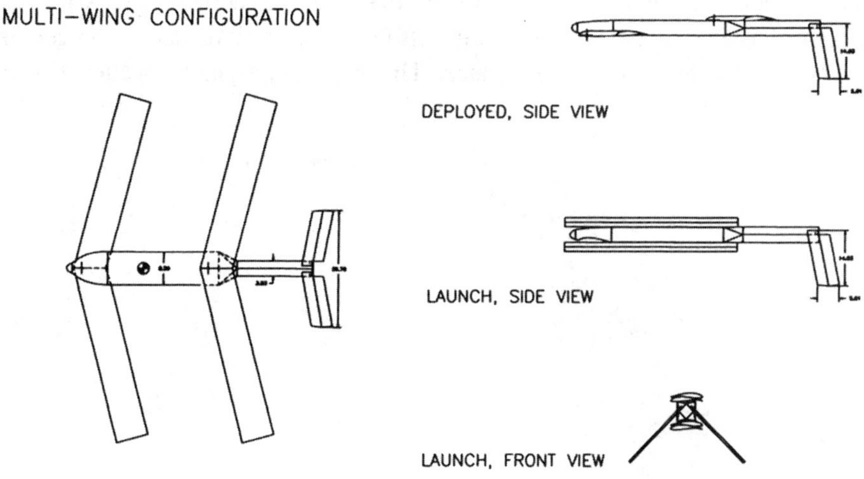

Fig. 2.36 Multi wing FLYRT ATD configuration.

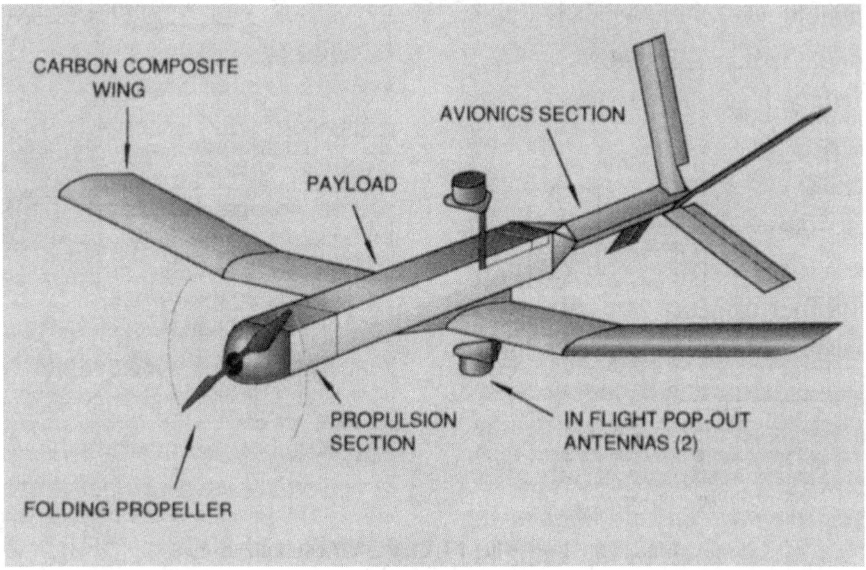

Fig. 2.37 FLYRT ATD low wing components.

The aircraft was designed to unfold after launch with a highly controlled number of steps (Fig. 2.38). The tails unfolded after clearing the launcher. The tails provided stability during launch, and their control surfaces were actuated to provide full three-axis control to command the aircraft toward an optimal launch trajectory. Each of the four tail surfaces was unfolded by torsional springs. They lay along a diamond-cross-section aft fuselage when stowed. The outer wing panels folded under the center wing panel prior to launch, and the center wing panel was rotated along the fuselage. After the spent booster separated from the aircraft via an explosive charge, the center wing pivoted and locked into place. The wing deployment occurred near

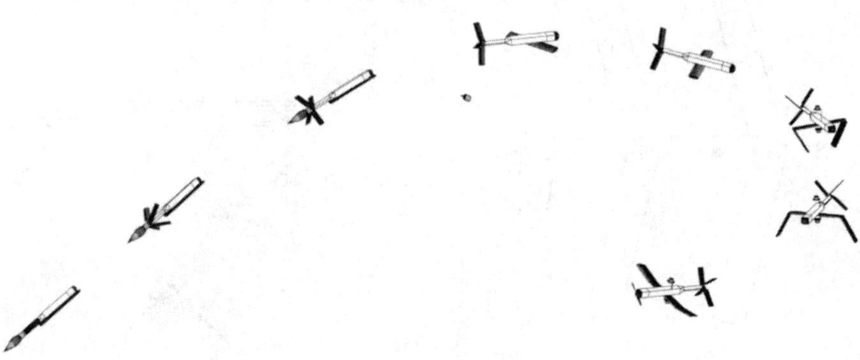

Fig. 2.38 FLYRT ATD deployment concept.

launch apogee to minimize loads. The pivot for the center wing was laterally asymmetric such that the pivot was about the quarter chord of the wing, but the wing could also occupy equal width along the fuselage when stowed. The asymmetric hinge also allowed inertial forces under acceleration to pivot the wing. Next, the outer wing panels unfolded about simple piano hinges and used aerodynamic forces to deploy and latch into position. The outer wing deployment approach was inspired by the success of the Variable Span LAURA and Venture vehicles (Chapter 4). The propeller folded along the fuselage for launch, and automatically unfolded when the motor started. Upper and lower payload antennas were each connected by a single arm that pivoted into place. Their deployment was spring driven.

Launch reliability is inversely proportional to the number of mechanisms, so the number of hinges and latches was minimized. A competing design need was to have a high-performance aircraft, which favored a larger airframe, which would be enabled by a greater number of folds. Therefore, there was a design trade-off between complexity and flight performance. In general, the performance of a folding aircraft is volume limited, not weight limited.

The launch sequence was carefully controlled. FLYRT followed a precise pitch profile. Angle of attack, roll, and yaw were commanded to open the wing. The yaw rate was controlled to ensure that both wings opened simultaneously. Initially the outer wing panels were to have potentiometers to measure opening position as a feedback loop for the yaw commands, as demonstrated in the LAURA and Venture projects, but this was later eliminated. The control system tended to zero sideslip and angle of attack, but it would yaw to the lagging wing to facilitate opening. The deployment took 4.5 to 5 seconds. FLYRT would begin its mission flight profile once the deployment was complete.

The FLYRT wing used a modified Wortmann FX 63-137 airfoil. This airfoil has a high lift coefficient, which enables low flight speed with a small wing. This airfoil also has an impressive endurance parameter ($C_L^{3/2}/C_D$), which reduces the power required for flight. The standard FX 63-137 has a very thin trailing edge cusp. NRL increased the trailing edge thickness to make the wing easier to manufacture and more rugged. The trailing edge thickness was approximately 1% of the chord length. Even with this modification, there was no measurable degradation to the airfoil performance. The thickness was added evenly to the upper and lower surface. Thanks to the advanced airfoil, FLYRT ATD had a maximum L/D of 20 without the antenna. The drag of the antenna reduced the L/D to 10 when deployed, due to a doubling of the drag.

All the control surfaces were in the X-tail. The FX 63-137 airfoil was not effective with flaps due to local separation on the trailing edge, so the wing had no control surfaces. The generous outboard panel dihedral provided effective dihedral effect. The tails provided yaw, which also produced a favorable rolling moment. The X-tails provided 3-DOF roll, pitch, and yaw control when the wings were stowed during launch under rocket power. The controls acted in pitch and yaw during winged flight. The control gains and scheduling were varied as needed throughout the launch sequence and flight.

The electric propulsion system was advanced for the time. The first propulsion system consisted of a 15-turn Astro Flight Cobalt 60 rare earth magnet brushed DC motor. The propeller had a 7-inch hub that cleared the fuselage, plus two folding 8.5-inch carbon blades. The overall system efficiency was 70%. The early flight tests revealed the need for a more powerful propulsion system. The brushed motor system was replaced by a custom-made, geared, nine-turn Aveox brushless motor with a gearbox. The 700-watt motor had 92% efficiency, and the overall system efficiency was 85%. This propulsion system produced 8.3 pounds of thrust at 54 knots. This was NRL's first use of a brushed neodymium-iron-boron magnet motor and may have been the first application of brushless motors to a Navy unmanned aircraft.

The Saft silver-zinc primary battery (nonrechargeable) achieved 125 watt-hours per kilogram specific energy under the flight discharge profile. Low-cost, commercially available rechargeable lithium polymer cells offer greater specific energy today, but this technology was not available during the FLYRT ATD program. The silver-zinc battery enabled the flight endurance while satisfying safety requirements. The battery powered all onboard systems, including the propulsion system, EW payload, automatic flight control system (AFCS), and communications. A single power bus was used.

NRL developed a custom digital autopilot for this program. Rugged and miniaturized GPS was not yet available. NRL's Optical Sciences Division developed an inexpensive, three-axis, fiber-optic laser rate gyro (Fig. 2.39). Vehicle attitude was estimated by integrating angular rates. The gyro had less than 1 degree of rate-integrated drift after a 20-minute flight, which was sufficient for the decoy mission. The position of FLYRT relative to the ship was more important than geolocation.

The fuselage was constructed of formed aluminum alloy sheet. The aluminum structure helped withstand the launch shock loads. It also helped with EMI protection of critical electronics in the presence of ship radars and the EW payload antennas. All avionics were housed within the metal fuselage. The forward fuselage had a rectangular cross section, which transitioned to a diamond cross section for the rear fuselage. The cross-section transition section was comprised of triangular facets. These facets generated vortices that helped keep the flow attached.

Fig. 2.39 FLYRT ATD fiber-optic laser rate gyro.

The wing was constructed of composite materials (Fig. 2.40). The outer skins and shear web were carbon fiber. High-density foam with high shear strength filled the structure. To get the necessary surface quality, strength, and manufacturing repeatability, the structure was built in a female mold. The FX 63-137 airfoil required tight tolerances. The tails were made of

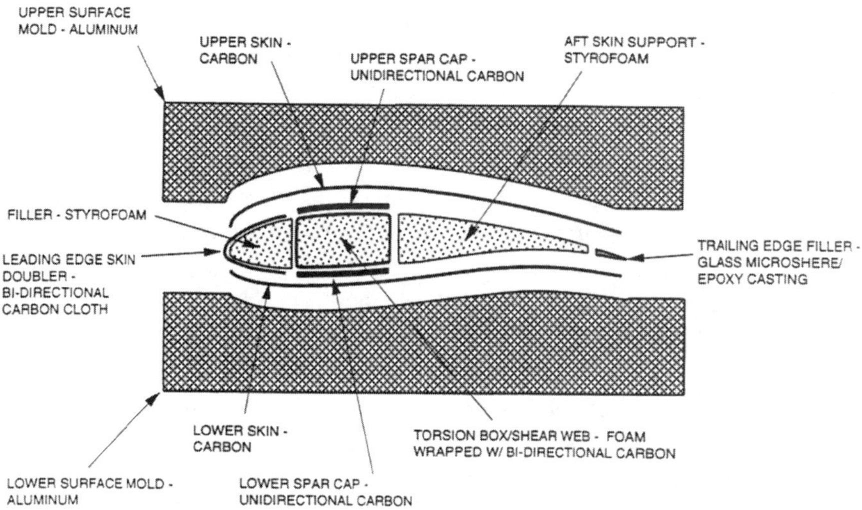

Fig. 2.40 FLYRT wing construction.

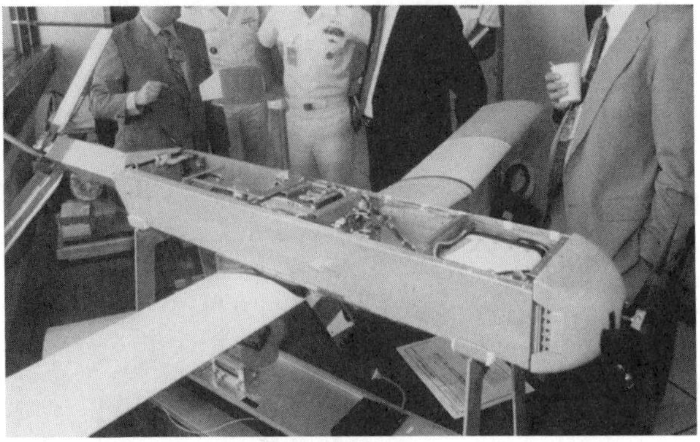

Fig. 2.41 Internal view of the FLYRT ATD fuselage.

carbon fiber over a balsa laminate to save weight. The extensive use of carbon fiber helped reduce airframe weight.

The electric motor and motor controller were located at the nose of the aircraft. The battery was in the forward fuselage ahead of the wing (Fig. 2.41). The EW payload was located aft of the battery. The laser rate gyro was housed in the fuselage transition region behind the wing. The autopilot boards were aft of that, in the rear fuselage, and the tail actuators were aft of the autopilot.

The control surfaces folded around the diamond fuselage cross section. A clever pushrod and linkage arrangement allowed the tails to unfold in a neutral position without adjusting the arm length. This difficult mechanical arrangement took an engineer one full month to design.

A number of 2/3-scale RC drop models of the selected low wing configuration were built. The models supported low wing and tail deployment tests and autopilot development, and provided qualitative validation of wind tunnel performance data. The drop models were built during the full-scale aircraft design phase, so lessons learned from the drop models could affect the final vehicle design. The drop models used rate gyros to improve damping. First, a fully unfolded FLYRT aircraft was dropped from NRL's NDM utility aircraft to evaluate handling qualities. Next, a drop model was released with the wing tips folded. After a few minor problems were resolved, the next drop model tested the pivoting center wing and unfolding outer panels. Finally, the aircraft was launched from a fully folded condition. The pilot sequenced the wing and tail deployment. The change from three-axis to two-axis tail control mixing was done using a toggle switch on the RC transmitter. For three-axis control, the right stick controlled roll and pitch, and the left stick controlled yaw and throttle. For two-axis control, the right stick controlled yaw and pitch, and the left stick controlled

Fig. 2.42 The NDM aircraft carrying a 2/3-scale FLYRT ATD drop model.

throttle. The RC controller manufacturer Airtronics assisted NRL in programming the radio control system.

Avionics testing was conducted using the NDM utility aircraft (Fig. 2.42). The airspeed sensors, attitude sensors, data logger, telemetry system, and autopilot boards were integrated into NDM, and autopilot modes were tested. Early EMI/RFI problems were encountered and resolved prior to testing on the FLYRT airframes. The SARA aircraft (Chapter 4) was used to test the electric propulsion system components and to support the battery and motor candidate downselection.

A 6-DOF flight simulation environment was used to develop the flight controls. The aerodynamic and propulsion data generated through the wind tunnel testing was applied. The simulation was updated based on flight test data to help refine the control laws. The deployment sequencing was evaluated in the simulator prior to testing.

A captive carry rig erected on a truck was used to evaluate the wing deployment prior to flight testing. This method was previously used on the LAURA program. The FLYRT was mounted forward and above the truck such that interference effects were minimized. The airframe was free to pivot in pitch, but it was constrained in roll. The truck traveled at the 65-mph design speed when the wing deployment was commanded. The test verified the wing deployment approach and the structural integrity.

The payload underwent extensive testing prior to the final demonstration. Brassboard payload prototypes were tested in an RF laboratory against threat simulators, and these helped to refine the payload design. Airborne captive carry tests were also conducted: The FLYRT aircraft was equipped with the payload and slung below a helicopter on a tether (Fig. 2.43). This provided a

Fig. 2.43 Vehicle-based captive carry rig test (left) and helicopter captive carry test (right).

representative tactical defense scenario, and it was tested in the vicinity of several Navy ships to test the defensive capability (Fig. 2.44).

The FLYRT flight test program was the most methodical yet conducted by NRL. The systems had detailed checklists. Flight test reports were generated immediately after the flight. The flight tests followed standard procedures influenced by the Society of Flight Test Engineers.

Table 2.11 shows a summary of the FLYRT ATD flight tests. Ten all-up flight test systems were serially assembled at approximately one-month

Fig. 2.44 FLYRT ATD in flight.

intervals. Each new system featured modifications based on lessons learned and data analyzed from each preceding test. A recovery system would have compromised the mission capability, so all flight test systems were expended (destroyed) in testing while preparing for the final demonstration. Two complete demonstration systems were assembled for the final demonstration.

Following the final demonstration flight, which ended in alighting onto the Chesapeake Bay and sinking, the FLYRT "flew" a long distance underwater. Like a sea glider, FLYRT gradually sank as its interior flooded, but with an L/D of 10. The rugged airframe was not damaged during the water impact, since it was designed to withstand a large launch shock. The CG did not change when the fuselage flooded. It traveled significantly farther than anticipated. Navy divers spent a considerable amount of time locating the aircraft, and it was recovered intact but badly damaged due to salt-water corrosion.

After the successful final demonstration (Figs. 2.45 and 2.46), the final report was written and the program ended. The NAVSEA sponsor felt that FLYRT ATD was a complete success. The last surviving aircraft was placed on display at NRL instead of being flight tested. The technology was proven, so losing the last expendable aircraft would serve no purpose. While FLYRT ATD would have been a valuable system for ship protection, the system did not transition to production or operations. FLYRT ATD was NRL's first Advanced Technology Demonstration program. It represented the culmination of two decades of technology investment that enabled a mission demonstration of a practical AED for the Fleet, first envisioned by Dr. Montgomery during the formation of the TEW Division in 1971. Prior to this program, NRL had primarily built small demonstrators and mock-ups to

Fig. 2.45 FLYRT ATD launch sequence.

Table 2.11 FLYRT ATD Flight Tests

Flight	Description
Fire 9	January 15, 1993. The purpose of the test was to demonstrate tail deployment and control in the boost phase. The wing was bolted in the stowed position, and the rocket booster was permanently mounted to the fuselage. The aircraft was ballasted for the launch configuration CG position. The tails opened after launch. The pilot commanded roll via radio control. The aircraft flew a ballistic trajectory into the ground as planned.
Fire 10	February 29, 1993. This test built upon Fire 9. In addition to testing booster operation and tail deployment again, this test was planned to verify umbilical separation and boost phase autopilot performance. The captive carry wing deployment tests were completed by the time of this test, so the wing deployment was also tested in this flight. Since the booster separation was not yet available, the CG was far aft of the flight position. The launch and tail deployment performed as planned. The wing deployment failed. The vehicle rolled to the right and yawed to the left when the center wing rotated into position. The left outboard wing panel latched into place first, which further increased the right roll rate. The right outboard wing broke away from the aircraft under the high loads. The aircraft rolled at a high rate and crashed. This behavior was not observed on the captive carry rig test because that model was constrained in roll. As a result of this test, the wing tip deployment was commanded separately from the center wing rotation. This permitted the yaw rate to damp out from the center wing rotation prior to deployment of the tips.
Fire 11	March 29, 1993. The purposes of this test were to evaluate autopilot performance, structural integrity of a fuselage modification, and the modified two-stage wing deployment sequencing, and to collect flight telemetry. The booster separator was not available. The test was successful in all objectives, including the wing deployment.
Fire 12	April 29, 1993. The purposes of this test were to evaluate the rocket motor separation system and survivability of the lightweight, thin-walled, U-channel fuselage. The model was configured as a ballasted RC aircraft, with the pilot controlling pitch and roll during launch. The wings were not configured for deployment. The test objectives were successfully achieved.
Fire 13	April 29, 1993. With the success of Fire 12, the Fire 13 airframe was tested on the same day. This test vehicle was configured with the motor separation unit, fully deploying airframe, autopilot, and electric propulsion. The rocket motor failed to separate. The CG was too far aft with the wings deployed, making the aircraft unstable in pitch. The aircraft was uncontrollable and it crashed. It was found that the bore rider alignment with the launcher umbilical arm caused the problem. The design was fixed for future flights.

Fire 14	May 18, 1993. The purposes of this test were to verify rocket motor separation, electric propulsion system, antenna deployment, and autopilot functionality, and to gather telemetry data. The rocket motor successfully separated. The deployment went as planned until the right wing tip deployment was delayed. The aircraft rolled inverted. The RC pilot overrode the autopilot but could not save the aircraft in time. The cause of the wing tip deployment failure was never determined, but the wing tip design was slightly modified as a precaution.
Fire 15	June 10, 1993. Fire 15 was a repeat of Fire 14, with the addition of updated autopilot software. As before, the launch was successful until the wing tips did not deploy simultaneously. The RC pilot successfully overrode the autopilot and was able to save the aircraft. The antennas deployed 65 seconds into the flight as programmed. The additional drag of the antennas caused the aircraft to descend at full power. The pilot landed the aircraft successfully, and onboard flight data was recovered. Following this flight, the NRL team conducted a drag reduction program that focused on all possible parasite drag sources. Key improvements included taping holes, adding internal bulkheads to reduce airflow through the fuselage, adding antenna splitter plates, and sealing all gaps with tape. The propulsion system power was modified to provide increased power.
Fire 16	July 27, 1993. This test vehicle included the three-axis, fiber-optic laser rate gyro and the upgraded brushless electric motor propulsion system. The antenna deployment was programmed to initiate 5 minutes into the flight to permit more time to gather flight test data. As with the previous two flights, the wing tips did not deploy properly. The RC pilot recovered the aircraft, and then returned control to the autopilot. The propulsion system appeared to lose power, so the RC pilot took over again. The vehicle stalled and crashed after 3 minutes of flight. The onboard data was recovered, but the telemetry system did not work. Data analysis revealed that the fiber-optic gyro (FOG) attitude estimation malfunctioned during launch and then recovered after motor separation. The loss of power was due to the autopilot adjusting the power setting for altitude control, which was not an error. The FOG was modified to better withstand the launch loads, and the autopilot was programmed to filter erroneous gyro data. From an investigation of flight data from several failed wing deployment tests, the development team concluded that the wing deployments occurred at too high an angle of attack. The stagnation point shifted to the lower wing, which generated sufficient force to prevent rotation. The autopilot's deployment program was modified to permit wing tip deployment at a lower angle of attack.

(*Continued.*)

Flight	Description
Fire 16B	August 13, 1993. This test repeated the objectives of Fire 16. This vehicle incorporated the modified deployment program. The launch was successful, and the wings successfully deployed. The modified propulsion system provided ample power for climb while the antennas were stowed. The autopilot entered an unstable phugoid mode while in the altitude hold loop. The RC pilot let the autopilot control the aircraft as long as possible in order to gather valuable test data. The pilot recovered the aircraft and then returned control to the autopilot. The antenna deployed on time, but it did not lock into place. There were several cycles of the autopilot stalling the aircraft followed by RC pilot recovery. The aircraft lost control during one of the stall recovery attempts. The flight had good onboard and telemetry data. Following this flight, the antenna deployment failure mode was corrected, propulsion power was augmented, and the autopilot software was updated.
Fire 16C	August 30, 1993. The purpose of this flight was to evaluate three autopilot control modes. The launch was uneventful. The RC pilot had to adjust the flight to correct for wind drift and for landing. Other than these minor manual override events, the autopilot had control over the flight. Two of the three autopilot modes were successful. The propulsion system provided sufficient power for climb. The antennas deployed and locked into place. The flight lasted for 12 minutes and 50 seconds. The flight had good onboard data and telemetry. This was the final land-based demonstration, and all test objectives were achieved.
Final Demo	September 9, 1993. This was the full system demonstration event over the Chesapeake Bay on NRL's Chesapeake Bay Detachment (CBD) Test Range. FLYRT was launched from a ship. The RC pilot was situated on a Boston Whaler downrange. The vehicle launched and the wings successfully deployed. About 10 seconds into the flight, the autopilot did not command the proper initial navigation maneuver. The RC pilot took over and flew the mission profile visually. The system properly engaged the threats. Despite the RC pilot override, the sponsor considered the demonstration to be a success, since complete autopilot functionality had already been demonstrated during the Fire 16C test.

Fig. 2.46 FLYRT ATD final demonstration launch.

mature the requisite technologies. The success of the FLYRT ATD and the experience gained by the development team paved the way for several additional NRL mission demonstrator UASs. Figure 2.47 shows a three-view drawing of FLYRT ATD, and Table 2.12 lists its technical characteristics.

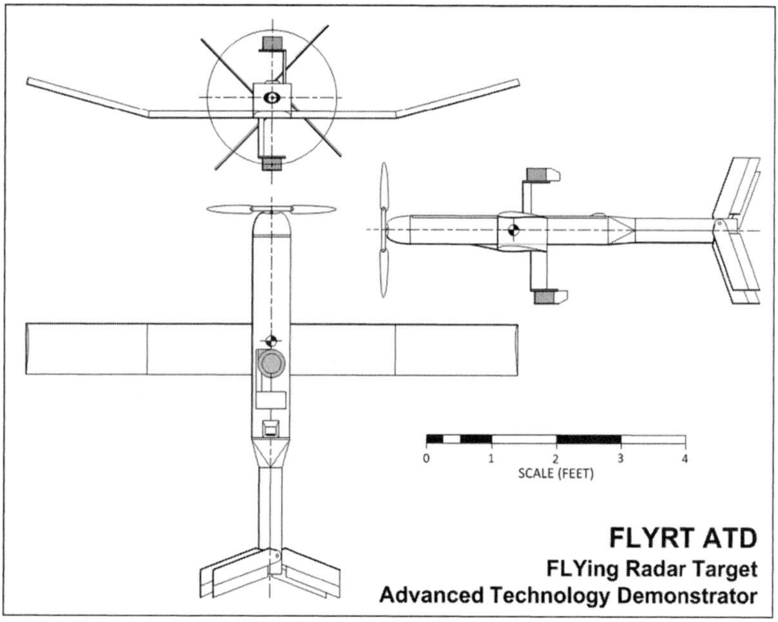

Fig. 2.47 FLYRT ATD three-view drawing.

Table 2.12 FLYRT ATD Technical Characteristics

Parameter	Value
Wing span	7.5 ft
Wing area	5.63 ft^2
Length	5.3 ft
Takeoff weight	72.0 lb (excluding booster)
	60.0 lb (with booster)
Empty weight	27.0 lb
Payload weight	25.0 lb
Battery weight	20.0 lb
Loiter speed	54 KTAS
Dash speed	70 KTAS
Endurance	>10 min
Max L/D	20 (antennas retracted)
	10.1 (antennas deployed)
Airfoils	Wortmann FX 63-137 wing flat plate tail
Launch method	Solid fuel rocket booster
Recovery method	Expendable
Propulsion type	Battery electric w/silver-zinc primary cells
Power	700 W

FLYRT ATD: Further Reading

Ailinger, K.G., Chaplin, H.E., Limparis, K.G., and Tayman, S.K., "Flying Radar Target (FLYRT) Technology Development," *1993 NRL Review*, Naval Research Laboratory, Washington, DC, 1993.

Banner, J.A., Davis, P.B., Peed, E.R., and Winchester, C.S., "SAFT America Lithium Sulfur Dioxide Battery (P/N 38303301) for FLYRT Application: Performance Discharge Test Report," Naval Surface Warfare Center, NSWCDD/TR-92/146, Aug. 1991.

Bovais, C.S., and Davidson, P.T., "Flight Testing the Flying Radar Target (FLYRT)," AIAA Paper 1994-2144, pp. 279–286, June 20–23, 1994.

Bovais, C.S., Haupt, M.E., and Toot, P.L., "Flight Testing with Sub-Scale Remotely Piloted Drop Models," AIAA Paper 92-4079-CP, 1992.

Bovais, C.S., Murphy, C.J., and Toot, P.L., "Flying Radar Target (FLYRT) Testing and Final Demonstration," *1994 NRL Review*, Naval Research Laboratory, Washington, DC, 1994.

Walden, R.B., and Page, G.S., "Shock and Vibration Testing of a Rocket-Boosted RF Decoy," Naval Research Laboratory, NRL/MR/5710-94-7486, Washington, DC, May 18, 1994.

TRANSONIC MISSILE ENGAGEMENT TARGET (TMET), 1994–1995

Fig. 2.48 TMET concept.

Transonic Missile Engagement Target (TMET) was a target system intended to replicate the flight behavior of a transonic anti-ship missile (Fig. 2.48). Ships would practice countermeasures against this simulated threat. The system was motivated by the need for more missile defense systems training for technical personnel and tactical planners against increasingly sophisticated threats. The training shortfall was primarily due to the high cost of existing threat simulation systems. Staging an operational training exercise with existing systems also posed logistics problems. TMET would be affordable due to its small size, which was enabled by its miniature TWT RF payload, small autopilot technology, new miniaturized turbojet engines, and NRL's experience with transonic low Reynolds number (LRN) aerodynamics. Because it would be launched from the ship, the training exercises could be conducted by the ship independent of other logistics.

TMET would be launched from the ship with a solid fuel rocket booster. A small turbojet with a diameter less than 5 inches provided propulsion after launch. The airframe reconfigured from a stowed geometry to the flight geometry during launch. Four fins rotated out of the fuselage to form an X-tail during the boost phase. The single-piece wing pivoted on top of the fuselage after the expended booster fell away. An intake for the turbojet extended from the lower fuselage. The overall fuselage length was 7 feet, including the booster.

TMET leveraged the avionics technologies developed for FLYRT ATD and SENDER (Chapter 5). It would use a FOG and GPS to provide the state

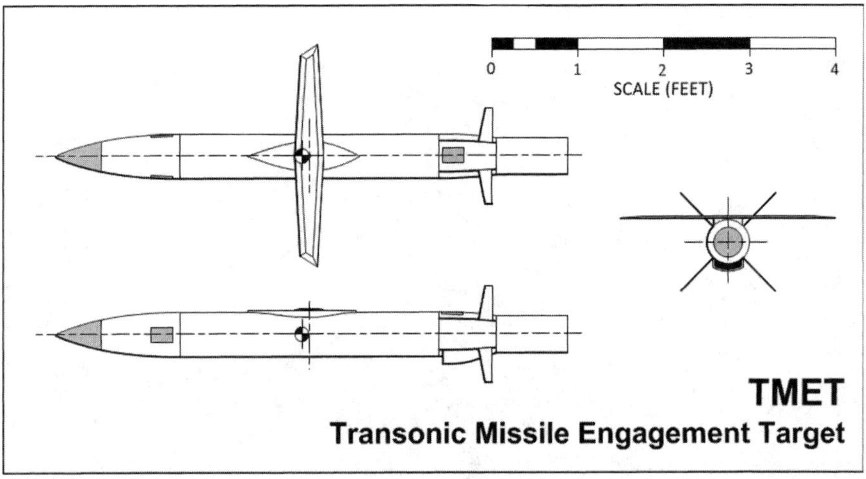

Fig. 2.49 TMET three-view drawing.

estimation. The avionics and payload would be powered by a permanent magnet alternator on the engine.

TMET would follow a simple flight trajectory. The vehicle would launch and fly several kilometers away from the ship. It would then perform a 180-degree 10-*g* turn and fly back toward the ship. Once on its inbound trajectory, the payload would activate to replicate a threat missile. The ship would either destroy the TMET, or TMET would run out of fuel and kinetic energy and scuttle before reaching the ship.

Engineering mock-ups of TMET were built, but the vehicle was not developed further, as sponsor needs changed. A TMET three-view drawing is shown in Fig. 2.49, and its technical characteristics are listed in Table 2.13.

Table 2.13 TMET Technical Characteristics

Parameter	Value
Wing span	2.5 ft
Length	7.0 ft (including booster)
Takeoff weight	49.0 lb
Empty weight	20.0 lb
Payload weight	5.0 lb
Fuel weight	16.0 lb
Cruise speed	Mach 0.78 at sea level
Endurance	1.2 min
Launch method	Solid fuel rocket launch
Recovery method	Expendable
Propulsion type	Turbojet

SUBMARINE-LAUNCHED TACTICAL AIRBORNE RECONNAISSANCE EXPENDABLE (STARE), 1995

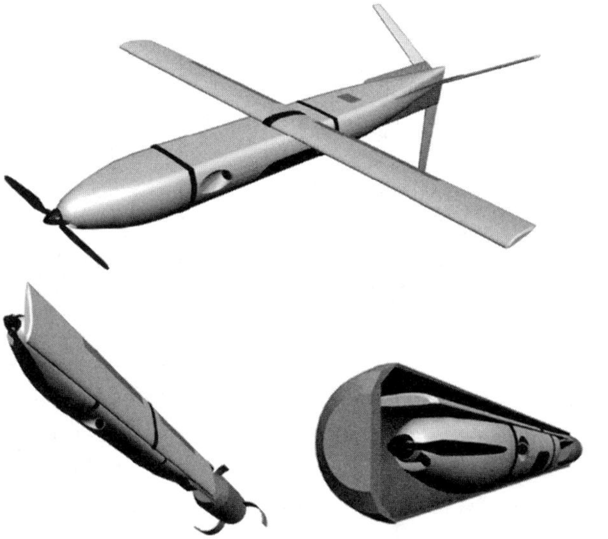

Fig. 2.50 STARE flight configuration (top), at initial launch (bottom left), and in the container (bottom right).

Over the course of two decades, NRL's Vehicle Research Section (VRS) considered the development of submarine-launched UASs and conducted many design studies. The compact size of small UASs and their capabilities for automated launch and autonomous flight make them well suited to be submarine-based systems. Initially, the submariner community was not interested in uses for UASs, but this position has changed as an increasing number of harbor security and counterterrorism missions have been developed to utilize the capabilities of the small UAS.

Submarine-launched Tactical Airborne Reconnaissance Expendable (STARE) was one of the first NRL detailed design studies for a submarine-launched small UAS (Fig. 2.50). It embodied the technologies of SENDER and the FLYRT ATD to provide a battery electric aircraft that could be deployed from a submerged submarine, float to the surface cocooned in a small buoy, self-deploy upon command from the buoy, and self-launch from the sea surface into the air to perform its misssion. STARE featured a fairly conventional airframe configuration. It was a high wing monoplane with a tractor propeller. The tail surfaces were configured into an inverted-Y layout, and each was all-moving, so there were no separate control surfaces. The tail surfaces provided full three-axis control during the ballistic launch phase before the wing was fully open. Following wing opening, the tail functions were reprogrammed to provide pitch and yaw control. The one-piece, no-dihedral wing had a simple constant chord planform and aileron control surfaces. The wing opened by rotating 90 degrees about a single central pivot atop the fuselage. For launching from the floating deployment buoy, STARE would utilize a solid fuel rocket booster that would be jettisoned after its fuel was spent, as successfully demonstrated two years earlier by the FLYRT ATD program.

The estimated performance characteristics for an unmanned aircraft of such dimunitive size as STARE were impressive. It was designed for a cruise airspeed of 100 knots for 30 minutes, providing a range of 50 nautical miles, all under battery electric propulsion. This level of performance and range was concurrently being demonstrated by the SENDER project. Unfortunately, for reasons lost to history, the development of STARE was not undertaken and no aircraft were built or tested.

Following STARE (Fig. 2.51, Table 2.14), NRL continued design studies for submarine-launched small UASs. During summer months, the VRS employed engineering students who were often given assignments to conduct preliminary design studies. These projects helped to train these prospective

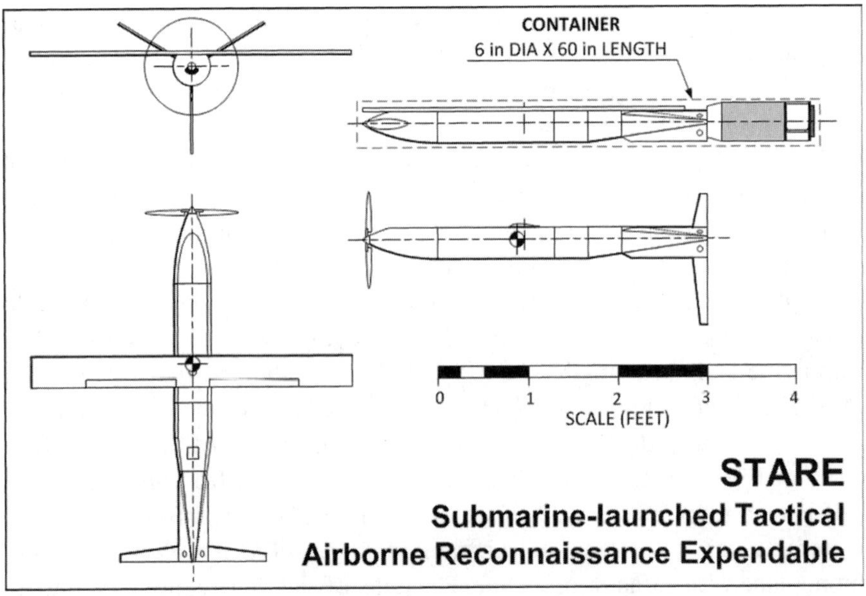

Fig. 2.51 STARE three-view drawing.

Table 2.14 STARE Technical Characteristics

Parameter	Value
Wing span	3.61 ft
Wing area	1.2 ft^2
Length	3.8 ft
Takeoff weight	25.0 lb
Empty weight	7.0 lb
Payload weight	6.0 lb
Battery weight	12.0 lb
Cruise speed	100 KTAS
Dash speed	100 KTAS
Stall speed	76 KTAS
Endurance	0.5 hr
Launch method	Rocket booster from floating container
Recovery method	Expendable
Propulsion type	Battery electric
Power	1.0 hp (750 W)

engineers in the engineering design process and sometimes resulted in innovative ideas that could be developed further. During summer 2001, the engineering students were grouped into two teams, each competing to develop a viable design concept for a submarine-launched small UAS. While their design studies were under way, the Chief of Naval Research, Rear Admiral Jay M. Cohen, visited NRL to be briefed on the status of the Dragon Eye program (Chapter 5), which was then in an advanced stage of development. RADM Cohen, one of the Fleet's top submarine commanders, was told about the student project. He enthusiatically listened to the team leaders describe their design studies and expressed his support by announcing an award for the winning team. RADM Cohen said that the first-place student team would be given the opportunity to spend one week aboard a Navy submarine. Then one of the students asked the admiral, "What about the second-place team?" RADM Cohen replied that the second-place team would get to spend *two* weeks aboard a Navy sub!

VERTICALLY LAUNCHED IMAGING INFRARED DECOY TECHNOLOGY (VLIIRDT), 2000

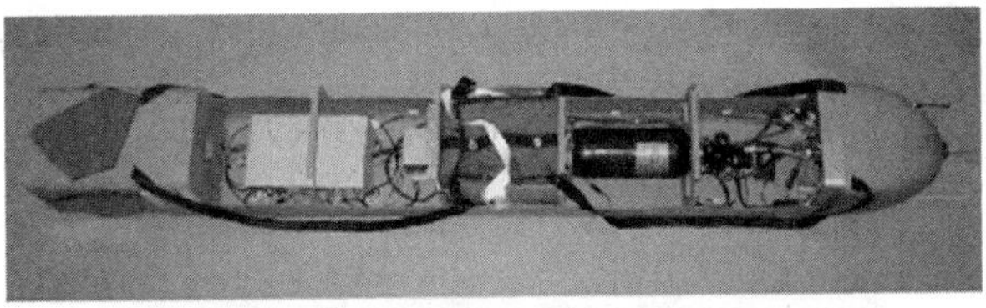

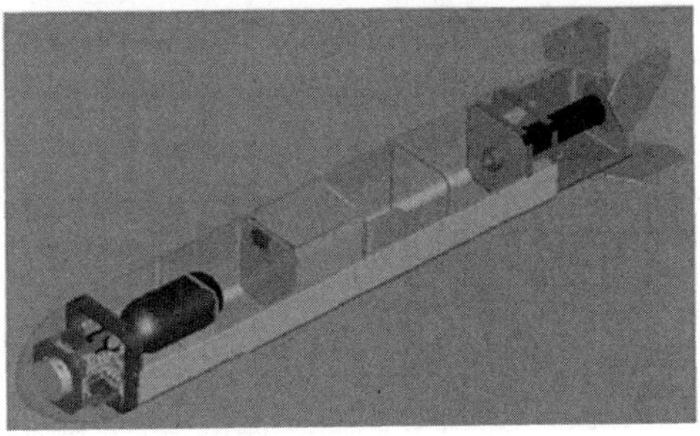

Fig. 2.52 Photograph (top) and concept drawing (bottom) of the VLIIRDT vehicle.

Vertically Launched Imaging Infrared Decoy Technology (VLIIRDT) was a vertically launched rocket that would fly a nonballistic, controlled trajectory from a ship to dispense an infrared (IR) chaff or obscurant cloud (Fig. 2.52). The mission was to provide countermeasures against heat-seeking anti-ship missiles. The VLIIRDT program was conducted to prove the technical feasibility of the concept.

VLIIRDT used multiple thrusters for propulsion and control. The main rocket was a modified version of an existing booster motor. The control thrusters ejected cold nitrogen gas from an off-the-shelf, high-pressure gas bottle. The valves were sequenced to provide control. The control thrusters fired in pairs located at the front of the body. There were lateral pairs for rolling and longitudinal pairs for pitch and yaw. The discrete firing gave the system bang-bang 3-DOF control.

The system flew a horizontal trajectory after thc desired altitude was reached. The autopilot commanded a pitch-over to horizontal flight. The square body generated lift, and the rear fins provided directional and longitudinal stability. The airframe consisted of an aluminum body with a composite nose and four deployable fins. After the main booster ceased firing, the vehicle gradually bled off airspeed and increased angle of attack. The aircraft descended ballistically into the water when there was insufficient aerodynamic lift to maintain level flight. The cold gas control system had

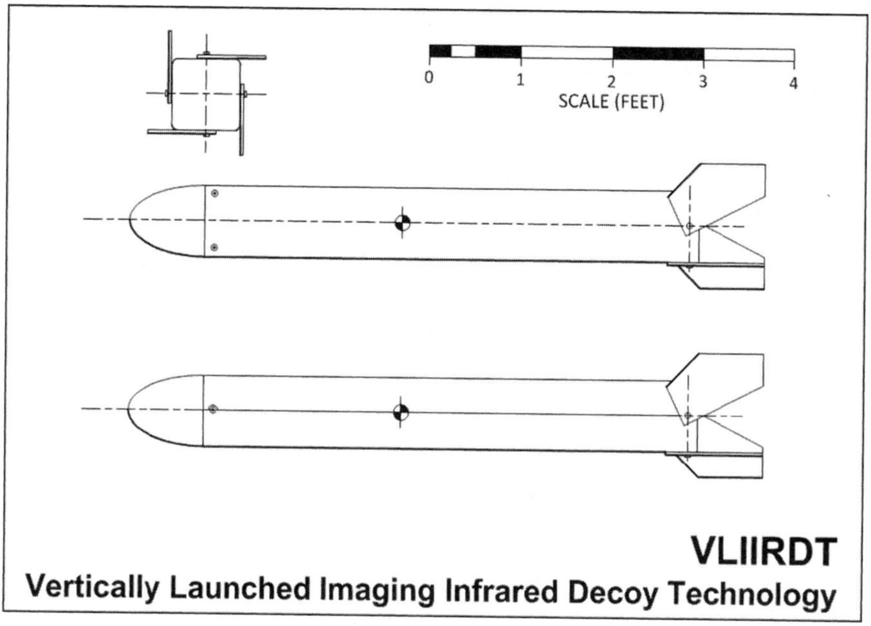

Fig. 2.53 VLIIRDT three-view drawing.

sufficient gas stored in the pressure tank to enable control for the duration of the horizontal flight.

The demonstration vehicle was equipped with a parachute to enable reuse. The parachute opened prematurely on the first flight, leading to a failure to fly until the attitude control propellant was spent. A second flight was fully successful and demonstrated the requisite technologies. Further development of this simple horizontal flight rocket has not yet occurred, and VLIIRDT remains a working prototype (Fig. 2.53, Table 2.15).

Table 2.15 VLIIRDT Technical Characteristics

Parameter	Value
Length	6.25 ft
Takeoff weight	45.0 lb
Endurance	1 min
Launch method	Solid propellant rocket launch
Recovery method	Expendable or parachute recovery
Propulsion type	Rocket

Chapter 3

Air-Launched Fixed Wing Aircraft

A broad range of military missions can be conducted with autonomous, air-launched, fixed wing unmanned aircraft. These missions include surveillance, reconnaissance, electronic warfare (EW), biological/chemical agent detection, and remote cargo resupply of ground forces. Design challenges for this class of unmanned aircraft systems (UAS) development include compact packaging, automatic unfolding while in flight, autonomous operation over broad airspeed and altitude ranges, affordable expendability, plus assurance that the UAS cannot damage the deploying aircraft.

Airborne Active Expendable Decoy–Forward Fired (AAED-FF), 1986–1990

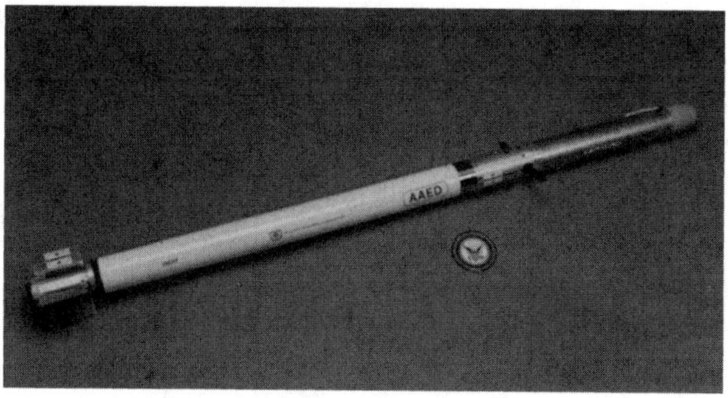

Fig. 3.1 AAED-FF.

The Airborne Active Expendable Decoy–Forward Fired (AAED-FF) was developed to be an air-launched EW decoy for aircraft defense against radar-guided missiles (Fig. 3.1). The payload was a radar repeater intended to spoof active missile seekers. This kind of payload receives the threat radar signal and transmits a signal that causes the missile to believe the decoy is as viable a target as the defending aircraft. NRL developed and evaluated both towed and forward-fired decoy versions under the AAED-FF program.

The towed decoy version was towed by a wire by the host aircraft. The advantage of a tethered system is that it can have long endurance. The fundamental disadvantages are that it can only fly in formation with the host aircraft, and it cannot perform independent maneuvers. This imposes limitations on the engagement envelope. NRL transitioned and supported the full development of the towed decoy, which was built by multiple industry teams and ultimately became the ALE-50 decoy.

NRL also developed the more technologically advanced forward-fired versions. These decoys were forward fired from the aircraft with a rocket motor. The decoy would fly at transonic speeds and was capable of maneuvering independently of the host aircraft. The booster was a modified version of the venerable 2.75-inch Zuni unguided rocket developed by the Naval Ordnance Test Station, China Lake, California. AAED-FF was capable of controlled, powered, nonballistic flight.

The controlled flight lasted for 30 to 40 seconds. The decoy cleared the aircraft, climbed, and turned to maneuver away. Pitch and roll were controlled by independently actuated small canard control surfaces. At its nominal altitude, AAED-FF began its flight at Mach 0.82 and increased its speed due to a significant reduction in mass as its solid propellant was burned, to end its powered flight at Mach 1.2. The propellant consumption created a large forward center of gravity (CG) shift that loaded the canard. The canard would reach its control limit near its stalling angle of attack at the end of the motor burn. The decoy then made a high-speed steep glide until ground impact.

The flight control system was an unconventional fluidic system that controlled the two-axis rate gyros for roll and pitch, control logic, and actuators. The compressed nitrogen fluid was stored in a high-pressure titanium toroidal tank, which was punctured at launch. The rate gyros worked by having the gas impinge upon an array of microphones that were sensitive to the fluid flow. The fluid followed its initial path when the aircraft was disturbed, but the sensor array was fixed to the aircraft. Motion was detected by the active elements of the microphone array. The control gains and maneuver scheduling logic was controlled by a simple fluid mechanical processor, which consisted of stacked welded wafers. The fluid vented past the gyros and logic device and then powered the actuators. The fluid-powered actuators drove the control surfaces for pitch and roll control. The roll actuation was torque-proportional, and the autopilot could command wings level in roll and pitch profiles. The maneuvers were all preprogrammed on the ground. McDonnell Aircraft was a pioneer of the fluidic control technologies for missiles. Garrett (Tempe, Arizona) was contracted to develop the autopilots for this program. Although this fluidic logic and control technology worked and was capable of performing under extraordinarily high G-forces, it has been superseded today by digital electronic flight control systems.

NRL built dozens of AAED-FF demonstrator systems and conducted flight test experiments. The AAED-FF demonstrators were launched from a rocket sled at Holloman Air Force Base, New Mexico. The systems were free-fired downrange, and then performed the mission profile. Initial tests had guidance and control problems, but these were resolved. Even successful flights resulted in the loss of the demonstrator, since it would impact the earth at Mach 1.2 at the conclusion of the flight.

An unusual oscillatory transonic low Reynolds number (LRN) phenomenon was encountered on the canards. The small canards had a 1-inch chord, and each side extended 4 inches from the body. It was discovered that a shock tripped the boundary layer on the canard. The boundary layer shed the separated wake and then the flow reattached. This change then moved the shock location by approximately a quarter inch. This repeated in an oscillatory manner at a few hundred cycles per second. The pitching moment changed by an order of magnitude during a cycle. Foch studied the problem in his master's degree thesis research. He discovered that the shock movement length was a function of Reynolds number and Mach number and its magnitude was inversely proportional to Reynolds number, so it was a particular problem for small chord wings under this transonic speed regime. The NRL team learned how to design tiny transonic airfoils to mitigate the shock boundary layer interaction, enabling the redesigned AAED-FF canards to provide effective control. Figure 3.2 shows a three-view drawing, and

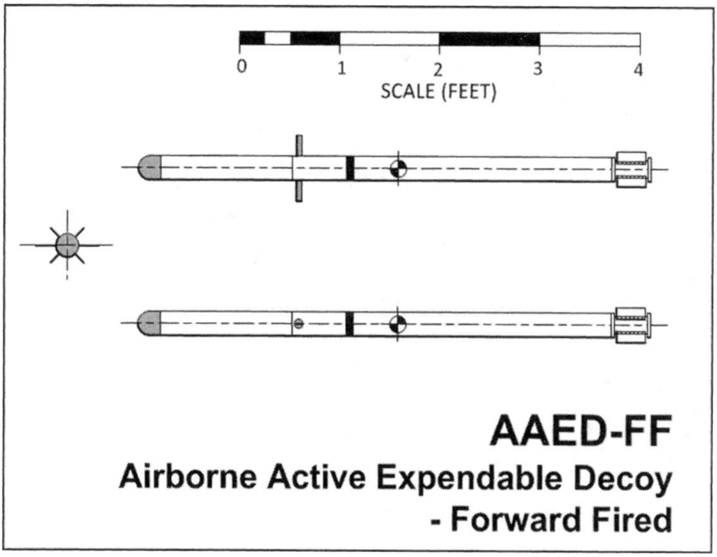

Fig. 3.2 AAED-FF three-view drawing.

Table 3.1 AAED-FF Technical Characteristics

Parameter	Value
Wing span	0.67 ft
Length	5.1 ft
Launch weight	25.3 lb
Empty weight	18.8 lb
Payload weight	7.0 lb
Cruise speed	Mach 0.82–1.2
Powered endurance	28 sec
Airfoils	Flat plate w/wedge leading edge
Launch method	Air launch
Recovery method	Expendable
Propulsion type	Solid fuel rocket

Table 3.1 lists the technical characteristics for this air-launched fixed wing aircraft.

The test program was considered a success, but the forward-fired system never transitioned to production due to limited funding. The towed version was produced instead, as the ALE-50, which is delivered by several vendors and has been widely employed by the U.S. Navy and Air Force since the 1990s.

PENDULUS, 1987

Fig. 3.3 Pendulus.

NRL was tasked to develop an air-launched small UAS that could carry a miniature magnetic anomaly detector (MAD) payload for submarine tracking. P-3 Orion aircraft performing antisubmarine surveillance are vulnerable to attack from the submarines if they make repeat passes. However, multiple detections are needed to determine a submarine's speed and bearing. The proposed operational concept was for a P-3 to make the first submarine detection and then launch Pendulus (Fig. 3.3). The UAS then would conduct a search pattern to locate the submarine and obtain the track. This mission required Pendulus to be rapidly air launched from the P-3. NRL was instructed to proceed under the assumption that the miniature MAD payload would be ready for integration into the airframe during the flight test phase.

There are limited options for air-launching an unmanned aircraft from the P-3. The alternatives are the bomb bay, sonobuoy dispenser, or door. NRL chose to use the A-size sonobuoy dispenser, which measures 4.8 inches in diameter and 36 inches in length. The limited volume of this dispenser was geometrically challenging, but NRL felt that it was possible to develop a suitable air vehicle that could package within the tight volume.

The Pendulus configuration was highly unconventional (Fig. 3.4). The fuselage consisted of a long tube structure that was oriented vertically. An inflatable streamlined fairing extended aft to reduce drag, since the circular

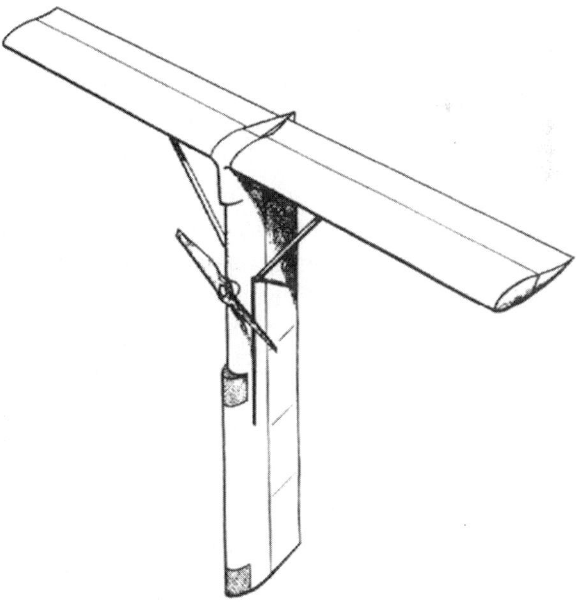

Fig. 3.4 Mission Pendulus design concept.

cross section fuselage would have had unacceptably high drag without remediation. The fairing was inflated with compressed nitrogen from a small bottle. The wings folded down the side when stowed, with a portion of the trailing edge folding behind the fuselage. The wings unfolded upward during deployment. The aircraft had the general appearance of a flying "T" when viewed from the front. The two-bladed propeller was recessed into a notched section on the front of the vertical fuselage. The thrust line passed through the CG, which was below the wing. The relatively heavy MAD payload was located at the bottom of the fuselage to facilitate sensing and to maximize pendulum stability.

The flight control system was originally controlled exclusively through weight shifting, also known as CG shifting. The fuselage could swing fore-aft to control pitch, and swing right or left for lateral-directional control. In many ways, this control system was similar to that of hang gliders, on which the pilot shifts his/her weight to control pitch and turning maneuvers. The pendulum stability approach was also similar to that of paragliders.

The T-wing configuration was inspired by the Spratt Controlwing. Spratt was a contemporary of the Wright brothers and had an active career for many decades. His designs featured monoplane or biplane wings elevated high above a fuselage. Spratt's designs flew successfully, but were not widely adopted. The pendulum stability enabled efficient cambered airfoils to be used on the wing, rather than reflexed airfoils that are typical of traditional flying wing designs. In the case of Pendulus, NRL used the RF 1165FB section. NRL engineers were inspired to attempt the development of this innovative configuration by articles about a Spratt float plane design in the June/July 1974 and May 1980 issues of the Experimental Aircraft Association's (EAA's) *Sport Aviation* magazine.

NRL engineers built several test models to prove the feasibility of the design and to evaluate its handling qualities. The first model was built up using balsa sticks and MonoKote covering (a heat-activated adhesive Mylar [polyester] film) in such a way that airfoils were approximated by faceted sections for the wing and fuselage. This model was built as a free-flight glider in 1.5 hours. The flights showed that the design had promise, and the experience encouraged the team to proceed with the T-wing configuration. The very rapid configuration evaluation was amazing in light of the reluctance of many unmanned aircraft development organizations to experiment with flight hardware. A very limited effort taught the developers much about the design before major development commenced.

Next, a series of full-scale radio controlled (RC) demonstrators were built. All the models were lighter than the mission aircraft. The first models did not fold and served to prove the handling characteristics of the deployed aircraft. The first RC model was a glider that was launched via a high-start glider

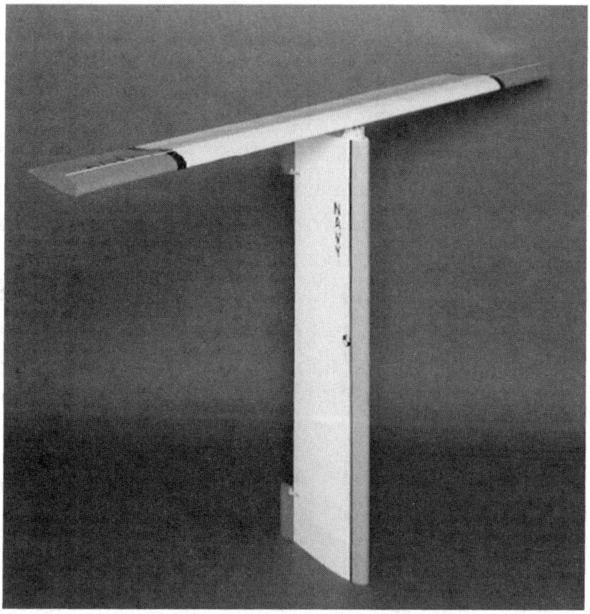

Fig. 3.5 Pendulus RC glider prototype.

launch system consisting of a bungee cord and kite string. The next model included the electric propulsion system. The final model series incorporated the folding mechanisms and served as drop models.

The RC models exhibited unusual flight dynamics (Fig. 3.5). One key problem was that the designs were too stable, such that maneuvering was difficult to initiate. Pendulus was spirally unstable with wings level, but spirally stable at a 20 degree bank. Once the aircraft was in the bank, it was difficult to get it out of the bank. The RC pilot had to continually remove the turn to return to wings-level flight.

The pendulum pitch control worked well enough for the purpose of changing speed. However, pitch oscillations occurred when high-rate pitch commands were applied. This led to transient pitch control reversals that damped out. An autopilot could have taken advantage of this phenomenon by initiating a quick pitch change by commanding the opposite input and then smoothing the maneuver; this autopilot mode was not implemented on the program.

The pendulum lateral-directional control approach was never quite satisfactory. To help fix the problem, two control surfaces were placed aft of the inflatable body fairing. At first, the two surfaces were actuated in opposite directions to serve as ailerons. The roll authority was not satisfactory. Next, both surfaces were actuated in the same direction to act as rudders. The rudder actuation worked as hoped, and satisfactory lateral control occurred through the design's natural yaw/roll coupling.

The envisioned mission aircraft was expendable, so recovery was not a design consideration, but the later flight tests adopted a net recovery technique to enable reuse of the demonstrators. Early flights simply landed on the ground, which routinely caused airframe damage, as the aircraft immediately and rapidly nosed over into the ground when the bottom of the fuselage first made contact. The high CG was a contributing factor.

Air deployment tests followed the flying qualities tests. Pendulus was dropped from a tower with the wings folded along the side. The drag and lift loads acting on the wings caused them to fold upwards. A simple one-way latch secured the wings in place. Once the wings deployed, the aircraft nosed over, gained speed, and pulled out of the dive. NRL also built an engineering mock-up of the aircraft that could fit within the sonobuoy dispenser. The model incorporated all the folding and inflating features.

To simplify the design and system operations, Pendulus employed battery electric propulsion. As discussed in the SARA program (1990–1994; see Chapter 4), electric propulsion technologies were still quite immature in 1987. Silver-zinc batteries and brushed rare earth electric motors could provide Pendulus with only 5 minutes of endurance. While short by today's standards, it was viewed as acceptable for the mission at that time.

Ultimately, the fully operational miniature MAD payload did not materialize, so the full program did not go forward despite the successes demonstrated by the air vehicle. The unfolding flying demonstrators were never launched from the sonobuoy dispensers. No high-speed aircraft launch tests were conducted under the program.

NRL proved that the T-wing configuration could be controllably flown and packaged. To the authors' knowledge, nobody has taken the configuration

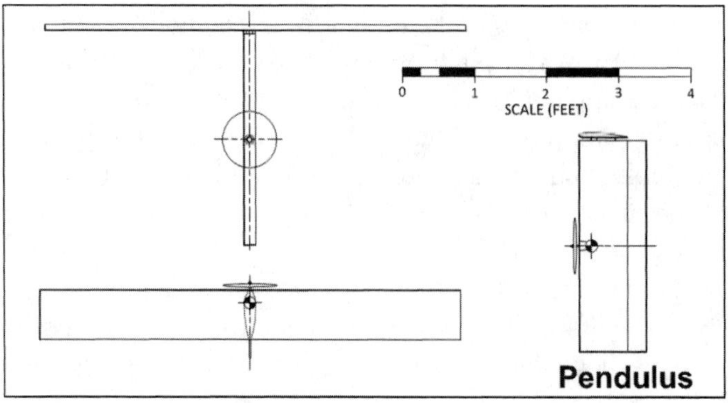

Fig. 3.6 Pendulus three-view drawing.

Table 3.2 RC Research Pendulus Technical Characteristics

Parameter	Value
Wing span	5.8 ft
Wing area	4.9 ft^2
Length	1.0 ft
Height	2.9 ft
Takeoff weight	3.5 lb
Payload weight	0.5 lb
Battery weight	1.2 lb
Loiter speed	28 KTAS
Stall speed	20 KTAS
Endurance	5 min
Max L/D	15.0
Airfoils	Wing: RF 1165FB
Launch method	Bungee
Recovery method	Conventional
Propulsion type	Battery electric
Power	100 W

further, despite its promise. Figure 3.6 shows a three-view drawing, and Table 3.2 lists the technical characteristics for this air-launched fixed wing aircraft.

Pendulus was perhaps ahead of its time. Today, the tandem wing ACR Coyote UAS (Advanced Ceramics Research Inc.) is launched from the P-3 sonobuoy dispenser and has been employed on P-3 hurricane research operations. The influence of the Pendulus program on the ACR system is unknown to the authors.

NRL RAPTOR, 1992

Fig. 3.7 Artist's concept of NRL Raptor in flight.

During the early 1990s, silicon solar panel technology was making great strides in performance. Thin-film amorphous silicon enabled very light membranes that could conceivably be used as a covering over a gossamer structure for a long-endurance, high-altitude unmanned aircraft. Consequently, the U.S. Air Force became interested in such an aircraft for reconnaissance applications. Its use was perceived as a supplement to a reconnaissance satellite; it could be rapidly deployed to an emerging situation to provide intelligence before a space asset could be put into place. The Strategic Defense Initiative Organization (SDIO) sponsored three studies for a high-altitude UAS called Raptor (Fig. 3.7).

Lawrence Livermore National Laboratory (LLNL) was sponsored by the SDIO to develop both solar-powered and liquid-fueled Raptors. LLNL awarded a contract to AeroVironment Inc. to develop the solar-powered version by refining and evolving that company's electrically powered, span-loaded, plank flying wing, the Pathfinder. The structure was based on the experience of Ray Morgan and Paul MacCready in designing pioneering human-powered aircraft and the world's first solar-powered manned aircraft. The Raptor/Pathfinder UAS was successfully developed, and its final evolution is known today as the Helios, which has set many world altitude and endurance records. Helios performed Environmental Research Aircraft and Sensor Technology (ERAST) science missions for NASA along with the earlier Pathfinder and Pathfinder Plus UASs. AeroVironment's superb flying wing designs all use the same Liebeck LA 2973A airfoil that NRL used for its FLYRT and CG Shift research UASs (Chapter 4).

LLNL awarded a contract to Scaled Composites LLC to develop a reciprocating-engine-powered, high-altitude Raptor. Its configuration was a conventional, tractor-propeller, low wing monoplane, which was a very conservative design for Scaled Composites founder and chief designer Burt Rutan. To reach its design altitude of 65,000 feet, a 100-horsepower class Rotax light aircraft engine was extensively modified to incorporate a two-stage turbocharger. Altitude chamber tests were successfully performed with this modified power plant at altitudes up to 70,000 feet. Like the AeroVironment aircraft, the Scaled Composites Raptor participated in the NASA ERAST program.

The third Raptor study sponsored by the SDIO was performed by NRL. Although NRL had no experience developing very large UASs, the Lab had extensive experience in LRN aerodynamics, advanced composite structures, autonomous flight, and solar-powered spacecraft, the requisite technologies for a Raptor. NRL also had more experience in autonomous flight and in-flight deployment than the other two Raptor developers, so the SDIO tasked NRL to utilize these technologies to investigate a second-generation operational capability with solar electric propulsion. Whereas the previous

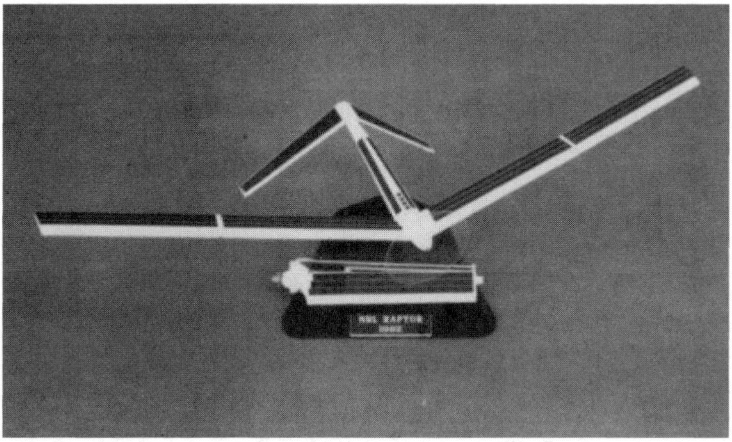

Fig. 3.8 NRL Raptor early design concept model.

two Raptor designs performed conventional runway takeoffs and the solar-powered version could only be launched under ideal low-wind conditions, NRL designed a Raptor that could be air launched from a C-141 or C-5A transport aircraft. This enabled the NRL Raptor to be deployed under all the weather conditions in which these manned transports could operate.

To provide the options of incorporating flexible-film solar panels or more rigid solar cells, Foch asked Dr. Robert Liebeck, aerodynamics expert and airfoil designer, if he could design an efficient airfoil with at least the aft 60% of the upper surface being flat. Liebeck accepted the challenge and developed a very efficient custom airfoil that was even tolerant of the boundary layer heating caused by the infrared absorption of the dark blue solar panels. Upon completing the airfoil design, Liebeck called Foch and jokingly thanked him for destroying his life's work! Prior to this airfoil, one of the signature aspects of Liebeck's airfoils was the cusped aft upper surface, designed to create an ideal Stratford pressure recovery without boundary layer separation. Unfortunately for Liebeck, but fortunately for Foch, this new flat-top airfoil performed better than many of Liebeck's cusped ones.

NRL studied several configurations that would maximize solar panel area, have the required high aerodynamic performance, and be reliably self-unfolding in flight (Fig. 3.8). NRL had learned a number of things about in-flight unfolding from many previous research projects: The reliability of in-flight unfolding is inversely proportional to the number of hinges and latches; telescoping is to be avoided; and some method of controlling the aircraft attitude throughout the unfolding process is necessary. The deployment event was the primary design driver in the NRL Raptor configuration. The final configuration selected for detail design is shown in Fig. 3.9.

Fig. 3.9 NRL Raptor configuration and deployment concept.

A controversial design aspect of the NRL Raptor was the propeller/airframe integration. To support compact packaging, it became desirable to set the propellers into the wing leading edge. There was concern that this installation might significantly reduce the installed propeller efficiency. Henry V. Borst, noted aerodynamics expert and former head of the Curtiss-Wright Propeller Department, was contacted, and his opinion was asked. After studying NRL drawings and calculations, Borst recommended that NRL proceed with the desired propeller installation and stated that the efficiency loss would be minimal.

To realize the desired flight performance, an extremely light structure for a cantilever wing was developed. NRL's weight estimates for the wing were severely criticized by the sponsor and the other Raptor design teams as being overly optimistic, since NRL was proposing the lightest aircraft structure ever constructed. NRL was confident in its calculations and its aircraft composite fabrication capability, so a sample wing panel was constructed with the new Liebeck airfoil (Fig. 3.10). This wing panel was carefully weighed and structurally load tested. Results exceeded the design estimates, and the NRL Raptor wing panel was the lightest cantilever wing panel yet constructed. When complete with its Tedlar (polyvinyl fluoride) film skin, its weight was slightly less than 0.1 pound per square foot!

Although the NRL Raptor had many advanced features, it was seen as having a much higher development risk than the other two Raptors being developed by established large aircraft manufacturers. Sponsorship to

Fig. 3.10 NRL Raptor wing panel prior to covering.

complete the detailed design and fabrication of a flying prototype was not forthcoming. Nevertheless, the NRL team enjoyed the design challenge and profited by the experience. Figure 3.11 shows a three-view drawing of the NRL Raptor.

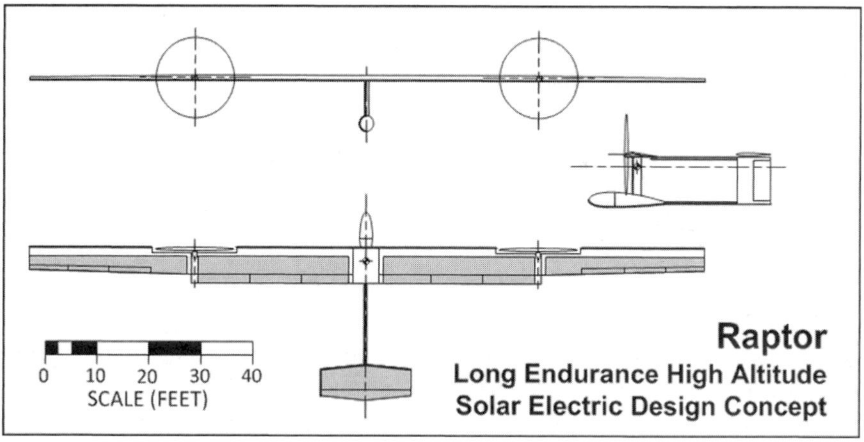

Fig. 3.11 NRL Raptor three-view drawing.

EXPENDABLE MULTI-MISSION MARINE AIRCRAFT (EMMA), 1999

Fig. 3.12 EMMA high-speed configuration (top) and low-speed configuration (bottom).

The Expendable Multi-mission Marine Aircraft (EMMA) was proposed to provide the capability for an F/A-18 aircraft to rapidly emplace a radar jammer deep within enemy territory in advance of an incoming air attack (Fig. 3.12). This type of EW mission is termed "Suppression of Enemy Air Defense," or SEAD. It places a higher value on the ingress speed of the UAS than do other EW missions. To take maximum advantage of the high launch speed and altitude of the deploying F/A-18, EMMA was designed with a unique flight profile that required more extreme airframe morphing between flight modes than previous NRL UASs, and a dual propulsion system.

Specifically, EMMA was designed to be launched at an altitude of 35,000 feet at speeds up to Mach 0.82. Upon clearing the aircraft, EMMA would establish a long-range powered glide at Mach 0.7 under turbojet propulsion until reaching a low altitude on the order of 1000 feet above ground level (AGL). This portion of the flight profile would provide range of 75 nautical miles during the first 10 minutes of flight. The turbojet propulsion unit (engine plus fuel tank) would then be jettisoned. EMMA would fly powerless at a constant low altitude, trading off its kinetic energy (via airspeed) to maintain altitude. When the airspeed dropped to 225 knots, the inner portion of the low-speed wing would deploy, which automatically converted the high-speed wing into a wing/fuselage fairing. On the inner portion of the low-speed wing, EMMA would maintain its low altitude as its airspeed was further reduced. Nearing its 165-knot cruise airspeed, the outer wing panels would then unfold and the electric propulsion system would be activated. It would use lithium primary batteries to power a brushless motor that turned a folding propeller. The total ingress range was estimated to be 100 nautical miles, flown over an elapsed time of 18 minutes. Upon reaching its target,

EMMA would activate its electronic payload and perform its EW mission for 30 minutes.

The performance capabilities proposed for EMMA were impressive to the potential users. All the requisite technologies were available, so the primary development risk was incorporating such extreme airframe and propulsion morphing into a reliable, yet affordably expendable system. Following the initial design phase, the program costs and schedule were estimated so NRL could produce a detailed proposal to the U.S. Marine Corps. On a very memorable day of the project, Dr. Montgomery and Mr. Foch were given the opportunity to present the proposal directly to the USMC Commandant at the Pentagon.

En route to the Pentagon, Montgomery asked Foch to provide a quick overview of the presentation. He was horrified to see that Foch had named the proposed project "EMMA." Yes, it was an excellent acronym, but

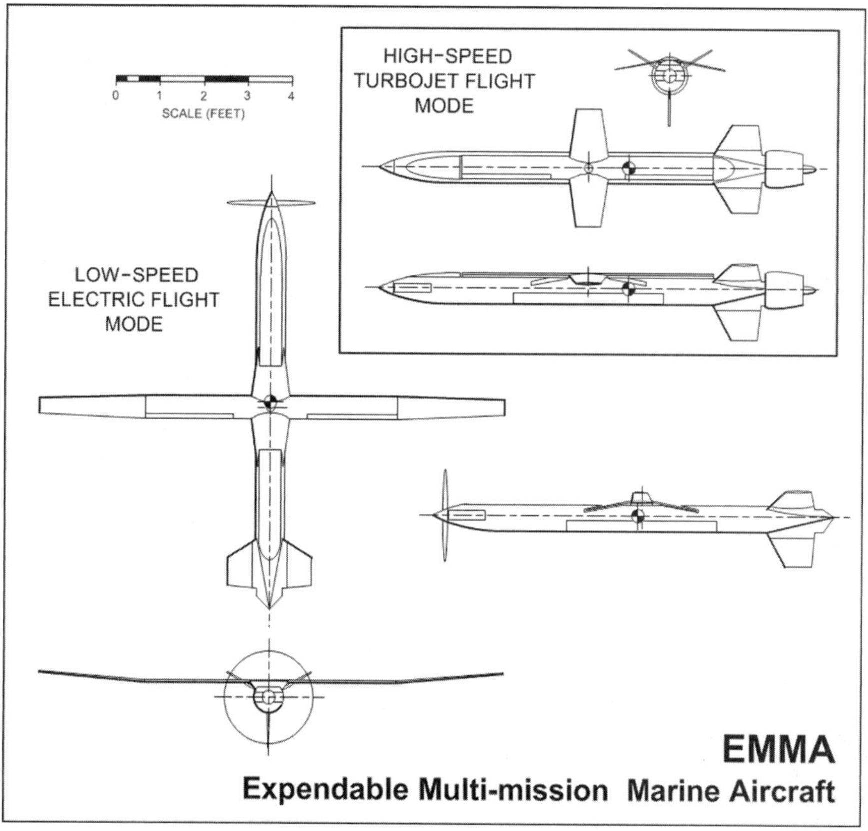

Fig. 3.13 EMMA three-view drawing.

Montgomery was certain that no Marine pilot would want to take an "EMMA" into battle! He was concerned that upon seeing the proposed name, the Marines would forever chide NRL about such a feminine name for a Marine weapon system. Unfortunately, there was no time to modify the slides and text. Montgomery was visibly displeased with Foch for the name choice. In the USMC Commandant's office, Foch put up the first slide as Montgomery cringed. The Commandant sat bolt upright, obviously taking great notice of the proposed name EMMA. Mongomery and Foch prepared for the expected fallout. Instead, the Commandant exclaimed, "I love the name; it is my daughter's name!" The rest of the presentation was uneventful and appeared well received. On the trip back to NRL from the Pentagon, Montgomery leaned over close to Foch and softly said, "You must be the luckiest man in the world."

For reasons lost to history, EMMA was never developed for the USMC, but it did lead to the ALICE program, discussed later in this chapter. EMMA would have been very technically challenging, but could have provided a mission capability that is still valid today. Figure 3.13 shows a three-view drawing, and Table 3.3 lists the technical characteristics for this air-launched fixed wing aircraft.

Table 3.3 EMMA Technical Characteristics

Parameter	Value
Wing span	2.5 ft (turbojet) 10.5 ft (electric)
Length	9.5 ft (turbojet) 8.75 ft (electric)
Takeoff weight	60.0 lb
Empty weight	30.0 lb
Payload weight	5.0 lb
Fuel weight	10.0 lb
Cruise speed	Mach 0.7 (turbojet) 165 KTAS (electric)
Endurance	48 min (10-min turbojet, 8-min glide, 30-min electric)
Ingress range	100 nautical miles in 18 min
Launch method	Air launch from F/A-18
Recovery method	Expendable
Propulsion type	Turbojet with JP-5 Battery electric with lithium primary cells

EXTENDER, 1999–2000

Fig. 3.14 Extender.

Extender was designed to demonstrate the feasibility of deploying a small unmanned aircraft from a Lockheed P-3 Orion aircraft to support radar site detection and geolocation missions (Fig. 3.14). Small electronic surveillance measure (ESM) payloads were under development at NRL for small unmanned aircraft. Thus equipped, the Extender and P-3 could cooperatively geolocate the target radar site from different locations. Extender could fly into hostile territory without putting a flight crew at risk, while the P-3 host aircraft safely stood off. The system was named Extender because it extends the capability of the P-3.

Obtaining a Naval Air Systems Command (NAVAIR) flight clearance to launch an unmanned aircraft from the P-3 or any other Fleet manned aircraft is a cost- and labor-intensive activity. It would have been necessary for all engineering data and testing to be reviewed and approved by a NAVAIR competency board, and the sponsor did not have sufficient funding to support a traditional flight clearance. A nontraditional approach that would result in the ability to safely deploy from the P-3 was needed.

NRL spoke with NAVAIR about options for air-launching Extender from the P-3. The traditional options are the torpedo bay, sonobuoy tubes, and wing stations. All of these would require the extensive NAVAIR flight clearance process. Foch was not satisfied, so he continued to talk with P-3 flight crews about other options. Finally, it was discovered that the P-3 could jettison a life raft out the side door to help a downed flight crew or a ship in distress. The P-3 would fly at low altitude and slow down below 175 knots. The flight crew member would hold door handles with both hands and kick the life raft container out the door with both feet. The life raft fell away

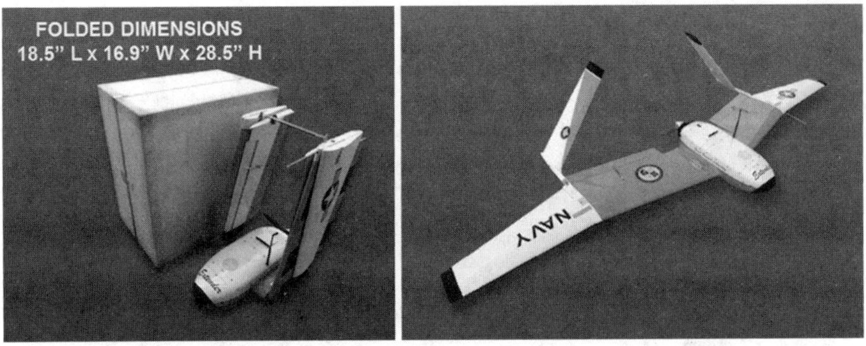

Fig. 3.15 Extender folded in life raft container (left) and fully deployed (right).

without interfering with the P-3. In further discussions with NAVAIR, it was concluded that if the Extender could fit within the life raft container and had identical dynamics, no new NAVAIR flight certification would be necessary, and the flight clearance for prototype testing would be relatively easy to obtain. The Extender was lighter than the folded life raft, but the container required the same ballistic coefficient, so appropriate ballast would be applied. There would be no data or power interfaces to the P-3. The flight crew would simply push a button on the container to activate Extender's avionics prior to jettisoning it from the P-3.

The Extender aircraft was designed to fit within the life raft container (Fig. 3.15). The internal container dimensions measured 20 inches long by 32 inches wide by 32 inches high, approximately the size of a large suitcase. The folded aircraft dimensions were 18.5 inches long by 16.9 inches wide by 28.5 inches high. The vertical box shape required the aircraft to have a stubby pod fuselage that would sit in the bottom of the box, and wings with two folds on each side. The inner hinge folded the inboard wing panel up, and the outboard wing panel folded down along the inboard wing panel. As various airframe configurations were considered, flying wing designs were the leading candidate, but the desired endurance performance could not be achieved for the size constraints and payload weight. The initial flying wing design had a low endurance parameter ($C_L^{3/2}/C_D$) due to the low trimmed lift coefficient; therefore, a displaced tail was added so the trim lift coefficient in loiter flight could be increased, with a corresponding higher endurance parameter. The displaced tail design had a down-loaded tail for trim, which enabled a more traditional cambered airfoil rather than a reflexed camber flying wing airfoil.

Extender had a unique configuration. The wings were cranked such that the inboard panels were swept back and the outboard panels were slightly swept forward. The displaced tails' booms were located at the end of the inboard panel so that the tail moment arm could be maximized with minimal tail boom length. This was critical for the folding geometric constraints. The forward sweep of the outboard panel maintained the relationship between the aircraft aerodynamic center and the required CG location for the appropriate static margin without ballast. The displaced tails were angled inward, so they acted as split inverted V-tail halves. The inboard wings had ailerons, and the displaced tails had ruddervators. The motor and propeller were on the rear of the pod fuselage in a pusher arrangement.

NRL designers were concerned about managing this complex deployment in free fall. The packaging approach left significant unused volume between the folded wings. It was decided that this volume would house a parafoil. The parafoil would slow the descent, stabilize the aircraft, and enable more time for the deployment (Fig. 3.16).

The original deployment concept had numerous steps. The box containing the Extender was kicked out the side door of the P-3 by the flight crew. The box opened and the folded airframe fell away. A small parafoil attached to the fuselage opened such that the Extender was oriented nose-first into the relative wind. Under the acceleration of gravity, the wing panels dropped about the inner hinge, and the inner wing locked onto the fuselage. Next, the outboard stabilizers unfolded about an angled boom hinge under aerodynamic forces. Next, the outer wing panels dropped and

Fig. 3.16 Extender original deployment sequence concept.

aerodynamically flew into the extended span position, using the same technique as FINDER (discussed next). Last, the parachute was cut and Extender began flight operations. The design had six simple hinges: four for the wings and two for the tails. The deployment sequence was controlled with an electronic timer. The geometry and inertias were arranged such that the deployment stages operated passively under gravity and aerodynamic forces. No springs were necessary.

Once the design was completed, an RC demonstrator with a nonfolding fixed wing was constructed to evaluate flight characteristics. It had a fixed tricycle landing gear to permit repeated flights from a runway. The first flight occurred on April 7, 1999, followed by the first autopilot flight on April 15, 1999. Initially, it was hoped that the wing control surfaces could be used as elevons without any tail control surfaces; the tails would simply act as passive stabilizers. However, the flight tests revealed that the wing elevon control surfaces were too close coupled. The next flights used wing ailerons and tail ruddervators, and the aircraft had good handling qualities. There were approximately three dozen Extender flight tests to evaluate handling qualities, stability, and control (Fig. 3.17).

Next, the deployment methods were evaluated in the Wright Patterson Air Force Base (WPAFB) vertical wind tunnel (Fig. 3.18). Extender was oriented with an angle of attack consistent with a 2:1 glide ratio associated with the parafoil glide. Due to the vertical oncoming wind, Extender had a nose-down orientation during the tests. The deployment worked as expected, and the tests confirmed that no springs were required. Multiple angles of attack were evaluated. One very significant result of these tests was the determination that the parafoil was not required for deployment. The Extender would descend bottom-first into the relative wind, and the aircraft could passively

Fig. 3.17 Extender RC demonstrator.

Fig. 3.18 Extender in the WPAFB vertical wind tunnel.

open itself. Eliminating the parachute would reduce altitude loss required for deployment. Furthermore, reliability would be improved and cost reduced by eliminating this additional equipment.

Two deployable Extender aircraft were built. The airframes were molded fiberglass and carbon fiber composite structures. Foam and fiberglass life raft containers were constructed to house the aircraft. Plans were prepared for the initial deployment tests to be made by dropping from a balloon. Prior to the deployment tests, flight tests with the deployable airframes were made to verify the operation of their systems. The first of these flight tests was performed on November 17, 1999.

The Extender system was ready for the balloon deployment tests, but an urgent project to support Fleet activities necessitated putting the program on hold. NRL was given permission by the sponsor to put the systems on the shelf for a year. The systems were completed, given corrosion protection, and stored in their containers. During the year that Extender sat on the shelf, sponsor interest waned and the project was never completed. Two nearly flight-worthy Extender systems are still in storage as of this writing.

The Extender configuration was the basis for NRL's Nexus unmanned aircraft proposal. Chapter 5 describes how Nexus influenced the Navy and Marine Corps' Small Tactical Unmanned Aircraft System (STUAS) and led to the Dragon Eye program. Figure 3.19 shows a three-view drawing, and Table 3.4 lists the technical characteristics for this air-launched fixed wing aircraft.

Extender was also used as the basis for a Mars Flyer proposal (Chapter 9). The Extender airframe was used as a subscale mock-up for the proposal. The Mars aircraft had a different propulsion system, among other key changes,

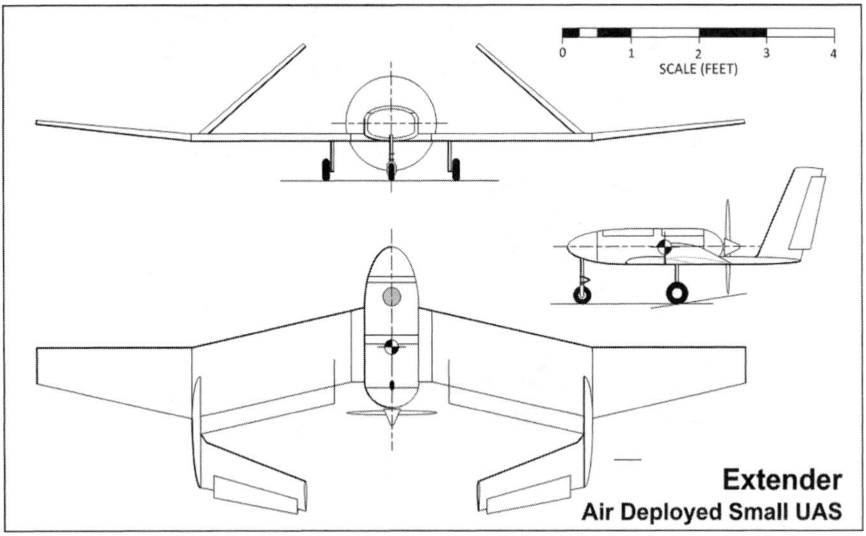

Fig. 3.19 Extender three-view drawing.

Table 3.4 Extender Technical Characteristics

Parameter	Value
Wing span	10.2 ft
Wing area	10.1 ft^2
Length	2.5 ft
Takeoff weight	31.0 lb
Empty weight	15.2 lb
Payload weight	7.0 lb
Battery weight	8.8 lb
Cruise speed	45 mph
Dash speed	85 mph
Landing speed	35 mph
Endurance	2.3 hr
Range	103 mi (one way)
Launch method	Air deployed (operational) Conventional (prototypes)
Recovery method	Expendable (operational) Conventional (prototypes)
Propulsion type	Battery electric, $LiSO_2$ batteries
Power	750 W (max) 272 W (cruise)

but Extender's aerodynamic configuration, packaging, folded geometry, and deployment approach were suitable for deployment from a Mars entry vehicle and flight of an exploration mission of the planet. One Extender airframe was painted in NASA markings and placed on display at the NASA Ames Research Center.

FLIGHT-INSERTABLE DETECTION EXPENDABLE FOR RECONNAISSANCE (FINDER), 1999–2010

Fig. 3.20 FINDER.

The Defense Threat Reduction Agency (DTRA) asked NRL if it was possible to apply autonomous, air-deployed unmanned aircraft technology for a new asset for chemical and biological weapons counterproliferation. This was part of the DTRA Counter Proliferation 2 Advanced Concept Technology Demonstration (CP 2 ACTD) for the European Command. The mission profile was (1) conduct an air strike on the chemical or biological weapons facility, (2) air-launch the Flight-Insertable Detection Expendable for Reconnaissance (FINDER), and (3) fly FINDER through a chemical or biological agent plume. FINDER would detect the agent in real time. The data would be transmitted back to the host aircraft, which would then relay it back to an exploitation station. FINDER measurements could be used for treaty verification purposes. As the requirements matured, FINDER also had to return a sample back to a laboratory. FINDER had to be affordably expendable, since it would be expended at the conclusion of the mission or destroyed after returning a sample (Fig. 3.20).

The early FINDER designs were envisioned as electric aircraft (Fig. 3.21). The unmanned aircraft would fit under the pylon of a tactical military aircraft

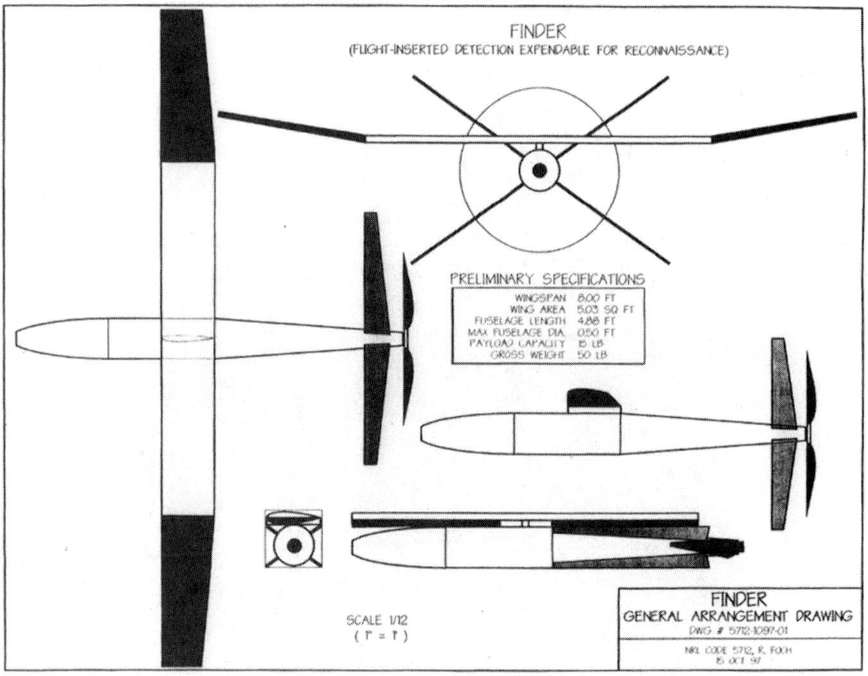

Fig. 3.21 Early FINDER concept.

or within an internal weapons bay. When stowed, the center wing section would pivot to align with the fuselage, and the outer wing panels would fold under the center wing section. In initial designs, the FINDER tails folded along the fuselage for packaging within an internal weapons bay, but later designs had fixed tails.

The CP 2 ACTD introduced new requirements. First, FINDER had to be air dropped from a General Atomics Predator A (MQ-1) unmanned aircraft. FINDER was thus considered to be a marsupial aircraft, since it was an unmanned aircraft deployed from a larger unmanned aircraft. The FINDER was to be controlled from the Predator ground control station. The two aircraft would cooperatively track a plume by sharing meteorological and chemical data. The second major requirement change was that the FINDER had to return to the base after performing its mission. The aircraft would carry a sample in addition to its real-time *in situ* detector. This requirement drove long range and endurance, which necessitated a gasoline engine rather than the original electric propulsion concept. FINDER would be incinerated after it returned the sample if contaminated by dangerous agents.

The FINDER airframe design was already well advanced by the time the requirements changed. Placing an engine in the tail would have been very

Fig. 3.22 Rendering of FINDER internal arrangement.

difficult, because the CG would shift too far aft. The solution was to place the engine inside the fuselage just aft of the wing and to add a shaft extension to the pusher prop at the rear of the fuselage (Fig. 3.22). Like every aircraft designer who has developed a shaft-driven propeller, the FINDER team initially had problems with torsional resonance and fatigue of the drive shaft, but these were fully resolved. Advanced polymer antivibration devices and universal fittings at each end of the shaft helped solve the problems.

No heavy fuel engines were then available at the desired power class. However, the Predator A used aviation gas, and suitable small engines that used this fuel type were available. The 3W 24i single-cylinder, two-stroke gas engine was selected to power the FINDER.

The payload was housed in the modular nose so that air samples would not be contaminated by engine exhaust. The primary payload was the “Spider,” an Army Engineering Research and Development Center (ERDC) chemical detection payload. NRL investigated biological sampler payloads, and camera payloads were used for other missions. The starter-generator provided up to 140 watts of payload power.

After completing the detailed design, NRL conducted wind tunnel testing. The FINDER system required a SEEK EAGLE flight clearance, since it was deploying from an Air Force asset. The wind tunnel tests gathered aerodynamic data for the FINDER at different stages of wing deployment. This helped with the simulation of FINDER in free fall. This testing and the simulation showed that FINDER would not hit the Predator during the deployment sequence (Fig. 3.23).

Next, the engineers had to work out the systems that interfaced to the Predator. The most expedient solution was to develop a custom wing pylon. This pylon included the communications link for the relay, and supplied power to heat the FINDER engine compartment, which enabled a reliable in-flight engine start.

Fig. 3.23 FINDER on the Predator wing pylon.

The air launch sequence had several steps (Fig. 3.24). First, FINDER was released from the Predator pylon. The tails did not need to fold when dropped in this manner. FINDER dropped on a ballistic trajectory, with the tails providing stabilization and attitude control. The center wing section rotated from its fore-aft stowed position to the flight position using an electrical actuator for rotation. Next, the outer wing panels dropped and aerodynamically flew into the extended span position. The wing deployment sequence was much like that of FLYRT ATD. Finally, the engine was started under battery power with the onboard starter-generator. There was a narrow time window for engine start as the aircraft descended toward the ground in a glide.

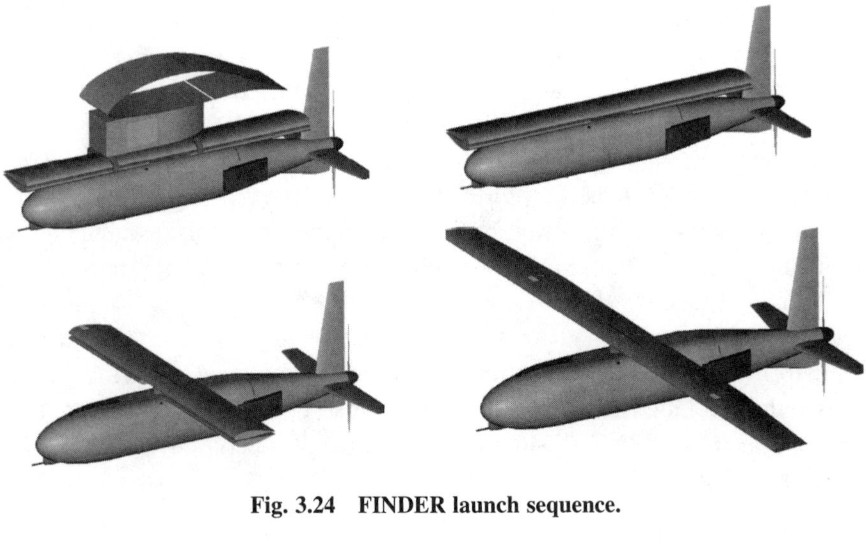

Fig. 3.24 FINDER launch sequence.

Fig. 3.25 FINDER aircraft with tricycle landing gear.

Initial flight testing took place from the ground using a catapult launcher and conventional wheeled landing using tricycle landing gear (Fig. 3.25). These initial flight tests proved the flight characteristics before proceeding to the more demanding air-launch tests, the first of which was performed on March 22, 2000. Tail wheel landing gear was used for the air-launch tests.

Following utility air-launch tests, the FINDER was fully integrated with the Predator. The first deployment from Predator was conducted in June 2002. Over the course of the program, 21 drops were conducted at altitudes of 15,000 to 20,000 feet. Every Predator launch was successful, and few problems were encountered (Fig. 3.26). The longest FINDER flight lasted 6.5 hours.

A later variation of the FINDER program was to support AC-130 Spectre gunships. The FINDER would be air launched from the gunship and fly down to lower altitude to obtain improved target imagery. The close-range surveillance provided improved resolution thanks to the reduced slant range. It also reduced image degradation caused by obscurants such as smoke, clouds, or dust. This program was called Spectre-FINDER.

Fig. 3.26 FINDER air launch from Predator.

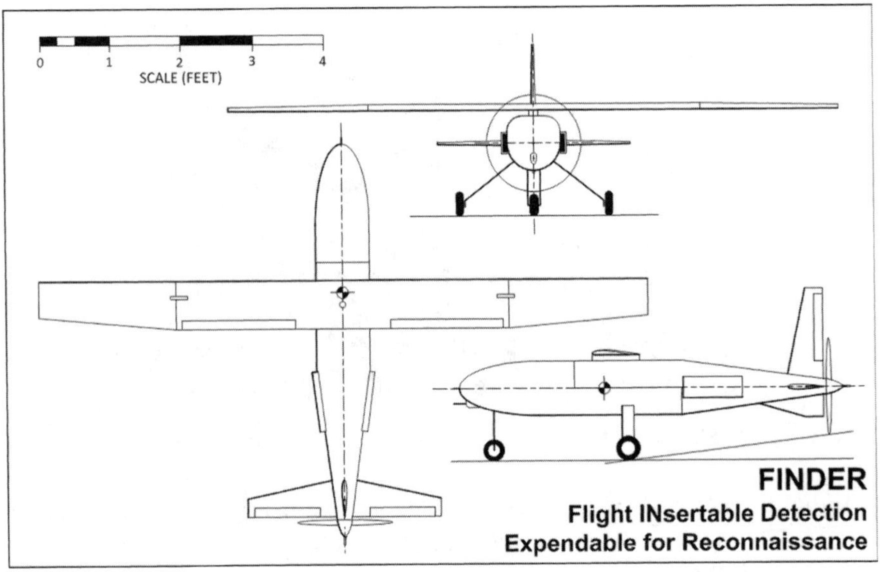

Fig. 3.27 FINDER three-view drawing.

The decade-long FINDER program reached all its goals and was fully successful. FINDER was a revolutionary capability. Approximately 20 residual airframes were used for advanced experimentation. Figure 3.27 shows a three-view drawing, and Table 3.5 lists the technical characteristics for this air-launched fixed wing aircraft.

Table 3.5 FINDER Technical Characteristics

Parameter	Value
Wing span	8.6 ft
Wing area	5.4 ft^2
Length	5.25 ft
Takeoff weight	50–59 lb
Payload weight	11.5 lb
Fuel capacity	2.8 gal
Loiter speed	70 mph
Cruise speed	80 mph
Dash speed	100 mph
Endurance	>5 hr
Max L/D	16
Launch method	Air launch or conventional takeoff
Recovery method	Expendable or conventional landing
Propulsion type	Gasoline reciprocating engine
Power	~2 hp

FINDER: FURTHER READING

Cross, A., "NRL's FINDER UAV: A Counterproliferation Asset," *2005 NRL Review*, Naval Research Laboratory, Washington, DC, 2005, pp. 203–205.

Kosmatka, J.B., and Panza, J., "Aeroelastic Analysis of the Predator Aircraft With Wing Mounted FINDER Mini-UAVs," AIAA Paper 2003-6501, delivered at the *2nd AIAA Unmanned Unlimited Systems, Technologies, and Operations Conference*, San Diego, CA, September 15–18, 2003.

AIR LAUNCHED INTEGRATED COUNTERMEASURE, EXPENDABLE (ALICE), 2001–2005

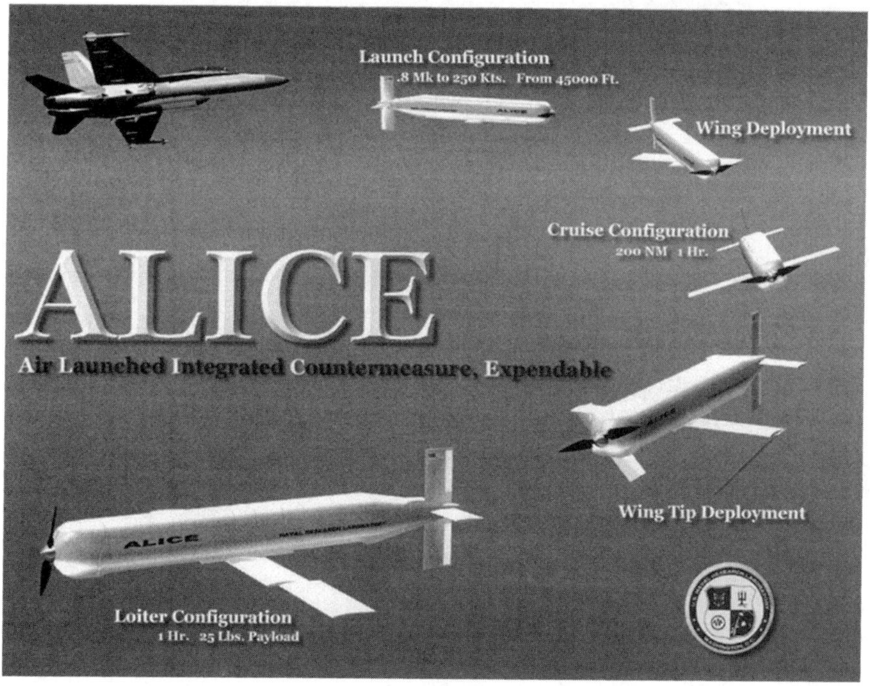

Fig. 3.28 Refined ALICE concept.

The purpose of the Air Launched Integrated Countermeasure, Expendable (ALICE) unmanned aircraft was to enable an unmanned aircraft to be air launched from a high-speed tactical manned aircraft for EW and other missions. The ALICE vehicle was designed to be reconfigurable to fly in both sustained high-speed cruise and efficient loiter (Fig. 3.28). The airframe geometry morphed in flight. ALICE would be launched from the host aircraft at Mach 0.8 and up to 45,000 feet, cruise at 165 knots for 200 nautical miles,

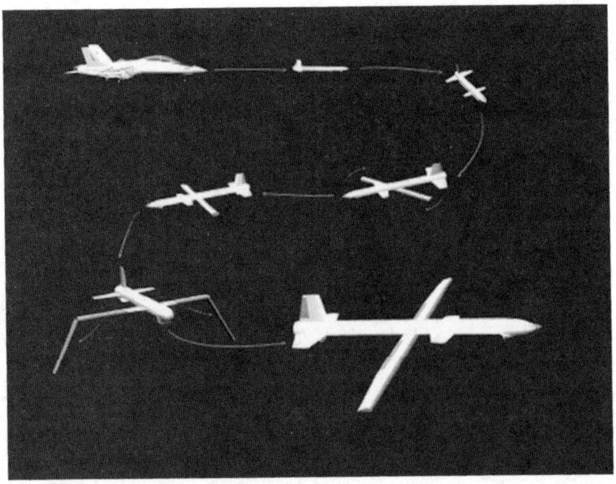

Fig. 3.29 Original ALICE concept.

loiter over its target at 65 knots for 2 hours, and then be expended. The mission payload weight was 25 pounds.

The deployment sequence was similar to that of the FLYRT ATD in many ways, except ALICE did not use a rocket booster. The tails deployed immediately after release from the host aircraft to provide stability. The vehicle would fall in a steep glide using body lift while decelerating. After ALICE slowed close to the cruise airspeed, the wing would deploy to the 165-knot cruise configuration. Once in the mission area, the outer wings would deploy to decrease the wing loading and to increase the aspect ratio for best endurance.

Like FLYRT ATD, the ALICE wing had three segments and the outer panels folded under the center panel (Fig. 3.29). The folded wing rotated to align with the fuselage when stowed. Unlike FLYRT ATD, the wing had an efficient, low-drag airfoil shape when the outer panels were folded. The final version of ALICE used a novel, staggered wing panel approach, in which the outboard panels were shifted aft from the center panel. When folded, the airfoils nested together in a manner that reduced the overall thickness and provided low profile drag at cruise lift coefficients. The outboard panels had control surfaces that protruded behind the center panel when folded. The control surfaces served as ailerons when the wing was folded for cruise and when fully deployed for loiter, though the deflection direction changed between the two configurations.

The propulsion system had to adapt to the large range of airspeeds. The proposed propulsion system consisted of a JP-8 heavy fuel rotary engine and a folding, variable-pitch propeller. The propeller folded along the fuselage when stowed and in the initial flight after launch. The folding mechanism was similar to those used by electric model aircraft folding propellers. Like electric motors, the rotary engine had low torque pulses compared to a reciprocating engine, which simplified the folding propeller design. The variable-pitch propeller allowed high pitch for high-speed cruise and more flat pitch for the low-speed loiter.

Wing section wind tunnel models were fabricated and tested to validate the performance of each airfoil and the nested (folded) pair. The controllable-pitch folding propeller and rotary engine dual-speed propulsion system was also wind tunnel tested.

ALICE did not proceed past the design stage. Its development team became preoccupied with developing the FINDER UAS, which did not make use of ALICE's novel technologies. The FINDER was air launched from a low-speed Predator unmanned aircraft and was optimized only for low-speed flight, so the dual-mode wing and propulsion system technologies pioneered in the ALICE design were not necessary. Figure 3.30 shows a three-view drawing, and Table 3.6 lists the technical characteristics for this air-launched fixed wing aircraft.

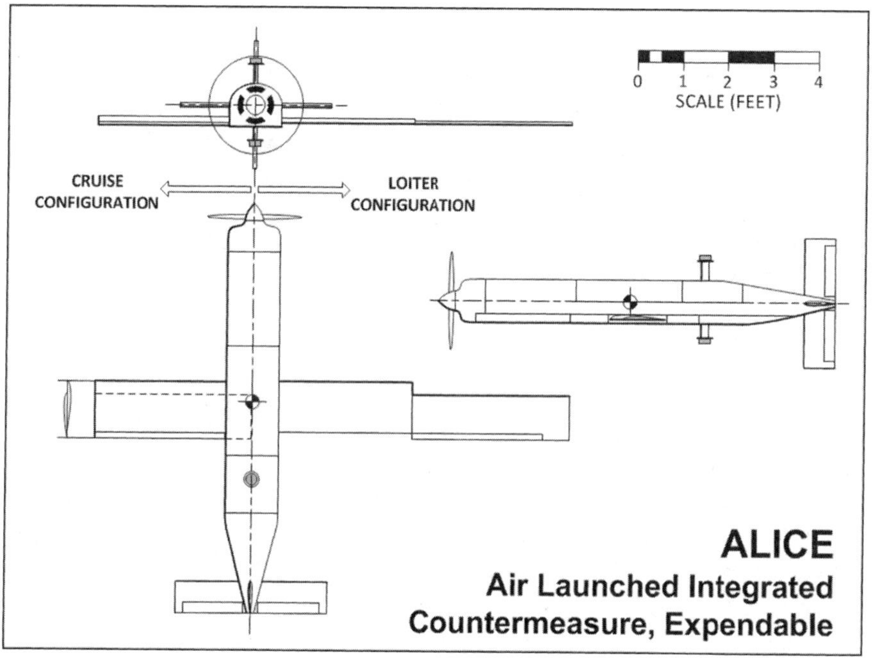

Fig. 3.30 ALICE three-view drawing.

Table 3.6 ALICE Technical Characteristics

Parameter	Value
Wing span	7.0 ft (cruise) 14.0 ft (loiter)
Length	8.7 ft
Takeoff weight	160.7 lb
Empty weight	136.7 lb
Payload weight	25.0 lb
Loiter speed	65 KTAS
Cruise speed	165 KTAS
Endurance	2 hr on station
Max L/D	19.9 (loiter)
Airfoils	Custom folding airfoils
Launch method	Air launch
Recovery method	Expendable
Propulsion type	JP-8 rotary engine
Power	18.0 hp

Duster, 2004

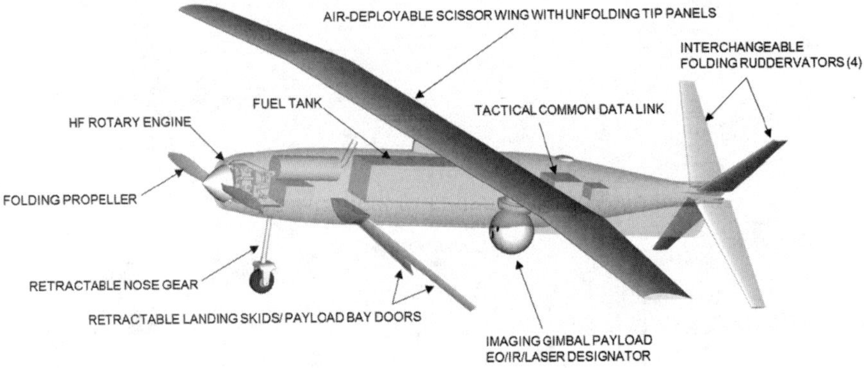

Fig. 3.31 Duster design features.

Duster was a concept for a larger follow-on to FINDER that could be launched from the P-3 Orion or the Multimission Maritime Aircraft (MMA), the planned P-3 replacement. Duster would be housed in an aerodynamic fairing such as a modified Sargent Fletcher external fuel tank pod, and launched from the P-3 or MMA wing pylons. In operations, it would extend the surveillance range of the P-3, the MMA, or the Navy's Broad Area Maritime Surveillance (BAMS) UAS. Missions would include electro-optical/infrared (EO/IR) surveillance, communications relay, biological/chemical agent detection, EW, and sensor emplacement. The primary mission was strike support, in which the Duster UAS could shorten the kill chain. The aircraft would be equipped with a high-performance EO/IR ball with laser designator. Duster could fly below clouds or operate over denied territory. Targeting information would be passed to the P-3 or MMA, which would then relay the information to a killer aircraft, such as an F-18. An advantage of high-performance, air-launched unmanned aircraft is that they can be brought to the area of interest more rapidly than land-based unmanned aircraft.

The Duster configuration was similar to that of FINDER in many respects. It featured a high-mounted pivoting wing with deployable outer panels (Fig. 3.31). The tails had an X arrangement, like the initial FINDER design, in which each tail surface folded along the fuselage and deployed individually. Each of the four tail surfaces was interchangeable. Unlike the FINDER's engine, Duster's engine was mounted in the nose of the aircraft with a tractor propeller to eliminate the need for a complex drive shaft. Duster's payload was mounted in the bottom and rear of the fuselage. The wing tips had moderate anhedral to counter Dutch roll from the parasol wing arrangement. Retractable nose gear and two skids could be deployed for ground recovery. Duster fit within an aerodynamic shell that resembled a drop tank. The propulsion system would be either a small turboprop or a recuperated turbogenerator hybrid electric system.

The Duster project did not go forward beyond a design study. The concept remains valid drawing, and this class of system has not been fielded. Figure 3.32 shows various views, and Table 3.7 lists the technical characteristics for this air-launched fixed wing aircraft.

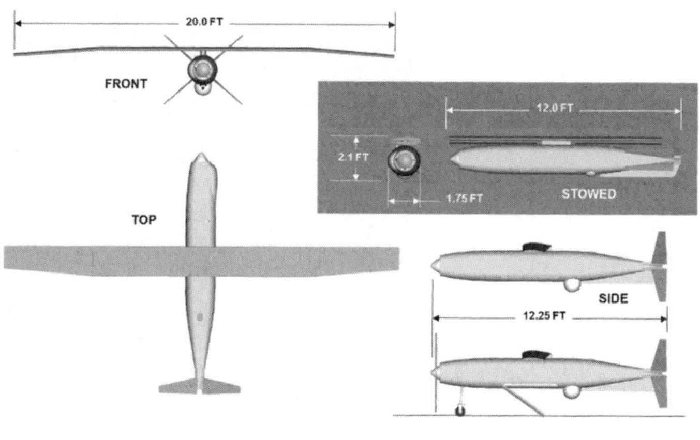

Fig. 3.32 Multiple Duster views.

Table 3.7 Duster Technical Characteristics

Parameter	Value
Wing span	20 ft
Wing area	28 ft^2
Length	12.25 ft
Takeoff weight	300.0 lb
Payload weight	75.0 lb
Fuel weight	75.0 lb
Loiter speed	58–62 KTAS
Stall speed	42 KTAS
Endurance	22 hr at 60 KTAS
Launch method	Air launch
Recovery method	Conventional landing with skids and wheels
Propulsion type	Heavy fuel rotary engine

DUSTER: FURTHER READING

Linne Von Berg, D.C., Howard, J.G., Kruer, M.R., and Lee, J.N., "Deployable Unmanned Systems for Targeting, Exploitation, and Reconnaissance (DUSTER)," *2005 NRL Review*, Naval Research Laboratory, Washington, DC, 2005, pp. 197–199.

AIR DEPLOYED LONG ENDURANCE RECONNAISSANCE (ADLER), 2005

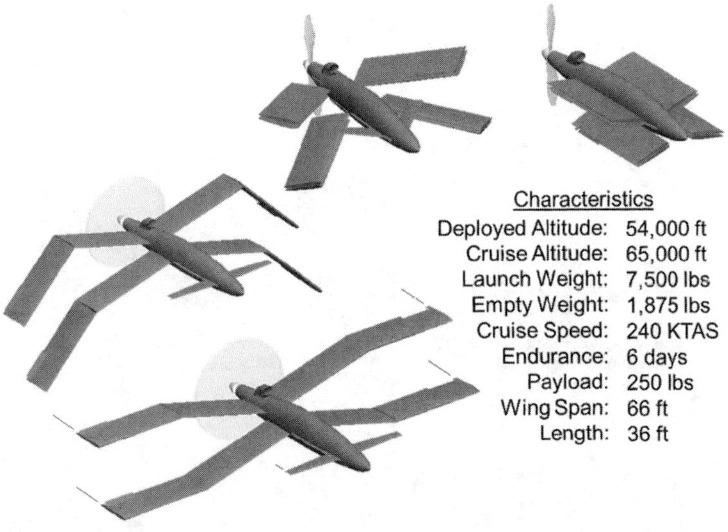

Fig. 3.33 ADLER deployment sequence and technical characteristics.

The Air Deployed Long Endurance Reconnaissance (ADLER) was a concept design for an unfolding, high-altitude aircraft developed for the Defense Advanced Research Projects Agency (DARPA). It could be ejected from the back of a C-17 cargo aircraft (Fig. 3.33). ADLER was designed to be dynamically stable while folded, so no stabilizing parachutes were necessary. The X-wing configuration had wings that folded along the fuselage when stowed and then pivoted open. Each wing had outboard panels that folded under the inboard panels. These outboard panels unfolded with autopilot-commanded aerodynamics to yield the full span, as on Venture (Chapter 4) and FLYRT ATD. A canard pitch control surface pivoted out of the fuselage. The large dihedral and anhedral of the rear wings provided sufficient side area for directional stability.

Two propulsion system alternatives were considered for ADLER. First was a turbocharged heavy fuel reciprocating engine for maximizing endurance. The second was a high-altitude turboprop engine for more rapid development with a proven technology engine, but lower flight endurance. Both would drive a large-diameter, folding propeller.

Unfortunately, during the design phase, the sponsor became enamored by the possibility of performing the same mission with a powered, high-altitude parafoil, and ADLER development was discontinued. Figure 3.34 shows a

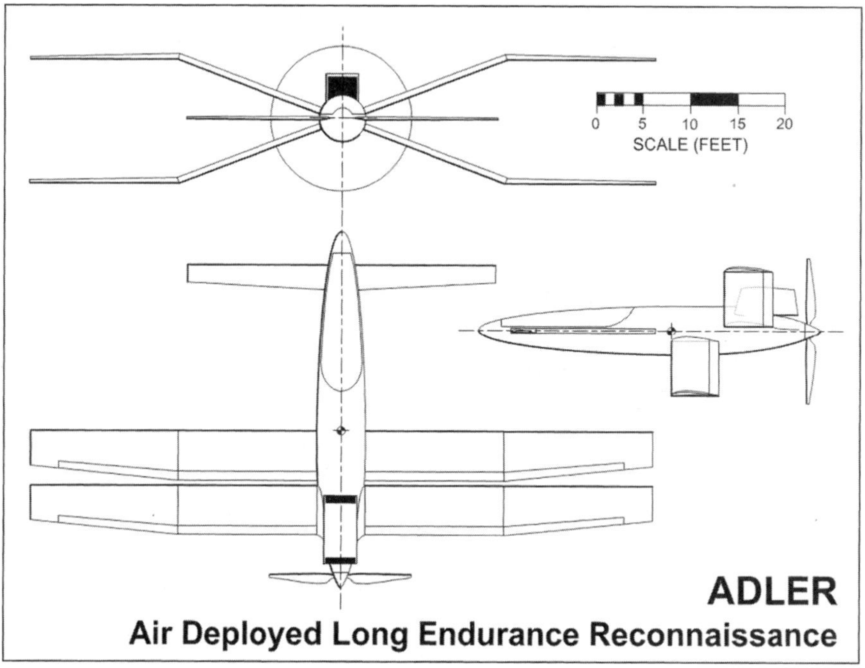

Fig. 3.34 ADLER three-view drawing.

Table 3.8 ADLER Technical Characteristics

Parameter	Value
Wing span	66 ft
Length	36 ft
Takeoff weight	7500.0 lb
Payload weight	250.0 lb
Fuel weight	5385.0 lb
Loiter speed	240 KTAS
Endurance	144 hr at 240 KTAS
Launch method	Air launch
Recovery method	Conventional landing with skids and wheels
Propulsion type	Heavy fuel reciprocating engine or turboprop

three-view drawing, and Table 3.8 lists the technical characteristics for this air-launched fixed wing aircraft.

Insertion of Cargo Expendable (ICE), 2005

Fig. 3.35 ICE conceptual design.

Insertion of Cargo Expendable (ICE) was designed to enable autonomous resupply to forces imbedded in an area where manned aircraft overflights would be too risky (Fig. 3.35). It was designed to be deployed from the aft cargo door of either C-130 or C-17 aircraft at all altitudes and airspeeds under which the cargo door might be opened.

The ICE airframe configuration was selected to be simple to deploy, compact to store, and stackable aboard the deploying aircraft (Fig. 3.36). The selected configuration was also scalable from a small UAS to a large tactical UAS. ICE incorporated a large payload-volume-to-stowed-volume ratio. The payload bay was centered on the aircraft CG, and so the best trimmed flight CG was maintained by symmetrically loading the payload bay, eliminating the need for weight and balance calculations prior to launch. To deploy, a lanyard was pulled to activate the preprogrammed automatic flight control system (AFCS), and the folded ICE was manually

Fig. 3.36 Top and bottom views of stowed ICE.

pushed out of the aft cargo door. To avoid airframe damage and to facilitate stacking, the stowed ICE rested on a custom, blow-molded polyethylene sled that fell away as ICE opened its wings when well clear of the host aircraft. Prior to wing deployment, ICE dropped as a stable, bomb-like shape. Two simple, spring-loaded piano hinges unfolded the wings. Elevon control surfaces provided two-axis roll/pitch control. Between each wing panel and the fuselage, a membrane trailing edge extension reduced wing loading, reduced landing speed, and increased pitch stability.

Potential sponsors considered funding the development of the ICE prototype, which from the start was designed to be easily produced by industry. Unfortunately, support for this interesting design never materialized. Figure 3.37 shows a three-view drawing, and Table 3.9 lists the technical characteristics for this air-launched fixed wing aircraft.

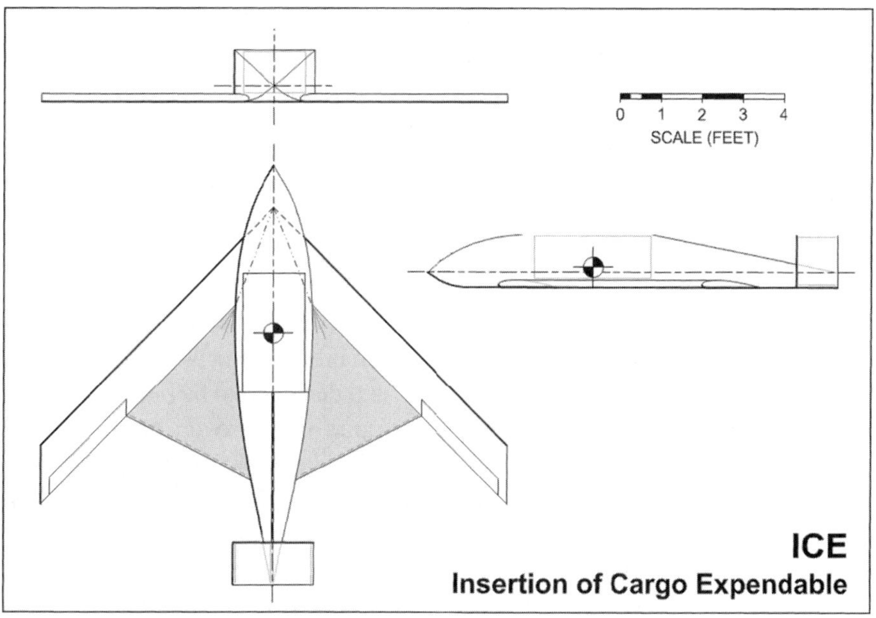

Fig. 3.37 ICE three-view drawing.

Table 3.9 ICE Technical Characteristics

Parameter	Value
Wing span	11.3 ft
Wing area	34.4 ft^2
Length	10.0 ft
Takeoff weight	370.0 lb
Payload weight	270.0 lb
Range	29 nmi from a 22,000-ft AGL launch
Glide speed	87 KTAS
Stall speed	41 KTAS
Endurance	20 min from a 22,000-ft AGL launch
Launch method	Air launch
Recovery method	Crash landing

Disposable Air Vehicle Experiment (DAVE), 2006

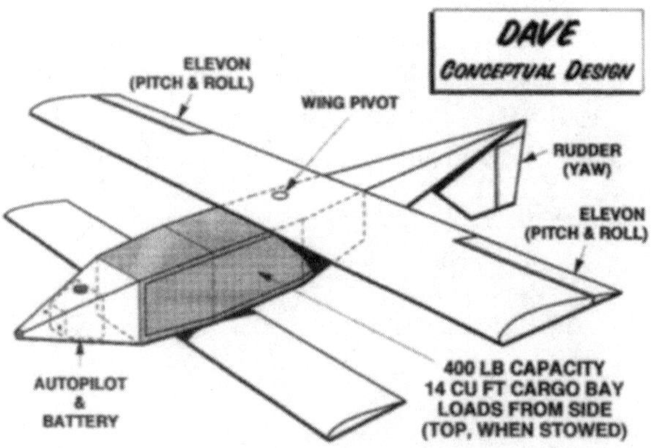

Fig. 3.38 DAVE concept.

The Disposable Air Vehicle Experiment (DAVE) was a conceptual design of a precision cargo resupply system for distributed ground forces (Fig. 3.38). It was approximately twice the size of ICE and designed to be built from non-aerospace materials. It used a simple airframe with proven pivoting rigid wings. It had an L/D of 10, which enabled a significant improvement in range over guided parafoil systems such as the Joint Precision Airdrop System (JPADS). In zero-wind conditions, DAVE would be able to fly 31 nautical miles from a 20,000-ft launch altitude. It could travel 20 nautical miles with a 25-knot headwind. The cruise speed was designed to range from 55 to 75 knots. The fuselage could carry 400 pounds of cargo in a 14-cubic-foot

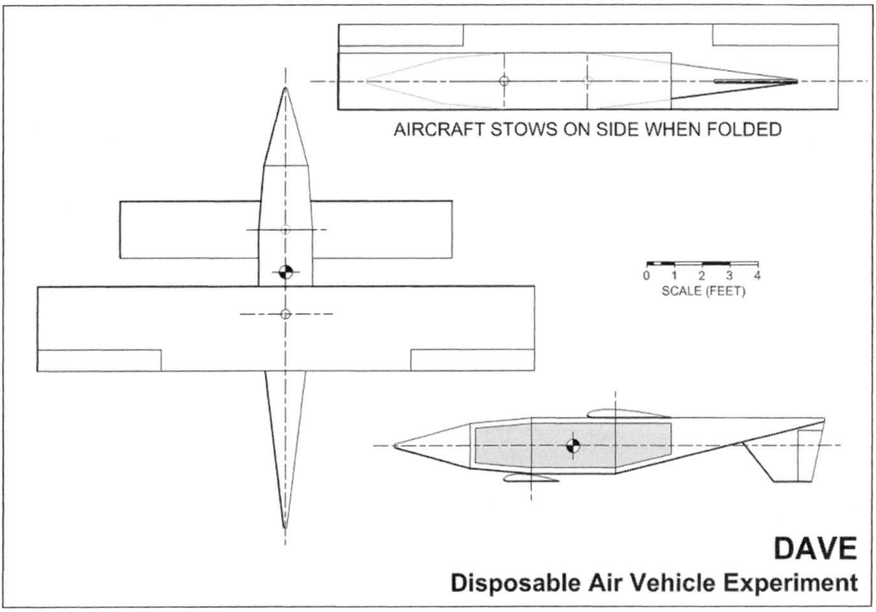

Fig. 3.39 DAVE three-view drawing.

volume. This volume was large for most missions, since only 6.4 cubic feet would be required for 400 pounds of water. Up to six DAVE vehicles could be stored in a single C-130. For launch, the DAVEs would be pushed off the rear cargo ramp of the C-130 or C-17. The wings would deploy after clearing the host aircraft. The payload was placed on the CG so that variable payload weight would not affect the aircraft balance. The airframe was to be

Table 3.10 DAVE Technical Characteristics

Parameter	Value
Wing span	18.0 ft
Wing area	78.0 ft^2
Length	15.5 ft
Takeoff weight	650.0 lb
Payload weight	400.0 lb
Glide range	31 nmi from a 20,000-ft AGL launch
Glide speed	65 KTAS
Stall speed	40 KTAS
Endurance	33 min from a 20,000-ft AGL launch
Launch method	Air launch
Recovery method	Crash landing

constructed of wood, paper, and polystyrene foam to minimize unit cost and to facilitate disposal. The simple controls consisted of elevons on the rear wings and a rudder.

Despite its overall simplicity, its performance advance over JPADS, and NRL's extensive experience with deployable, expendable, autonomous aircraft, sponsorship was not forthcoming to develop this promising concept. Figure 3.39 shows a three-view drawing, and Table 3.10 lists the technical characteristics for this air-launched fixed wing aircraft.

Chapter 4

Research Vehicles

Since the opening of the NRL, researchers at the laboratory have conducted experiments to support technology development. Experimental research provides hand-on experience in the application of concepts and is often the most cost-effective and expeditious method to determine the feasibility and practicality of a new idea. Not surprisingly, when the newly formed Tactical Electronic Warfare (TEW) Division identified the need for self-launching autonomous air platforms for Fleet applications and recognized that existing technologies did not adequately support this need, long-term unmanned aircraft systems (UAS) flight research efforts were initiated, and they continue to the present day.

The goals for developing NRL research vehicles, which are different from mission demonstrators, were to conduct exploratory development of new technologies and to collect engineering data. The technologies explored by NRL UAS researchers included mission payloads and their integration into airframes; advanced aerodynamics; new composite structural materials and fabrication techniques; new and improved forms of aircraft propulsion; new airframe configurations; autonomous launch, flight, and recovery; optimized subsystem integration and function fusion; and affordable expendability.

These test vehicles were predominantly flown via radio control by highly skilled UAS test pilots and were often the most challenging, frustrating, and rewarding projects as they paved the way for new paradigms and revolutionary systems. Often, the new technologies demonstrated were 10 to 20 years ahead of production systems and provided an exciting glimpse into the future. And the lessons learned during these projects were invaluable for enhancing the collective knowledge of the development team and the professional experience of each individual team member.

SHEARWING, 1980

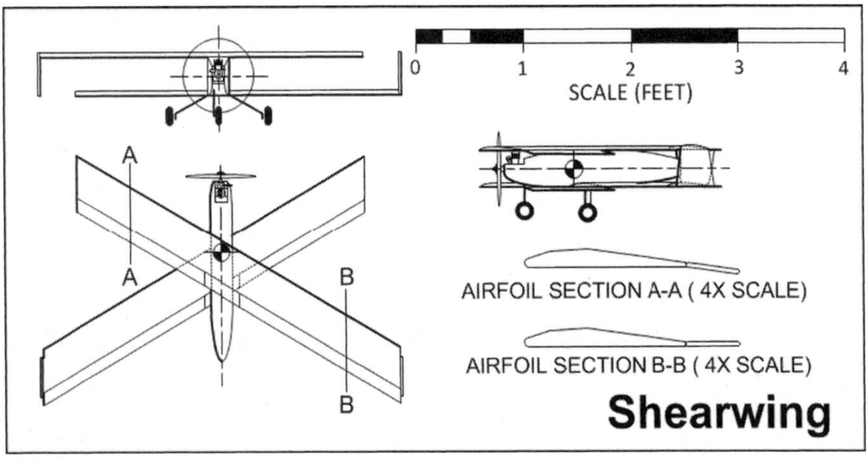

Fig. 4.1 Shearwing three-view drawing.

Foch developed the small X-wing configuration demonstrator pictured in Fig. 4.1 as part of Radio Frequency (RF) decoy research, to determine how well this kind of configuration would fly. It was a configuration that could be made to easily fold and unfold to meet RF decoy packaging needs. Shearwing had wings above and below the fuselage, and each wing was centered about the fuselage such that one half was swept forward while the other half was swept aft. Each wing was straight from tip to tip. The rear tip of each wing had a winglet. The lower wing had an upward-facing winglet, while the upper wing had a downward-facing winglet. The Shearwing design influenced the Experimental Fuel Cell (XFC) design and its U.S. patent nearly 30 years later (2006–2013; see Chapter 10). The primary configuration difference between Shearwing and XFC is that XFC has folding span extensions on the forward part of each wing.

The Shearwing fuselage was a pod configuration with a tractor engine. The aircraft had a fixed tricycle landing gear. The entire wing had camber flaps that could be adjusted for trim. The forward swept portion of each wing had positive camber. The aft swept wing halves had upward trim on the camber flaps and they were actuated as elevons. The camber flap positions were adjustable on the ground to aid in the flight research. Shearwing also used a ribless wing construction technique (see Appendix B).

This small demonstrator used a glow-fueled, single-cylinder, 0.09-cubic-inch, two-stroke engine. When Foch flew the aircraft at the Navy's Webster Outlying Field, he became disoriented due to the aircraft's unusual

Table 4.1 Shearwing Technical Characteristics

Parameter	Value
Wing span	3.33 ft
Wing area	3 ft^2 (reference)
Length	2.25 ft
Takeoff weight	2.5 lb
Empty weight	2.25 lb
Fuel weight	0.25 lb
Dash speed	40 KTAS (est)
Stall speed	20 KTAS (calc)
Airfoils	Faceted
Launch method	Conventional
Recovery method	Conventional
Propulsion type	Enya 09, 0.098-cu.-in., 2-stroke engine
Power	0.2 hp

appearance and had great difficulty leveling the wings. This was due to an inherent feature of the configuration: It appeared the same whether right side up or inverted. In addition, since the wings had low camber airfoils, the aircraft behaved similarly both right side up and inverted. The aircraft crashed in the St. Mary's River next to the airfield. One of the engineers volunteered to recover the crashed aircraft, which was still afloat in the shallows near the shore, if Foch would buy him lunch. Having no access to a small boat, the engineer waded into the river to the aircraft, but lost both of his shoes in the mucky bottom. He successfully recovered the Shearwing, so as compensation for his efforts, as well as his shoes, Foch bought him lunch for an entire week.

A wind tunnel model was constructed and tested at low speed at the University of Notre Dame wind tunnel. The tests revealed aerodynamic performance comparable to that of a more conventional aircraft, and there was less roll-pitch coupling, caused by the right-left asymmetry, than expected. The configuration provides a parabolic longitudinal cross-sectional area distribution considered ideal by Dr. Richard Whitcomb (NASA Langley) for transonic flight because it reduces transonic wave drag. This feature could make the configuration applicable for use as a cruise missile design. However, the transonic X-wing configuration has not yet been evaluated further by NRL. Table 4.1 lists the technical characteristics for this research vehicle.

LOW ALTITUDE/AIRSPEED UNMANNED RESEARCH AIRCRAFT (LAURA), 1986–1990

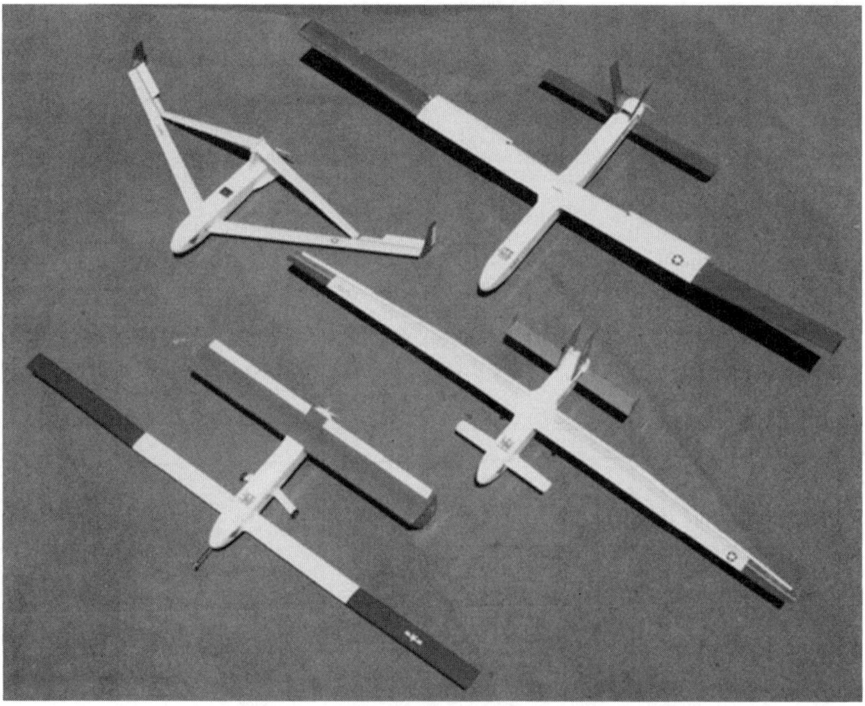

Fig. 4.2 The four LAURA demonstrators.

By 1986, NRL had pushed the development of low Reynolds number (LRN) aerodynamics for efficient airfoils and low-speed airplane design for over a decade. The goals of the Low Altitude/Airspeed Unmanned Research Aircraft (LAURA) program were to provide a practical demonstration of these technology advances (Fig. 4.2) and to collect real-world data to compare theoretical performance with that which could be attained with a practical military system. To exploit these technologies in the most expeditious way for the Navy, it was decided to develop a research test bed that was capable of performing a relevant unmanned aircraft electronic warfare (EW) mission. The Office of Naval Research (ONR), the program sponsor, agreed that using Fleet notional requirements would facilitate direct technology transition to operations. After due consideration, it was decided that the performance goal would be to maximize endurance while flying at ship-like speeds. Although very low-speed flight is taxing with regard to stability, control, and gust response, its one aerodynamic advantage is that the power required to fly is proportional to the cube of velocity. Therefore a very low-speed, aerodynamically efficient aircraft could fly with a small power plant and have long flight endurance on a limited amount of fuel. Without the need for a large fuel load, one could then design a relatively small UAS with a large mission payload mass fraction, as needed for EW platforms.

When airplane designers have attempted to develop the most aerodynamically efficient aircraft possible, one fundamental question has plagued them since the dawn of aviation: "What is the optimal aircraft configuration?" Sailplanes are undoubtedly the most efficient airplane designs, and these designs have gravitated toward conventional monoplanes with high aspect ratio wings, minimum wetted area fuselages, and small tail surfaces. However, sailplanes do not rapidly unfold from a shipboard container; do not carry bulky, power-consuming military payloads; are not adorned with antennas, radomes, and other drag-making accessories; and do not have to be affordably expendable. So, with various UAS and manned airplane designers favoring unusual configurations as being optimal for the stated Navy application, it was decided that the LAURA program would develop multiple prototypes of different configurations in order to assess the performance trade-off when an aircraft is optimized for Navy unmanned EW aircraft requirements. It would be instructive to learn which configuration might have the most potential for a practical Navy aircraft.

ONR sponsored NRL to lead a program with multiple contractors. The effort included flying aircraft with LRN airfoils, acquiring flight test data, and demonstrating long endurance at ship-like speeds.

Recent advances in LRN airfoils showed great promise. Relative to high Reynolds number airfoils used in an off-design LRN condition, purpose-designed airfoils could increase the maximum section lift coefficient by 50% and reduce the loiter drag coefficient by about 20%. However, many such airfoils had thin cusped trailing edges that were difficult to fabricate. Also, these airfoils often demanded good surface tolerances and good surface finishes in order to function as designed. ONR wanted to know how such airfoils performed under operational conditions. NRL tasked contractors with achieving high aerodynamic performance but did not specify manufacturing surface tolerances. If aircraft producers employed challenging airfoils with tight tolerances, they had to find a way to produce them and make them work in the flight test program.

The NRL Vehicle Research Section (VRS) was concerned with ship-launched EW countermeasures missions in which the unmanned aircraft flew at ship-like speeds to generate an effective false radar target. LAURA's specified minimum airspeed was 20 to 25 knots, and the required dash speed was 50 knots. The aircraft were required to have a climb rate of 600 feet per minute or more at sea level. The mission endurance was at least 24 hours. The stability and control characteristics had to satisfy FAA general aviation requirements.

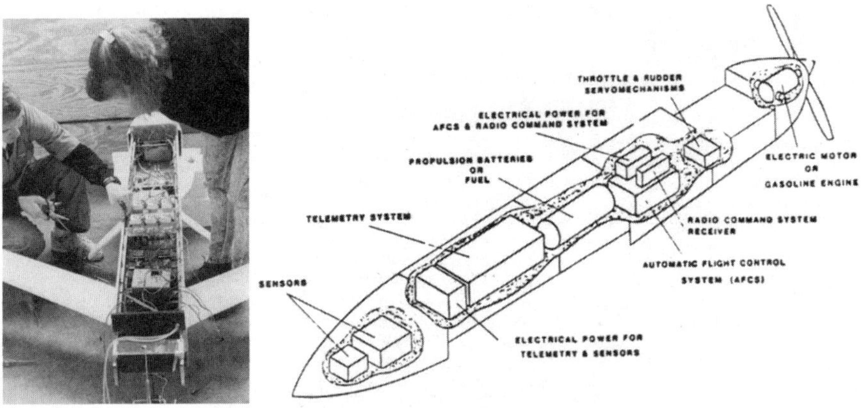

Fig. 4.3 LAURA fuselage internal arrangement.

The gust response behavior of these low-speed unmanned aircraft was not well understood. Some airfoils operating at low Reynolds numbers experienced post-stall hysteresis. After stall, the airfoil would plunge through an angle of attack less than the stall angle of attack before the separated flow abruptly reattached. If a gust caused a stall, the aircraft would oscillate vertically and possibly crash at low altitude. Such bobbing and unpredictable behavior were obviously undesirable for practical aircraft, and these issues would need to be addressed in the LAURA designs.

NRL developed a common modular fuselage and propulsion system to provide to each contractor to reduce duplication between contractors and to enable a fair comparison between configurations (Figs. 4.3 and 4.4). Various wings and tails could be attached to this fuselage. Structural load-bearing hatches were positioned at various points along the upper and lower surfaces and at various positions along the length. The engine was located in the back of the fuselage. Tricycle landing gear was on the fuselage underside. An actual EW mission would use a catapult launcher and the aircraft would be expended at the conclusion of the mission, but a landing gear was used in the test program to enable recovery of the aircraft for multiple flights. The flight data system and telemetry were housed in the nose. The core fuselage, nose section, hatches, and propulsion were provided to each contractor as government-furnished equipment (GFE).

The aircraft had to package within a shipboard stowable box of specified dimensions, but, aside from the defined fuselage, there were no other geometric limitations. The purpose of this was to give the designers freedom to fully exploit the configuration design potential. NRL did not specify the wing span, wing area, or wing loading. Fixing the wing area among all aircraft would have favored the conventional tail aft configuration, for example.

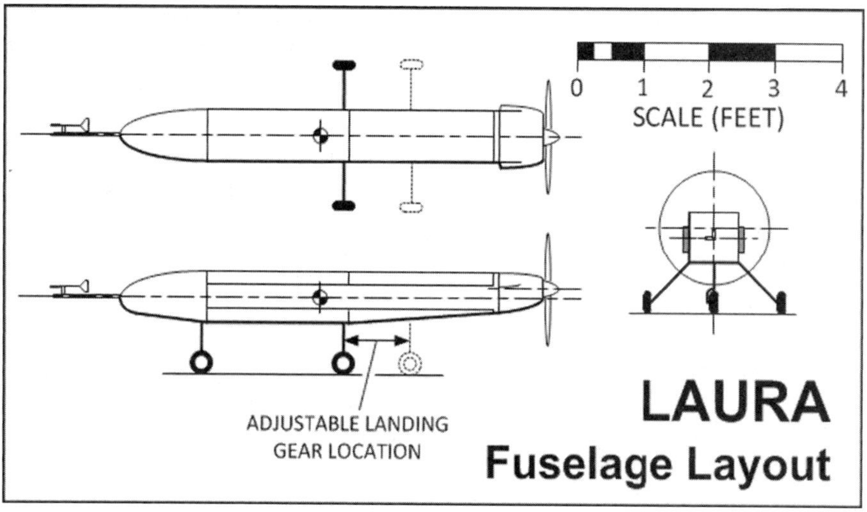

Fig. 4.4 LAURA fuselage three-view drawing.

Designers could choose how the various aircraft components would fit within the container. The box volume was generous, with the fuselage consuming only 25% of the available volume. The box internal length was equal to the fuselage length. The box internal width was equal to twice the fuselage width, and the box internal height was equal to twice the fuselage height. The aircraft had a total weight limit: The gross weight goal was 50 pounds, and the combined wings and tails had a goal weight limit of 18 pounds.

NRL sent out a request for proposals to industry, government organizations, and national labs. These groups were to build four different configurations. NRL wanted diverse responses, so the bidders were encouraged to propose unconventional designs. One government lab and three contractors were selected (Table 4.2). Each vehicle design represented a team's best attempt to satisfy the mission, packaging, and stability and control requirements. At the beginning of the program, each designer was confident that its design would prove superior.

Table 4.2 Four LAURA Configurations and Designers

Vehicle Name	Designer
LAURA Three Surface	NASA Langley and AS&M Inc.
LAURA Joined Wing	ACA Industries
LAURA Tandem Wing	Locus Inc.
LAURA Variable Span	Advanced Aeromechanisms

The program had three phases. The first phase was the detail design phase for all four configurations, which included a rigorous design review. The second phase was a wind tunnel test campaign for the top three designs, downselected from the design review. The third phase was a flight test program for the best one or two designs. In the end, all the designs looked promising and there was no clear factor to justify elimination of any of the four systems, so there was no downselect; all four designs were carried forward through wind tunnel testing into the flight test program. ONR provided additional funding to allow the full demonstration program for all vehicles.

All four teams tested their designs in the University of Maryland Glenn L. Martin 8-foot by 10-foot low-speed wind tunnel (Fig. 4.5). Each team was given one week (40 hours) of tunnel schedule to gather flight data in the first tunnel entry. A second tunnel entry was allowed to remedy any issues and to collect additional data. The wind tunnel models had removable hatches just like the full-size aircraft, so various configurations could use the same body. The models were built at 1/3 scale, with wingspans ranging from 5 to 8 feet. The tunnel was run at three times the flight speed to produce the same corrected Reynolds number. NRL fabricated the fuselage from aluminum

Fig. 4.5 LAURA wind tunnel models.

with removable hatches. The nose and cowl were made from mahogany hardwood. Aluminum fuselage hatches were provided as GFE to the contractors to enable them to interface their wings and tails to the fuselage. The contractor-fabricated model wings and tails were constructed of fiberglass or aluminum.

All four configurations were tested in the same wind tunnel back to back, which provided a major advancement for LRN flight regime research. Approximately 60 to 80 hours of wind tunnel test data were collected. These data included drag polars, forces and moments from alpha and beta sweeps, stability derivatives, and control curves. Processing the wind tunnel results required several months of data reduction. As can be seen in Fig. 4.6, each aircraft configuration had an excellent lift-to-drag ratio (L/D) range for such small, low-speed aircraft. The wind tunnel test data were published in a 1986 Royal Aeronautical Society paper.

Next, the full-scale flight demonstrators were built. The aircraft had spans ranging from 16 to 26 feet. The aircraft included the NRL-fabricated GFE fuselage, complete with its power plant and nose-mounted telemetry and air data systems. NRL installed the main landing gear to the fuselage at the locations provided by the contractor for each of their unique designs. Spare

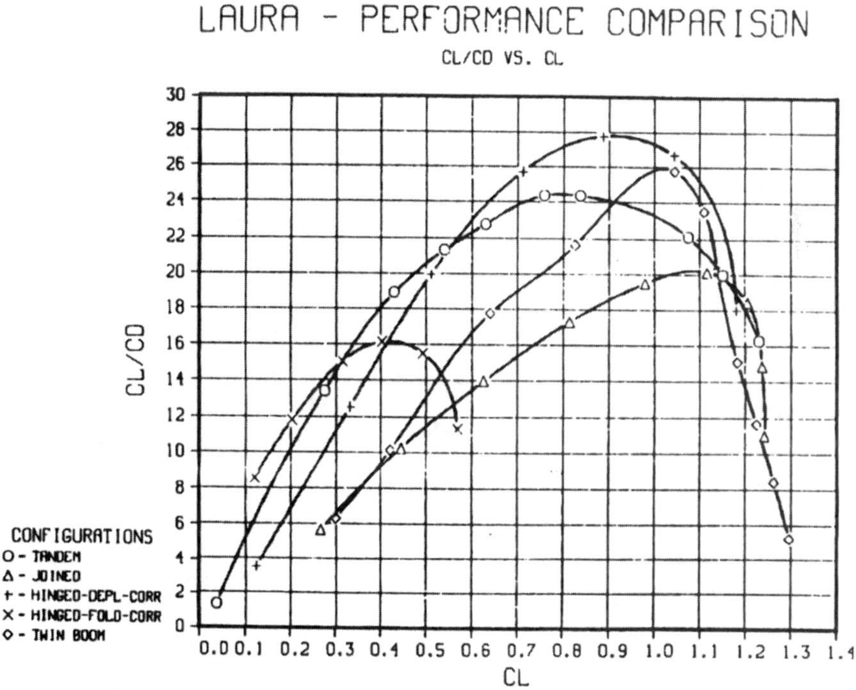

Fig. 4.6 LAURA wind tunnel test results: lift-to-drag ratios.

Fig. 4.7 LAURA Variable Span on the captive carry rig.

fuselage structural components were fabricated to support replacement due to damage.

As a crucial part of pre-flight checkout, each aircraft was tested on a captive carry rig (Fig. 4.7). The rig was cantilevered in front of a truck that could drive at the flight airspeed range of 25 to 40 knots. The aircraft was suspended on the rig but allowed to rotate in pitch so that the control system could be exercised and stall behavior evaluated. The test also allowed evaluation of the engine, cooling system, avionics, and telemetry system.

The flight test program commenced once the aircraft successfully completed the captive carry rig tests. All four aircraft were flown by NRL pilots using radio control. None of the aircraft had artificial stability augmentation. Some aircraft underwent design modifications through the flight test program to remedy handling qualities issues. Flying aircraft with very light wing loadings and at high lift coefficients was challenging for certain flight tasks and especially for landing under less than calm conditions. Prior to flying each design, the test pilots trained on a six-degree-of-freedom (6-DOF) flight simulator. As this was a flight research program, the aircraft were instrumented and significant flight test data were obtained.

LAURA was NRL's most expensive UAS program to date. This multiyear research program involved four large unmanned aircraft with extensive wind tunnel, ground, and flight testing. The lessons learned paved the way for many future NRL systems. NRL demonstrated that LRN unmanned aircraft could achieve long endurance and fly at ship-like speeds. These aircraft flew on normal days, when wind gusts could exceed 15 knots. NRL published the flight test and wind tunnel data. The characteristics of multiple aircraft configurations were evaluated against a single specification and equivalent technology, which provided a basis for direct comparisons in this new flight regime. This research showed that there was no clear aerodynamic advantage between the configurations tested. However, there were different attributes in packaging efficiency and in stability and control, and so different configurations may have advantages for particular applications. The four configurations are described individually in the pages ahead.

LAURA Three Surface

Fig. 4.8 LAURA Three Surface.

NASA Langley Research Center and AS&M Inc. initially developed a high aspect ratio twin-boom conventional design that used advanced airfoils. It used a custom LRN 1010 laminar flow airfoil developed for this project by Dr. Werner Pfenninger. The wings were tapered. The design emphasized aerodynamic performance. After wind tunnel testing, this design evolved into a three-surface configuration by Shahid Siddiqi of AS&M (Fig. 4.8).

The wind tunnel tests yielded disappointing results of the initial twin-boom design. The aerodynamic performance was evaluated in a series of tests. First, the wing and body were tested in an image plane. Next, the tails were added. Finally, the full aircraft was evaluated, including the angle of attack and sideslip sweeps. The untrimmed lift and drag characteristics of the wing and body were encouraging. The wings required trip strips to prevent laminar separation bubbles. The drag of the complete configuration when trimmed was unexpectedly high. The tail booms and tail boom juncture interference drag produced significant drag. However, the biggest contributor was the trim drag caused by the tail download. The airfoils had a high nose-down pitching moment that had to be countered by the tail. In addition, the designers were extremely conservative in providing an exceptionally large static stability margin with a forward center of gravity (CG) location. The resulting large tail size needed for trim made the aircraft very stable, but it

Fig. 4.9 Original twin-boom configuration wind tunnel model.

also added considerable drag-producing surface area (Fig. 4.9). These combined effects reduced the aircraft performance, and the design was considered inferior to the other three LAURAs and was going to be dropped from the flight research phase.

However, because of the excellent performance of the wing/body combination measured in the wind tunnel tests, NASA decided to redesign the aircraft. The new design was a three-surface configuration with the original wing. Now there was a horizontal tail at the rear of the fuselage and a canard ahead of the wing. The twin booms were eliminated and the canard and horizontal stabilizer attached to the fuselage. Two vertical tails with rudders projected from the fuselage for directional stability and yaw control. NASA conducted computational simulations and wind tunnel tests at the Langley Research Center using their own funds to evaluate the modified design. The aerodynamic performance was much improved. The simulation showed that the revised design would be short coupled longitudinally. This raised some concern but appeared acceptable for flight research by skilled UAS test pilots. NRL decided to proceed with this design for the flight test program. The LAURA program was restructured, and ONR provided the requested funding to build the prototype LAURA Three Surface and to conduct flight testing.

Unfortunately, the flight test aircraft had a fabrication shortfall. There were insufficient funds to build a molded composite wing with carefully controlled surface tolerances. The wing as built used a carbon fiber spar and a polystyrene foam core with thin mahogany wood veneer sheeting. The vacuum bagging process for epoxying the skin to the core created pockmarks and bubbles that degraded the surface quality. Also, the airfoil section had variation along the span due to fabrication errors. The program did not have the resources to build a new wing, so the flawed wing was used for the test program. Unfortunately, it was discovered during flight testing that the advanced airfoil was not tolerant of these poor tolerances.

The captive carry rig tests revealed that the LAURA Three Surface had "soft" directional control, as expected due to its short tail arm. Additional fin area was added to remedy the problem. The captive carry tests confirmed that the aircraft was sensitive in pitch, also as predicted by the simulation. Prior to the flight, the pilots conducted at least six hours of flight simulator time to learn how to control pitch attitude on this close-coupled aircraft. With the system made ready and the pilots trained, this interesting configuration was cleared for flight.

The LAURA Three Surface made a single flight. The aircraft rolled down the runway, rotated, and began the initial climb. The aircraft appeared to begin to stall shortly after liftoff, and the right wing began to drop. It did not immediately respond to roll or yaw commands. The pilot reduced the pitch to reduce the angle of attack and to build up airspeed. The lateral-directional control was regained. The aircraft was able to continue forward flight at a nose-level pitch attitude with no climb rate despite having what appeared to be a stalled wing, and the test pilot felt the aircraft was trapped flying on the back side of the lift/drag curve. The flow did not reattach despite the moderate angle of attack, apparently due to an LRN stall hysteresis effect. The pilot attempted to turn the aircraft back for an immediate landing. However, the aircraft experienced a tip stall that resulted in a sharp roll rate to the right. The aircraft rolled over about 180 degrees, and the nose then fell through into a vertical dive. With only 100 feet of altitude, the pilot was

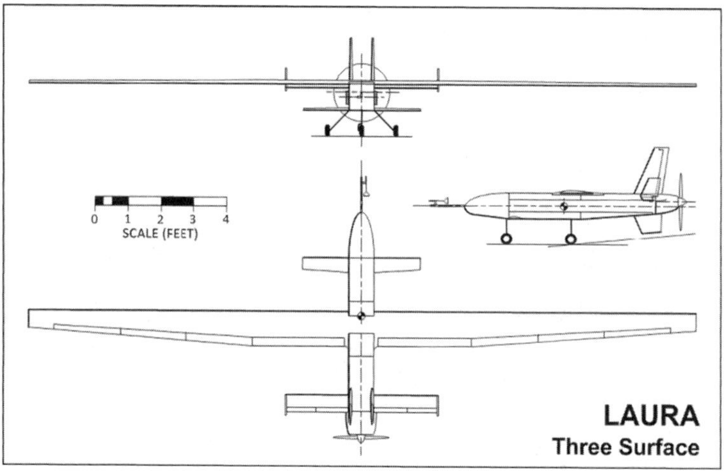

Fig. 4.10 LAURA Three Surface three-view drawing.

Table 4.3 LAURA Three Surface Technical Characteristics

Parameter	Value
Wing span	22.0 ft
Wing area	19.0 ft^2
Length	6.67 ft (three surface)
Takeoff weight	56.4 lb
Max L/D	25.79 (twin boom)
$C_{L,\max}$	1.29
Airfoils	LRN 1010 on wing
	NACA 0010 or 0012 for horizontal tail
Launch method	Conventional
Recovery method	Conventional
Propulsion type	Reciprocating gas engine
Power required	0.138 hp

unable to recover the aircraft, and it impacted the ground nearly vertically. Surprisingly, there was little damage to the aircraft thanks to the soft ground. The expensive telemetry and data acquisition systems were housed in the frangible Kevlar nose, and these were destroyed. The pilot observed that the aircraft had marginal directional control due to the short moment arm of the vertical tails and believed the aircraft was flying in a post-stall mode throughout the flight. The aircraft was repaired after the crash but never flown again due to funding constraints that prevented fabrication of an improved wing. Figure 4.10 shows a three-view drawing, and Table 4.3 lists its technical characteristics.

LAURA Joined Wing

Fig. 4.11 LAURA Joined Wing.

The LAURA Joined Wing demonstrator was developed by ACA Industries Inc. of Torrance, California (Fig. 4.11). ACA's president and chief designer, Julian Wolkovitch, was a pioneer of the joined wing configuration. Joined wings can be thought of as a form of close-coupled tandem wings. However, because of the rather small rear wing chosen for the LAURA Joined Wing, this design could be considered to be a high aspect ratio flying wing with a lift strut.

This joined wing design had a swept main wing with dihedral that was attached to the lower fuselage. The front wing was highly cambered. The aft fuselage had a vertical stabilizer that also served as a post for the rear wing. The rear wing was unswept with anhedral to join the main wing at mid-span and the top of the vertical stabilizer. The vertical stabilizer had a rudder. Elevons on the outboard portion of the main wings provided roll and pitch control.

The wind tunnel tests revealed that the aircraft had a good L/D but undesirable post-stall pitch-up (also known as an unstable pitch break). The rear wing stalled before the front wing, causing the nose-up pitch. This behavior could lead to end-over-end tumbling or a nonrecoverable flat spin. The solution to the stall pitch-up problem was the addition of vortilon turbulators on the rear wing. Vortilons are angled flat plates on the wing leading edge that align with the flow during normal flight conditions but that generate vortices at high angles of attack. The vortilons were applied to the model in the second wind tunnel entry, and they fixed the problem. The elevon chord was also increased.

The wind tunnel tests also revealed that the elevon control surface response was highly nonlinear with control deflection. Moderate control surface movement did not generate a control effect. Large control effects would then be seen at large deflections. The elevons were flat plates extending from the trailing edge of the main wing. The flow remained attached when the elevons were aligned with the rear camber of the highly cambered and cusped Wortmann FX 63-137 main wing airfoil, but the bottom surface separated with moderate upward control deflections.

The yaw control using the rudder was ineffective, or soft, and the directional stability was weak. Winglets were added to the main wing's tips to improve the directional stability. These were located aft of the CG such that there was a yaw moment arm. Additional sideslip data were obtained in the second wind tunnel entry to test the winglets.

The captive carry rig tests revealed that the elevator authority was still marginal, despite the control surface enlargement from the wind tunnel tests. The elevon chord was enlarged by 80%. Two horizontal anti-stall strakes were added to the lower rear fuselage as additional protection against unstable pitch break behavior. These strakes would produce vortices at high angles of attack that provide a restoring nose-down pitching moment. The static margin was found to be too small, making the aircraft excessively sensitive in pitch. The CG was moved forward to increase the static margin. At this point, the stability and control characteristics were deemed acceptable for flight, and the pilots planned to avoid intentionally stalling the aircraft.

Next, the program began flight testing. On the first flight, the aircraft climbed out without incident. The nonlinear pitch response was observed as expected. There was little control authority until about two-thirds deflection, when, suddenly, a large response was generated. The pilot had to pitch the aircraft down quickly at times to prevent excessive angle of attack when the elevons became effective at large deflections. This gave the pilot the feel of a control reversal, though this was caused by local separation on the control surface and not aeroelastic elevon reversal. A very large down elevator control deflection was required to trim the aircraft. Pitch doublet maneuvers showed adequate pitch damping at cruise flight speeds, with degrading damping at the lower-speed end of the flight envelope.

The rudder response was even poorer than expected. Full rudder deflection provided no yaw authority, but its drag did raise the nose. Rudder deflection created large interference drag with the rear wing. The drag was above the CG, so a nose-up pitching moment resulted. The aircraft had good yaw damping. The aircraft could be turned using the elevon differential deflection. However, the elevons generated significant adverse yaw with differential deflection. The dihedral effect of the configuration generated a restoring rolling moment that acted opposite of the roll command.

Landing on the first flight was challenging. The pitch control problems made it difficult to not hit the nose gear with excessive force. The aircraft had a tendency to over-rotate or under-rotate. When the aircraft over-rotated, it would pitch up, then pitch down and strike the nose gear. The high L/D and light wing loading caused the aircraft to float down the runway. The first flare occurred too high, resulting in a nose gear strike. The landing gear was too stiff for the aircraft weight, and also too springy. The energy stored in the landing gear acted as a spring, which caused the aircraft to bounce significantly with limited damping. At or slightly below stall speed (which can be reached when flying close to the ground due to ground effect), the aircraft had very little pitch damping due to the close-coupled design. The aircraft bounced multiple times on landing. There was insufficient weight on the nose gear, so steering during taxi was challenging. The outboard portion of the main wing sustained localized spar damage due to the inertial loads caused by the bouncing.

The aircraft was modified after the first flight. The elevon chord was increased to improve the pitch characteristics, but this had marginal influence. The elevon neutral trim position was adjusted so that the pilot would not need to hold down elevator for normal flight. The yaw control problems were addressed by greatly increasing the winglet size. The winglets were flat plates constructed of laminated balsa and carbon fiber. The winglets had drag rudders, which were split control surfaces that generated drag when deflected. The drag acting at the wing tip generated a yawing moment.

The second flight result revealed that the design modifications improved the handling characteristics. The lateral-directional handling qualities were similar to those of a general aviation aircraft. The pitch control was still soft near neutral and sensitive near the deflection limits. This was acceptable for a test aircraft but undesirable for a man-in-the-loop piloted operational system. The aircraft was landed successfully despite a downward wind gust during the flare.

The third flight evaluated the lateral-directional handling characteristics. The test confirmed that the elevons produced excessive adverse yaw. The aircraft was flown without incident.

During the final flight test of the LAURA Joined Wing, the aircraft had a hard nosewheel strike on landing. This caused the aircraft to bounce and over-rotate. The rear wing stalled and the aircraft attained a 40- to 45-degree nose-up attitude at an altitude of 5 feet. The aircraft dropped and hit the wing tips. Post-flight X-ray images of the wing revealed hundreds of cracks in the carbon fiber wings. The aircraft was considered unrepairable, and it never flew again.

The nonlinear elevon control response was never solved in this program. A solution would have been to apply a reduced camber airfoil on the main wing outboard of the wing intersection. Large camber was not necessary at the tips, which operated at a lower lift coefficient than the root. The nonlinear behavior was associated with the large camber due to separation effects. A production aircraft could be designed to avoid this problem.

The LAURA joined wing proved that a joined wing, low-speed aircraft was viable but that additional modifications were necessary. The test pilots

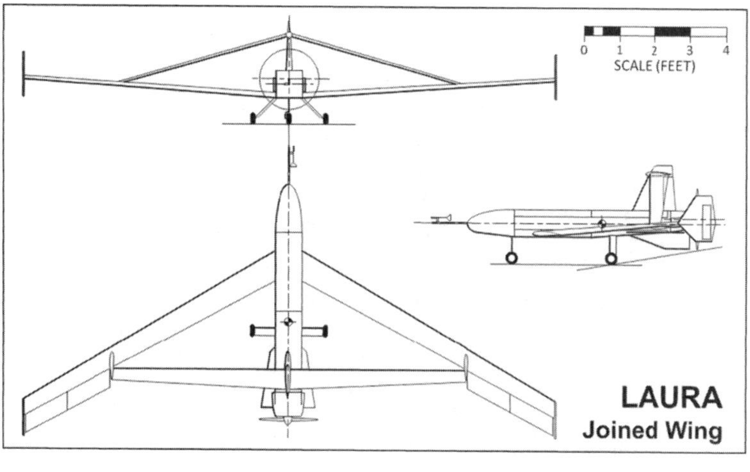

Fig. 4.12 LAURA Joined Wing three-view drawing.

Table 4.4 LAURA Joined Wing Technical Characteristics

Parameter	Value
Wing span	15.8 ft
Wing area	14.4 ft^2
Length	6.67 ft
Takeoff weight	50.3 lb
Loiter speed	26–28 KTAS
Stall speed	25 KTAS
Endurance	18 min (demonstrated)
Max L/D	20.13
$C_{L,\max}$	1.24
Airfoils	Wortmann FX 63-137 (front wing)
	Eppler E201 (rear wing)
Launch method	Conventional
Recovery method	Conventional
Propulsion type	Reciprocating gas engine
Power required	0.140 hp

could control the aircraft adequately. This aircraft had the least wing area of the four designs, so it had higher cruise and dash speeds. The high wing loading would be well suited to a range mission. The mission aircraft would be catapult launched and expendable, so the landing gear problems would not be a factor. Figure 4.12 shows a three-view drawing of this research vehicle, and Table 4.4 lists its technical characteristics.

LAURA TANDEM WING

Fig. 4.13 LAURA Tandem Wing.

Locus Inc. was awarded the contract for the LAURA Tandem Wing aircraft (Fig. 4.13). Richard Eklund was the lead designer from Locus, and he also designed the earlier LODED tandem wing unmanned aircraft (Chapter 2). Several lessons learned from the LODED program were applied to this research aircraft to create an improved design.

The front and aft wings had equal area to make it a true tandem wing. To have a positive static margin, the front wing carried twice the lift coefficient of the rear wing and it produced two-thirds of the total lift. The front wing had a higher aspect ratio to reduce induced drag, since it does most of the lifting. The rear wing had a lower aspect ratio to provide a lighter structure. The rear wing had twice the chord and half the span of the front wing, so the aspect ratio was one-quarter that of the front wing. The front wing used the highly cambered Wortmann FX 63-137 airfoil, and the rear wing used a thick, flat-bottom airfoil. Downward-facing winglets extended from the tips of the rear wing for directional stability.

The LAURA Tandem Wing had an unusual flight control arrangement. Pitch control was originally achieved by rotating the entire front wing in pitch. As with LODED, the wing pitch was actuated with an irreversible worm gear. The rear wing originally had aileron control surfaces that deflected differentially. The wing pitch actuator had development problems, so the pitch and roll control arrangement was changed to elevons only using the rear wing control surfaces, and the front wing was to be actuated to trim the aircraft in pitch for minimum drag. Yaw control was originally provided by a ventral rudder on the lower fuselage. This was an all-moving vertical control surface that rotated along a vertical axis. The rudder and ailerons were actuated with conventional rotary servos. To improve the yaw control response, split drag rudders were added to the winglets after the first flight.

Like LODED, the LAURA Tandem Wing tended to be tail heavy due to the distribution of components within the airframe. The rear structure had to be lightened to achieve the desired CG. The rear wing was built exceptionally lightweight, so the aircraft could achieve the desired longitudinal CG without nose ballast. It was constructed of a carbon shell with the supporting foam removed, plus ribs and spars. The resulting structure was quite light, but strong enough for static ground tests. The front wing had female molded carbon fiber skins with foam core.

The wind tunnel tests showed reasonable stall behavior. The front wing stalled before the rear wing, so there was not an unstable pitch break. However, the aircraft had a tendency to stall toward a tip of the front wing, which would cause a roll at stall. The low aspect ratio and lightly loaded rear wing was resistant to stall.

The rear wing structurally failed in a captive carry test. The failure occurred when the test rig heaved as the truck ran over a bump. In an attempt to make the rear wing light, the structure was slightly too weak for the flight loads in combination with an inertial shock such as might be caused by a sharp gust. The rear wing was strengthened, and nose ballast was added to balance out the extra weight in the tail. The structure did not break in a follow-on captive carry test, and the LAURA Tandem Wing was cleared for flight. This experience proved the value of the captive carry rig testing by identifying a problem that could have caused a crash due to in-flight structural failure.

Next, the program entered flight testing. The aircraft made a normal takeoff on the first flight. It rotated and had a good climb out. Large down-elevator deflection was required to keep a reasonable climb attitude. The aircraft trimmed out well in pitch. The flight test team cheered, but the pilot was concerned. He commanded first gentle then moderate roll commands to initiate a turn to head back to the takeoff area, but there was no response from the aircraft whatsoever, and it continued to maintain a constant heading. Even with full, stick hard-over roll and yaw commands, there was no response, and the aircraft continued to climb straight away from the takeoff point. Then due to a gust, the aircraft began to turn on its own through a 270-degree heading change. When the pilot attempted to level the wings, the aircraft did not respond. The rudder and ailerons did not have any effect on initiating or arresting turn rate!

Although there was no roll response when the pilot commanded full left roll, the flight crew could observe the ailerons moving as commanded. This control response problem was not observed during the captive carry tests because the captive carry rig had no freedom to roll.

To keep the aircraft from leaving the test range, the pilot stalled the aircraft in an attempt to have a minimum-impact crash on the airfield. He closed the throttle, he eased the control stick fully back, and the aircraft stalled. As the aircraft entered this power-off stall, the left wing dropped and the aircraft began a turn. The pilot recovered from the stall, and the aircraft was in a 30-degree bank. The aircraft continued level banked flight until the heading changed by about 60 degrees and its natural stability leveled its wings. The series of stalls and recoveries continued until a 180-degree turn was completed so that the aircraft returned to the runway. Over several tense minutes, the maneuvers were systematically completed again and again, so that the aircraft could align with the runway. The only turn authority occurred in stall and stall recovery. On landing, a gust caused the aircraft to roll to the right. The pilot had to stall the aircraft once again to correct the heading. The aircraft hit the runway with a high rate of descent. However, the aircraft survived with only minor damage to the landing gear, rear wing, and winglet.

Several modifications were made to the aircraft prior to the second flight. Post-flight investigations revealed that the structure near the rear wing

Fig. 4.14 LAURA Tandem Wing in flight.

mountings had been crushing during flight, allowing the wing to rock laterally on the fuselage. The wing rocking produced enough yaw to counter the aileron rolling command thanks to the roll-yaw coupling. The winglet size was increased. This eliminated the winglet ground clearance during rotation, so wheels were added to the bottom of the winglet. Split drag rudders were added to the winglets for improved yaw control.

The handling qualities were improved on the second flight, and while the Tandem Wing's turning characteristics were just acceptable for test flying, they would need improvement on a mission aircraft (Fig. 4.14). The drag rudders produced a reasonable yaw response. The ailerons generated adverse yaw that limited their effectiveness. Differential ailerons (deflection is larger up than down) were not possible because they would generate a coupled pitching moment. The maximum lateral control was approximately 10 degrees per second. The aircraft was capable of turning with large control deflections of the drag rudders and ailerons. Control mixing of the ailerons and drag rudders slightly improved the turning effectiveness. The pilot commanded approximately 3/4 down elevator to trim in pitch for the entire flight, which probably indicated that the rear wing incidence was too low, or the forward wing incidence too high.

The LAURA Tandem Wing had a third and final test flight. The forward wing incidence was reduced so that the pilot would not need to hold elevator beyond the trim limits. It was discovered that the lateral-directional control improved with increased airspeed.

The aircraft made three flights with no crashes. It could be trimmed at different airspeeds. The tandem wing aircraft met its performance objectives.

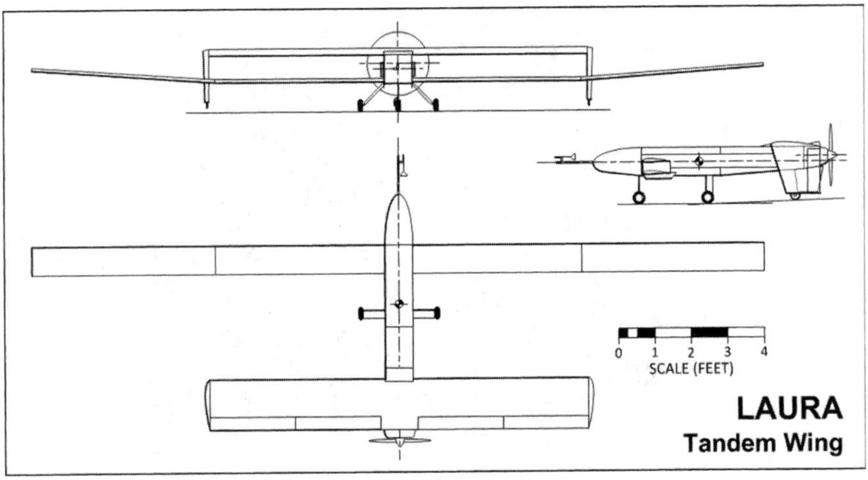

Fig. 4.15 LAURA Tandem Wing three-view drawing.

The pitch control was excellent, though the lateral-directional control was never ideal. Landing could be difficult in gusty conditions due to the roll rate limitations, but it certainly appeared that a mission-capable aircraft could be developed with acceptable lateral-directional control characteristics. Figure 4.15 shows a three-view drawing of this research vehicle, and Table 4.5 lists its technical characteristics.

Table 4.5 LAURA Tandem Wing Technical Characteristics

Parameter	Value
Wing span	20 ft
Wing area	30.5 ft^2
Length	6.67 ft
Takeoff weight	61.0 lb
Loiter speed	26 KTAS
Max L/D	24.36
$C_{L,\max}$	1.23
Airfoils	Wortmann FX 63-137 on front wing Eppler 205 on rear wing
Launch method	Conventional
Recovery method	Conventional
Propulsion type	2-stroke reciprocating engine
Power required	0.126 hp

LAURA VARIABLE SPAN

Fig. 4.16 LAURA Variable Span with wings retracted (top) and deployed (bottom).

The LAURA Variable Span was developed by Advanced Aeromechanisms of La Jolla, California (Fig. 4.16). Alberto Alvarez-Calderon was its designer. This aircraft had a conventional tail-aft configuration but an in-flight variable-span wing invented by its designer. The benefit of the variable span was that the wing area and span loading could be adjusted for dash or loiter conditions. The variable-span approach was also known as the "hinged wing."

This design attached a second fuselage section to the NRL standard fuselage to yield a single fuselage of about one and one-half times the standard length. The motivation was to provide a longer tail moment arm to help accommodate the pitching moment change due to the variable span. The tail loads changed significantly between the two span states. The other design teams protested the fuselage modification, but NRL decided that the rules permitted the box volume to be used in this manner. As it turned out, the other design teams could not have used the extra fuselage to advantage. The horizontal tail was sized to be zero loaded (zero lift coefficient) for the full-span, maximum-endurance configuration.

The variable span worked by aerodynamically deploying the wing tips from the stowed position to the full-span position. The stowed airfoils joined together to form a single, thick, semi-symmetrical airfoil. The top airfoil of the inboard wing had a leading edge cusp to help seal the gap between the top wing and folded lower wing leading edges. This cusp also enhanced the maximum lift coefficient when the wings were deployed due to increased camber. This clever cusp was informally named the "Calderoneron" by Foch in honor of the designer. The inboard and outboard wings used Eppler airfoils that were modified to be flat near the leading edge in order to facilitate folding. The approach to unfolding was to release the folded wing, which would drop under gravity and aerodynamic downward lifting forces. No torsional springs were required. As it dropped, the lower wing would rotate about a horizontal hinge at the tip of the upper wing. The angular momentum plus the aerodynamic forces were sufficient to complete the rotation to the horizontal, full-span position. The wings would then latch. This design used ailerons on the outboard wings to facilitate the deployment. The ailerons would control the rotation rate so that the wings were not overstressed and the final latching was smooth with minimal shock. The rotation position of each wing was measured by a potentiometer. A control system on each wing managed the deployment via an outer wing aileron servo, ensuring synchronization. Processors on autopilot boards hosted the control software. The control system could also enable the wings to fly back to the stowed position if the wing latch was configured for release. Such an electronic latch would be a simple engineering problem, but it was not part of the LAURA program. The wing deployment was tested on the captive carry rig but not in flight during the LAURA program due to funding limitations that prevented improvement of the wing deployment autopilot. The aircraft did fly with the wings deployed and with the wings folded.

The wind tunnel testing showed that the aircraft exceeded its aerodynamic performance requirements and had good stability characteristics in the folded and full-span configurations. This was not surprising for a conventional design with robust modified Eppler airfoils. The designer elected to conduct additional experimentation with the extra wind tunnel time. He installed a new fuselage that aligned with the wing flow but had constant depth. The fuselage had the appearance of an inch worm or banana. Despite the unusual appearance, the L/D increased by about 4%. However, per NRL requirements, the standard LAURA straight fuselage was used for the flight test program.

The flight vehicle was evaluated in thc captive carry rig. The wings were successfully deployed. A wing flap servo failed during the tests, leading to a quick repair. Following the tests, the weight grew due to local strengthening, repairs, and additional avionics. The modified aircraft's gross weight was higher than originally estimated, so some of the structural margin was eliminated. The structure was designed for +4/−2 g with a 50% margin, so the aircraft was cleared for flight with restrictions. The structure was found to be very flexible in the captive carry tests, which placed limitations on the flight envelope. NRL engineers were concerned about flutter in the flight test program. The stall behavior was found to be benign with ample stall warning. The captive carry tests showed no critical problems, and the aircraft was cleared for flight.

The airframe was designed to meet the design requirements with little additional margin and no growth provisions. The structure passed the static testing on the ground, as required. The aircraft had flight envelope restrictions in a dive, it could not fly at full throttle in level flight, and it could not withstand gusts while turning with a large bank angle. These limitations were not the result of design flaws. Rather, the designer followed the rules and specifications and sought weight reduction to maximize performance.

The aircraft first flew in the full-span configuration. The telemetry did not work while the aircraft was on the ground. The aircraft was placed on top of a truck and the telemetry then worked. The flight test team decided to proceed with the flight despite the risk that the telemetry system might not function properly. Telemetry was important not only for seeing the complete set of flight data, but also because the airspeed could be read to the test pilot to provide a feedback loop with the pilot. The aircraft rolled along the runway, and the pilot rotated the aircraft based on visual feedback. The pilot held full-up elevator to enable takeoff at the minimum unstick speed. The aircraft climbed at a steep angle to keep the airspeed low due to the flight envelope restrictions. There was little concern about stall, because the outboard wing panels had washout, and the inboard camber flaps were locked down in the takeoff position. As shown in both wind tunnel and captive rig testing, the inboard wing would stall first, so no tip stall behavior was expected. The controls were programmed for coupled rudder-ailerons to counter adverse yaw from the ailerons. The vehicle had positive longitudinal, lateral, and directional stability with high damping about all three axes. The inherent spiral stability allowed the aircraft to fly for several minutes without pilot inputs.

The landing approach was challenging for the pilot because the descent rate with the engine at idle was very low in the full-span configuration. The aircraft had a high L/D and light wing loading. A long approach was required to get the aircraft on the runway. The pilot had to perform 360-degree turns near the runway to lose sufficient altitude for landing. The aircraft floated above the runway in ground effect for a long distance. The second flight also flew with the wings deployed.

The wings-folded configuration was evaluated in the third flight. The wings were latched in the closed position. The inboard camber flaps were deployed to the maximum lift setting, and they were configured to act as ailerons. The shorter-span wing was stronger, so the flight envelope had higher speed limits. The folded wing had less dihedral than the full-span wing. Computer simulations showed possible problems with spiral divergence, so small winglets were added to provide more dihedral effect.

Despite the shorter span, the aircraft still had a light wing loading. Climb performance and overall handling characteristics were very good.

The LAURA Variable Span had two full-flying rudders. Each vertical stabilizer rotated entirely to provide yaw control. There were no traditional rudder control surfaces. One of the rudders failed on the third flight. The right rudder began to flutter due to a servo failure. Signal noise caused the servo to command an excessive rudder deflection, resulting in excessive load that stripped the servo gears. The flutter caused the rudder main spar to break. The redundancy of having twin vertical tails permitted the pilot to execute a safe landing.

The purpose of the fourth and final flight was to measure fuel consumption. The aircraft was flown with a fuel quantity appropriate for one hour of flight. The troubled data acquisition system was removed and replaced with a large battery to power the avionics for the long-duration flight. Pilots took shifts at 20-minute intervals. At 55 minutes into the flight, the aircraft experienced interference that caused an uncommanded pitch-up and stall. The interference was likely caused by microwave radar equipment testing at the test facility. The aircraft was recovered from its stall back to normal flight, and the pilot brought the aircraft in for a landing. The engine quit on final approach, but the aircraft made a successful landing. Fortuitously, the engine stopping at the end of the flight due to fuel starvation facilitated a good estimate of fuel consumption.

The wing deployment autopilot experienced problems, so no in-flight unfolding was attempted in the LAURA program. The wing was not strong enough to handle an uncontrolled deployment. Alvarez-Calderon continued to develop the wing deployment technology on his own. The wing and tail were reused on the lighter-weight Venture project to successfully demonstrate the in-flight wing deployment (1994–1995, discussed later in this chapter).

NRL proposed a transatlantic demonstration using this aircraft to ONR. However, the LAURA aircraft were designed for an EW mission, and at the time, this capability for an EW platform was considered sensitive information. So, ONR elected to keep the technological capability of having a transatlantic small tactical UAS solely within the Navy. It is worth pointing out that this capability existed about 10 years ahead of the famous Insitu Aerosonde transatlantic crossing, which was sponsored by ONR! The LAURA Variable Span had the aerodynamic performance to fly 60 hours on a gallon of fuel. The only technological barrier at the time was the engine. More fuel was lost dripping out of the carburetor intake than was burned in the engine. A solution would have been to develop a miniature fuel injection system. Figure 4.17 shows a three-view drawing of this research vehicle, and Table 4.6 lists its technical characteristics.

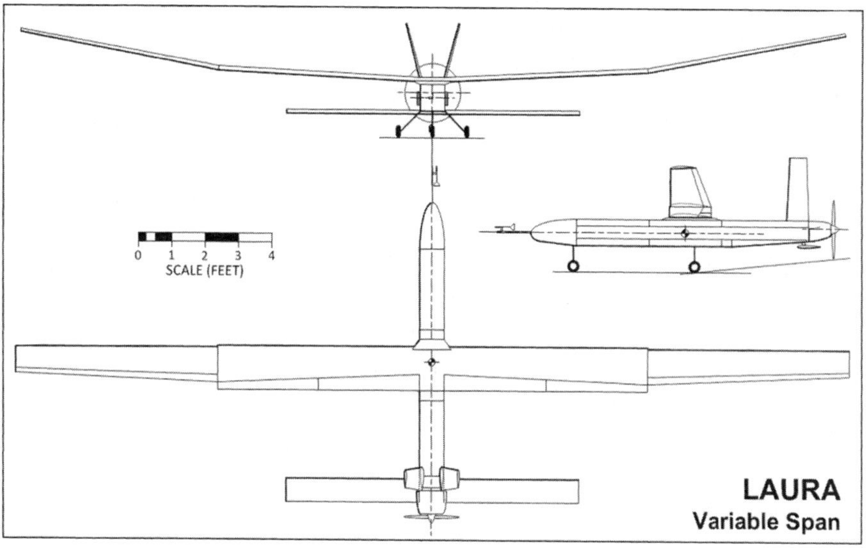

Fig. 4.17 LAURA Variable Span three-view drawing.

Table 4.6 LAURA Variable Span Technical Characteristics

Parameter	Value
Wing span	12.9 ft (retracted)
	24.6 ft (deployed)
Wing area	38.2 ft^2 (reference)
Length	10.1 ft
Takeoff weight	63.0 lb
Fuel capacity	1 gal for a transatlantic mission
Loiter speed	22 KTAS
Endurance	55 min (fuel tank limited)
	60 hr (with 1 gallon)
Max L/D	16.13 (folded)
	27.76 (deployed)
$C_{L,\max}$	0.57 (folded)
	1.16 (deployed)
Airfoils	Modified Eppler 193 (wing)
	NACA 0010 (horizontal tail)
Launch method	Conventional
Recovery method	Conventional
Propulsion type	Reciprocating gas engine
Power required	0.236 hp (folded)
	0.094 hp (deployed)

LAURA: FURTHER READING

Foch, R.J., "Test Pilot's Notes on Flying The Low Altitude/Airspeed Unmanned Research Aircraft (LAURA)," AIAA 92-4078, presented at the 6th AIAA Biennial Flight Test Conference, Hilton Head, SC, Aug. 24–26, 1992.

Foch, R.J., and Ailinger, K.G., "Low Reynolds Number Long Endurance Aircraft Design," AIAA Paper 92-1263, presented at the 1992 Aerospace Design Conference, Irvine, CA, Feb. 3–6, 1992.

Foch, R.J., and Toot, P.L., "Development of the Low Altitude/Airspeed Unmanned Research Aircraft (LAURA)," *1989 NRL Review*, Naval Research Laboratory, Washington, DC, 1989, pp. 137–138.

Foch, R.J., and Toot, P.L., "Flight Testing Navy Low Reynolds Number (LRN) Unmanned Aircraft," *Lecture Notes in Engineering, Low Reynolds Number Aerodynamics, Proceedings of the Conference*, Springer-Verlag, Berlin, 1989.

Foch, R.J., and Toot, P.L., "Low Reynolds Number Aerodynamics for Electronic Warfare," *1991 NRL Review*, Naval Research Laboratory, Washington, DC, 1991.

Foch, R.J., and Wyatt, R.E., "Low Altitude/Airspeed Unmanned Research Aircraft (LAURA) Preliminary Development," *Aerodynamics at Low Reynolds Numbers, Proceedings*, Vol. III, Royal Aeronautical Society, London, Oct. 1986.

PLICATE WING ASW UAS, 1987

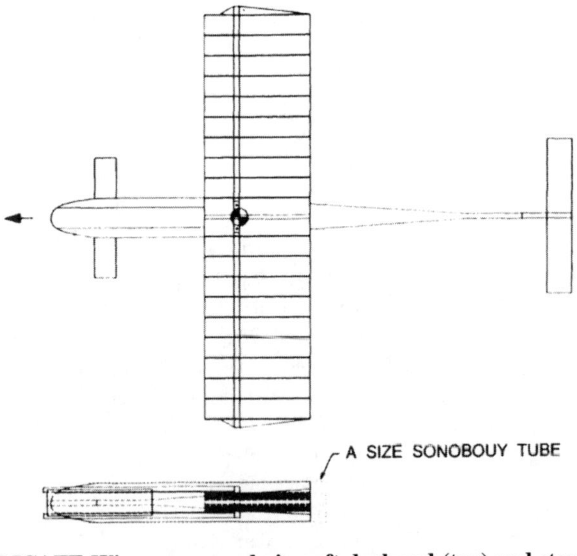

Fig. 4.18 PLICATE Wing unmanned aircraft deployed (top) and stowed (bottom).

The PLICATE Wing unmanned aircraft investigated a design to fit within the tight volume of an A-size sonobuoy tube, approximately 4.8 inches in diameter and 36 inches long (Fig. 4.18). Its design was an alternative to Pendulus (Chapter 3). The wing skins were a flexible polymer membrane. The internal structure was a series of sliding ribs over a box spar. The ribs slid from the wing tips to the wing root as the wing deployed. The membrane collapsed around the ribs to provide a taut surface. When folded, the ribs and spars fit within a fuselage recess. This configuration allowed a much larger wing area to be packaged within a constrained volume. The PLICATE Wing was a cross between a rigid folding wing and an inflatable structure. This was thought to be an optimum balance between mass and volume limitations.

NRL built engineering mock-ups that could be unfolded. However, all the readily available wing skin fabrics developed wrinkles when stored. The performance advantages of the lighter wing loading would be offset by the poor airfoil performance due to the surface quality. Therefore, NRL did not proceed past the engineering mock-up stage. A fresh evaluation with modern materials might yield a viable design.

SEMI-RIGID DELTA WING, 1987

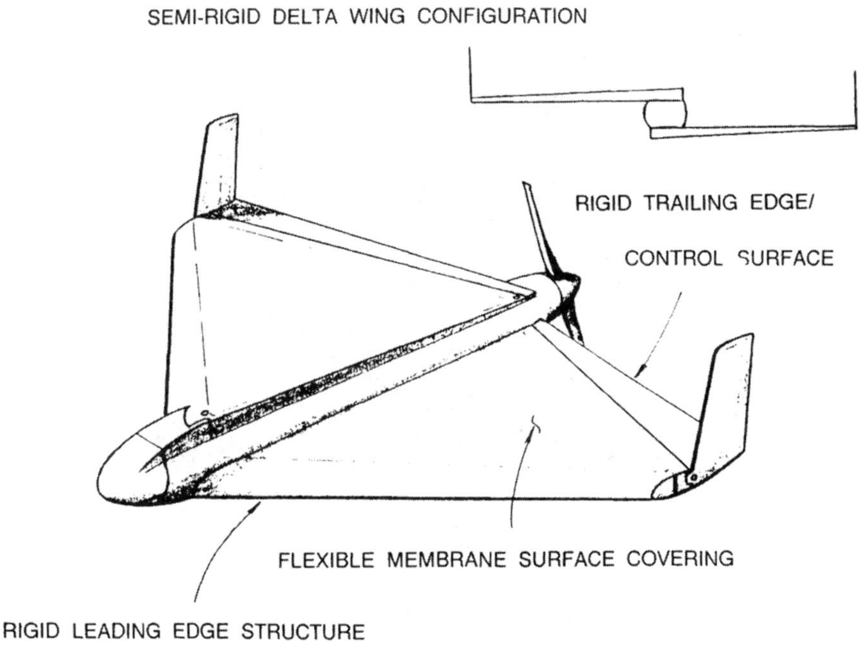

Fig. 4.19 Semi-Rigid Delta Wing concept.

The Semi-Rigid Delta Wing was a design concept for a folding, fabric wing aircraft (Fig. 4.19). The design was developed after an investigation of the Fairchild Sailwing AED, which had inconsistent flight performance. The Delta Wing design offered a better membrane wing that solved the Sailwing problems of trailing edge thickness and twist. The leading edge consisted of a hollow C-section. The wing fabric collapsed within the leading edge. The trailing edge consisted of a box spar with rigid control surfaces. At deployment, the trailing edge slid down the fuselage to form a triangle consisting of the fuselage, trailing edge, and leading edge. The resulting planform was a delta wing. This action pulled the membrane over the upper and lower surfaces in a uniform manner. The control surfaces were uniform and rigidly held, so there was no undesired twist. The arrangement provided full control over the wing geometry.

NRL built an engineering mock-up that could show the full wing deployment. The rigid leading edge provided good shape control over this critical region. Any wrinkles from the skin were not in a critical area. Delta wing aircraft operate at a low lift coefficient, and minor wrinkles are less important. The design never progressed beyond the mock-up stage. This design would be useful when a large wing area is required for a small packaging volume.

Flying Wing Research Testbed (FLYRT), 1987–1988

Fig. 4.20 FLYRT.

The Flying Wing Research Testbed (FLYRT) was a research vehicle designed to explore plank flying wing characteristics (Fig. 4.20). The later Dragon Eye system (Chapter 5) inherited many features from FLYRT. Note that the name FLYRT was later adopted by a much larger offboard countermeasure program, the Flying Radar Target Advanced Technology Demonstrator (FLYRT ATD), though the two programs share little in common.

The FLYRT research aircraft had a constant chord wing with a span of 4 feet and an aspect ratio of 4. The fuselage had a vertical tail mounted aft of the wing. The power plant was an O.S. 45 single-cylinder two-stroke driving a tractor propeller at the front of the fuselage. A fixed tricycle landing gear enabled conventional takeoff and landing on a runway.

FLYRT used the Liebeck LA 2573A flying wing airfoil. This airfoil was intended for higher Reynolds number applications, and this was NRL's first experience with it for this vehicle class. The airfoil worked better than expected. The stall characteristics were benign, with no abrupt separation. The operating lift coefficient was 0.68, which was impressive for a flying wing airfoil. The pitching moment remained positive at the low Reynolds numbers. The only restriction was that the airfoil required a trip strip to force transition when the chord Reynolds number was below 280,000. This was discovered through wind tunnel tests. The airfoil's Stratford recovery required that the boundary layer transition from laminar to turbulent at 30% of the chord before encountering the adverse pressure gradient on the upper surface. Automotive striping tape served as the trip strip. This modification provided good aerodynamic performance down to a Reynolds number of 100,000. The penalties for not applying the trip strip were decreased maximum lift coefficient and higher drag.

The FLYRT prototype was designed and built by two summer interns. The wings were constructed of foam cores with balsa sheeting. The airframe was painted with epoxy paint.

The only problem with the aircraft was that it was unstable in Dutch roll at high speeds. The vertical tail was tall, due in part to its high aspect ratio. This vertical tail added negative $C_{l,\beta}$ (dihedral effect). This problem went away at low speeds. At higher angles of attack, the center of pressure of the tail was lower relative to the CG, so the tail contribution to dihedral effect was reduced. The wings had slight geometric dihedral as well. Based on these lessons, the later Dragon Eye design had a tapered, low aspect ratio vertical tail and no geometric dihedral.

After several flights revealed that suitable handling characteristics were attained, the wing tips were modified. New plug-in wing tip extensions were applied, which were highly swept and tapered. This configuration is known as a C.P. Van Dam wing tip, shown in Fig. 4.21. The modified aircraft showed enhanced performance without degrading handling qualities. The wing tips improved stalling characteristics.

FLYRT had a successful flight test program of approximately 50 flights. This was NRL's first experience with a plank flying wing configuration. The design was enabled by the tripped Liebeck airfoil. Figure 4.22 shows a three-view drawing of this research vehicle, and Table 4.7 lists its technical characteristics.

Fig. 4.21 FLYRT with C.P. Van Dam wing tip modification.

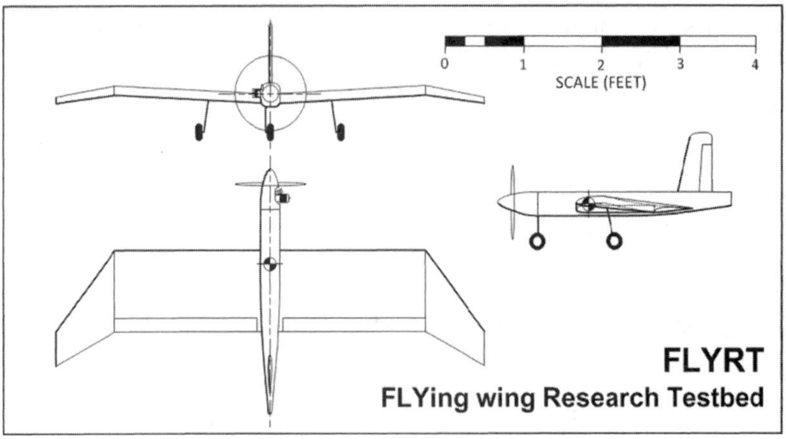

Fig. 4.22 FLYRT three-view drawing.

Table 4.7 FLYRT Technical Characteristics

Parameter	Value
Wing span	4 ft
Wing area	4 ft^2
Length	3.0 ft
Takeoff weight	5.5 lb
Dash speed	70 KTAS
Stall speed	25 KTAS (est)
Endurance	20 min
Airfoils	Liebeck LA 2573A
Launch method	Conventional
Recovery method	Conventional
Propulsion type	O.S. 45 FSR, 0.045-cu.-in., 2-stroke engine
Power	1.2 hp

SHADOW DISPLACED TAIL (DT), 1988–1991

Fig. 4.23 Shadow DT.

The objective of the Shadow Displaced Tail (DT) program was to develop a long-endurance aircraft that also had low Radar Cross Section (RCS) (Fig. 4.23). Most low-RCS aircraft have a flying wing configuration and are propelled by jet engines. Such aircraft perform well for range-optimized missions, as the lift coefficient is low. Long-endurance aircraft, on the other hand, try to fly at the best endurance parameter ($C_L^{3/2}/C_D$). The 3/2 exponent on the C_L term favors a high lift coefficient. Flying wing designs are generally unable to generate high C_L values when compared to conventional configurations. This is caused by the need to trim the pitching moment with positive static margin using only the wing. Trim is generally accomplished via reflexed airfoils and wing washout.

The solution to the problem of combining long endurance and low RCS was to modify the flying wing with a down-loaded tail placed behind the wing tips. The term "displaced tail" was coined by Foch to describe configurations in which the tails are laterally displaced to the wing tips (Fig. 4.24). A displaced tail surface provides directional and longitudinal stability and pitch trimming to enable traditional cambered, negative pitching moment airfoils on the wing. The configuration is capable of generating much higher trimmed lift coefficients than a flying wing, and the endurance parameter is therefore higher. The angled displaced tail has a lower radar signature than a traditional tail, and the overall aircraft is more flying-wing-like with respect to RCS. The configuration was inspired by World War II German fighter design concepts by Blohm und Voss. The German designs had a tapered tail surface extending out and down from a short boom that extended from each wing tip; the surface had a large anhedral angle.

Another advantage of the displaced tail over a flying wing is that less wing area is required. The lift coefficient for best endurance parameter is lower for a flying wing. A displaced tail flies at a higher lift coefficient, so less wing area is

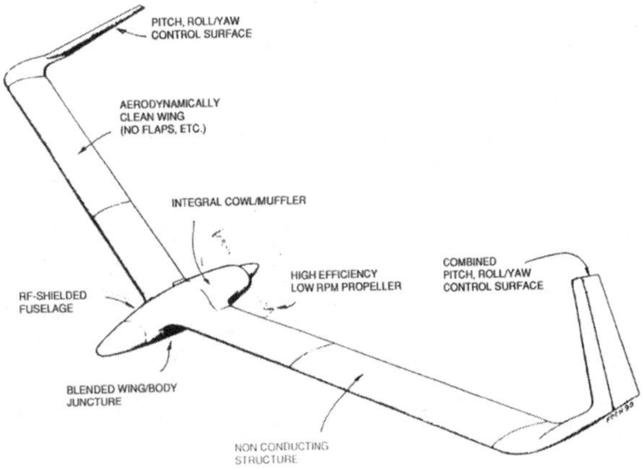

Fig. 4.24 Shadow DT features.

required for a given flight speed. The reduced vehicle size was expected to have RCS benefits that would offset the impacts of the extra tail surfaces.

The Shadow DT wing was swept to provide a moment arm for the tails and for RCS purposes (Fig. 4.25). The wing sweep was approximately 14 degrees, which had an arctangent of 0.25. The convenient arctangent made prototype construction slightly easier. The sweep was kept relatively low to prevent aerodynamic performance losses. Later during NRL's XFC program, it was discovered that a moderate amount of wing sweep can have beneficial boundary layer effects for low-speed, LRN aircraft. However, this was not understood at the time. More wing sweep would have provided additional tail moment arm, which would have reduced tail size and improved dynamic damping.

The program began by flow visualization wind tunnel testing of a model with adjustable winglets that could be canted inward or outward and upward or downward. Only the downward canted outward and inward canted upward positions provided yaw moments in the same direction as the commanded roll. A positive pitching moment airfoil was used for initial tests. The outward canted Blohm und Voss tail configuration was tried first. This arrangement showed poor high angle of attack characteristics. At high angles of attack, the wing tip vortex stalled the stabilizer control surface. The aircraft lost the ability to control pitch. Also, there was a tendency for one stabilizer to stall first, which caused the aircraft to roll over. The outward canted displaced tail had a larger span than a wing alone, but the download on the tail eliminated the benefit of the span augmentation for induced drag. The Blohm und Voss displaced tail was a poor layout for low-speed aircraft.

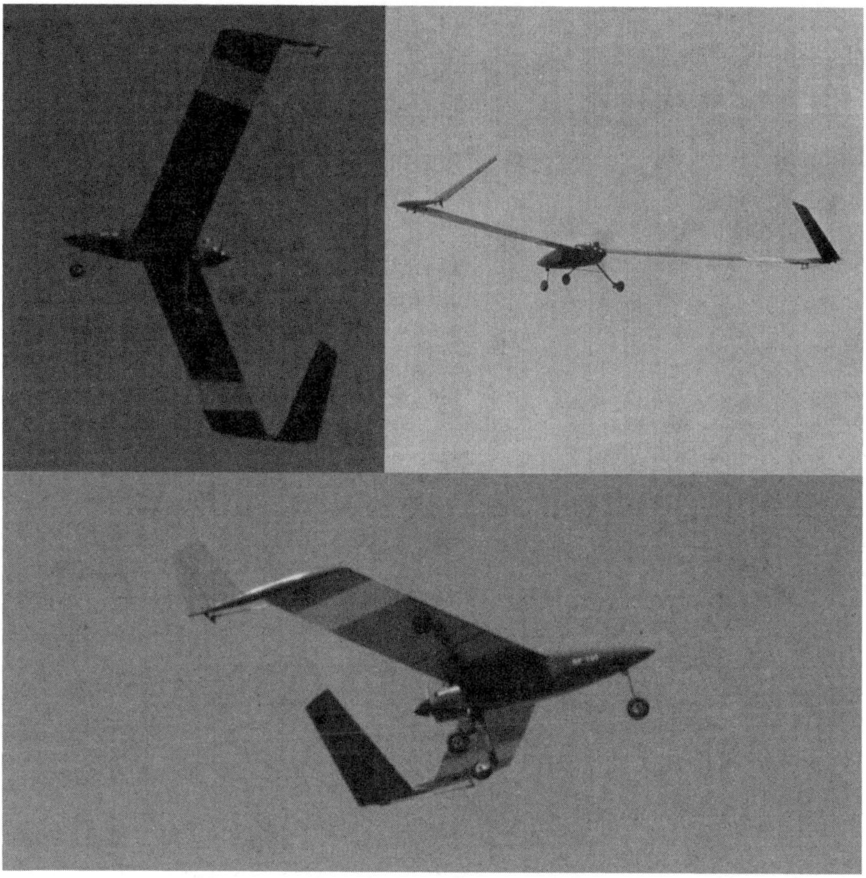

Fig. 4.25 Shadow DT in flight.

The wind tunnel model was reconfigured with the displaced tail surfaces canted inward and upward. The flow field improved significantly. The tail operated with a down force, so the lower side of the tail had the lowest pressure. This kept the wing tip flow attached at high angles of attack, which caused the stall to occur further inboard on the wing. The wing tip lift coefficient was higher than for a flying wing because of this effect. The stall characteristics were much more benign than those of a typical swept wing. No wing washout or strakes were necessary for good stall behavior. From earlier wind tunnel tests, the incidence angle of the tails was adjusted to provide the required download at the endurance flight condition with cambered airfoils. The wind tunnel tests also led to the addition of a small strake at the leading edge of the root of the tail to prevent local separation. The tails had a favorable interaction with the wing tip vortex that reduced induced drag much like a winglet.

The only negative impact of the inward canted tails was that they had an anhedral effect (or positive $C_{l,\beta}$), which reduced roll stability. The solution was to give the wing some dihedral. The additional dihedral plus the high wing position of this first research aircraft design gave the aircraft a thrust line below the CG. Changes in throttle resulted in pitch changes that required pilot control compensation. This undesirable handling characteristic could be easily remedied by raising the thrust line.

The longitudinal stability and control characteristics of the Shadow DT were similar to those of a close-coupled tail conventional configuration, as shown in Fig. 4.26. The pitch inertia was low, and there was low pitch rate damping. The aircraft was very responsive to pitch commands.

The tail control surfaces acted as elevons. That is, they provided both pitch and roll inputs. The lateral-directional handling characteristics were not optimized for a broad speed range. One possible change would be to add wing ailerons and then use the tails as pitch and yaw effectors. This control arrangement was not developed or tested until the NRL Extender project.

Having the tails at the wingtips could create structural problems for larger aircraft. The offset mass of the tail could generate aeroelastic problems, but the tails probably provide a flutter damping benefit. The Shadow DT demonstrator was built with a large wing torsional rigidity. The control surfaces were also balanced to prevent flutter. Figure 4.27 shows a three-view drawing of this research vehicle, and Table 4.8 lists its technical characteristics.

The Shadow DT program was successful at developing a new airplane configuration. The RC test aircraft flew more than 50 flights. Lessons learned from the program helped inform the designs of the CG Shift Displaced Tail aircraft, Extender, and the Mars flyers.

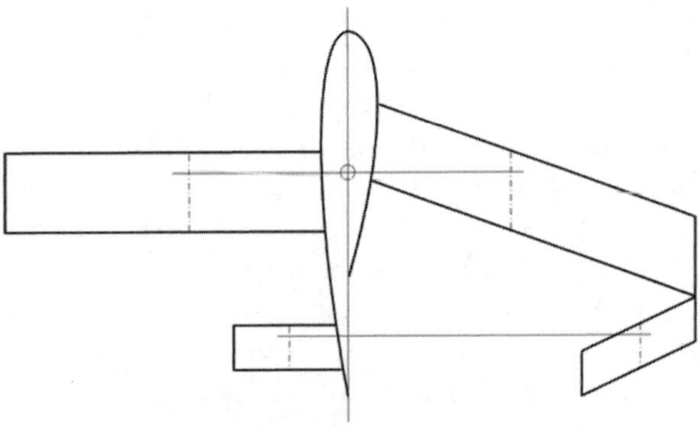

Fig. 4.26 Comparison of the displaced tail to an equivalent conventional configuration.

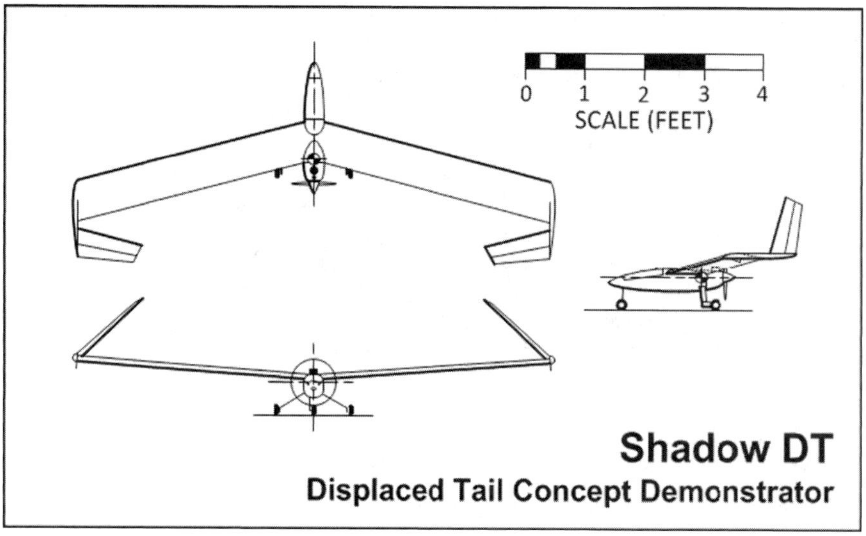

Fig. 4.27 Shadow DT three-view drawing.

Table 4.8 Shadow DT Technical Characteristics

Parameter	Value
Wing span	8.17 ft
Wing area	5.34 ft^2
Length	3.25 ft
Takeoff weight	4.75 lb
Empty weight	4.25 lb
Payload weight	N/A
Loiter speed	27 KTAS (est)
Dash speed	45 KTAS (est)
Stall speed	20 KTAS (est)
Endurance	0.4 hr
Max L/D	18.0 (calc)
Airfoils	Wing: Foch 10% t/c flat bottom Tail: flat plate
Launch method	Conventional
Recovery method	Conventional
Propulsion type	O.S. 25 FSR, 0.25-cu.-in., 2-stroke engine
Power	0.65 hp

SEMI-AUTONOMOUS RESEARCH AIRCRAFT (SARA), 1990–1994

Fig. 4.28 SARA.

The Semi-Autonomous Research Aircraft (SARA) was an electric propulsion test bed (Fig. 4.28). New battery and motor technologies that could make unmanned aircraft electric propulsion more practical were emerging. In particular, $LiSO_2$ primary batteries (nonrechargeable) offered much higher specific energy than NiCd batteries. These batteries and rare earth samarium cobalt brushed motors could offer extended-endurance electric flight.

Electric propulsion has several advantages over reciprocating engine propulsion. First, starting the electric motor is as simple as closing the circuit, while reciprocating engines require either ground starting or complex onboard starting systems. Electric propulsion systems can also be stored for a long time without maintenance.

$LiSO_2$ batteries had a critical temperature limit. The battery had to be kept under 70°C or the cells could short. SARA was instrumented with temperature sensors to monitor the cooling effectiveness. NRL gained substantial experience with battery and motor cooling through this program.

SARA had a low wing mounted to the bottom of the fuselage. The three-panel wing had 12 degrees of dihedral on the outer panels. A V-tail was used for simplicity. Another motivation for using a V-tail was that many deployable EW decoy concepts used V-tail and X-tail arrangements, so SARA could expand NRL's experience with this tail arrangement. The ruddervator control surfaces took up 40% of the V-tail chord. The motor was mounted in the nose of the vehicle with a tractor propeller. Tricycle landing gear was added to simplify launch and recovery operations, since there was sufficient power available. The controls were rudder, elevator, and throttle. The fuselage and tails were built-up balsa structures. The wing was constructed of foam core with balsa skins. The entire aircraft was covered with MonoKote plastic film.

The electric motor was an AstroFlight Cobalt 40 brushed motor. It produced approximately 400 watts at 28 volts and 14 amps. The motor spun a 12-inch-diameter (12×8) folding fixed-pitch propeller. The motor rotational rate was 8000 rpm for takeoff and 5400 rpm for cruise. The battery pack for initial flight testing was a 20-cell, 24-volt, 1.2-amp-hour NiCd pack.

SARA had a productive flight test program, and one flight on a relatively calm day achieved a flight endurance of 2 hours and 47 minutes. This was a record for NRL, and was possibly a world record for battery electric flight at the time. The vehicle was found to be climb power-critical on a warm day due to battery temperature limitations, but experimentation with cooling helped solve this problem. The vehicle demonstrated low-noise operation of efficient electric aircraft. The one drawback of SARA was a low climb rate of only 400 feet per minute at sea level.

The testing proved that the performance advantages of $LiSO_2$ batteries were impressive. The problems were that these primary batteries were not rechargeable, had a low power density, and were expensive. However, expendable decoys did not need rechargeable batteries. These batteries had a specific energy of approximately 125 watt-hours per kilogram when drained slowly over many hours. NRL showed that faster discharging over the duration of a flight yielded only slightly lower specific energy. Note that today's inexpensive lithium polymer rechargeable batteries achieve similar specific energy values. SARA proved that it was practical to build electrically powered small tactical unmanned aircraft capable of performing missions of 2 to 3 hours' endurance.

SARA had significant initial electromagnetic interference (EMI) problems related to the electric propulsion system. The first power-on ground tests revealed major EMI issues. The servos would "chatter," and the motor experienced power surges at a range of 100 feet. There was complete control signal loss at 500 feet separation. Spectrum analysis tests revealed that the EMI noise peaked near 72 MHz, which was in the RC radio frequency range used in tests. Several steps were taken to solve the EMI problems. A common electrical ground was used for the entire aircraft. The motor, motor controller, and battery were wrapped in a Monel (nickel alloy) sock, which was an anti-magnetic-field bag. Servo leads, servo cases, and the receiver were shielded with copper tape. Optical isolators were applied between the servo and the receiver. The motor and the motor controller were given separate power sources. It took several months to resolve these problems. Ground range checks still showed EMI-related range limitations, but these got better when the aircraft was airborne. SARA was never flown more than a few thousand feet away from the pilot. Figure 4.29 shows a three-view drawing of this research vehicle, and Table 4.9 lists its technical characteristics.

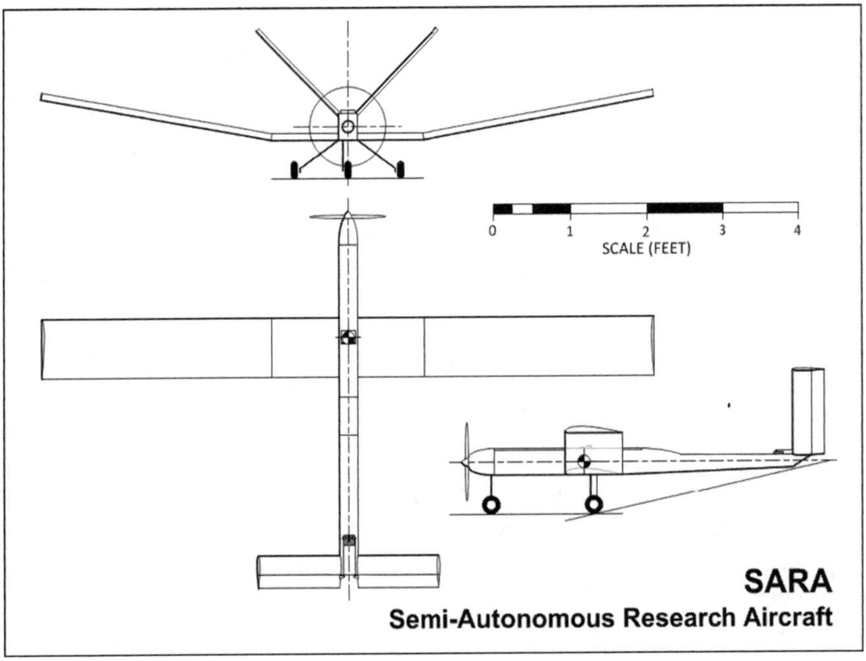

Fig. 4.29 SARA three-view drawing.

Table 4.9 SARA Technical Characteristics

Parameter	Value
Wing span	8 ft
Wing area	6 ft^2
Length	4.66 ft
Takeoff weight	10.3 lb (est)
Battery weight	3.25 lb
L/D	11.5
Loiter speed	25 KTAS (est)
Stall speed	19 KTAS
Endurance	2 hr 47 min
Airfoils	Wing: RF 1165FB
	Tail: NACA 0009
Launch method	Conventional wheeled takeoff
Recovery method	Conventional wheeled landing
Propulsion type	Lithium primary battery electric
	AstroFlight Cobalt 40 brushed motor
Power	400 W

Many of the EMI issues were related to the motor technology. The brushed motors had spark-producing brushes that created strong radio frequency emissions. Today's brushless motors produce much lower EMI. Also, new 2.4 GHz spread spectrum radio systems are much more robust and noise-tolerant than the 72 MHz FM radios available during the SARA program.

CG Shift Demonstrators, 1992–1995

In the early 1990s, NRL UAS designers began to consider the potential of augmenting the control and stability of small aircraft by shifting the aircraft's CG, instead of using aerodynamic control surfaces. Potential advantages of weight/CG shift control included a simpler, less expensive airframe; control power at much lower airspeeds and while unfolding in flight; lower aerodynamic drag; and lower radar cross section (stealthier). Conceptually, CG shift control could be applied in the same way as it is on hang gliders—beginning with the first gliders by Otto Lilienthal in the late 1890s and continuing to this day on flex-wing gliders—by moving the main payload mass both fore/aft and laterally about the center of pressure. There are two methods of accomplishing this for a streamlined UAS. First is to move a massive internal item, such as the battery or electronic payload, within the airframe. In this approach, the entire airframe is in one piece, but there must be sufficient internal volume to move the mass. Excess internal volume increases the frontal area and the drag, but some configurations, such as a low aspect ratio flying wing, inherently carry excess internal volume and are candidates for this approach. The second method for CG shift is to move the wing fore, aft, and laterally relative to the fuselage. Both methods are conceptually simple, but in practice, a number of technical issues emerged and design trade-offs were required.

To address the technical issues and design trade-offs, NRL performed 6-DOF dynamic flight simulations, and awarded a contract to McDonnell Douglas to develop detailed control architecture and stability and control algorithms for small UAS control using CG shift. The primary technical issues were electrical power requirements and actuator size needed for adequate control; trade-offs between inherent stability and control authority; rate damping; and lateral-directional control.

Longitudinal (pitch) control is a relatively simple matter of moving the CG forward or aft relative to the longitudinal center of pressure, which causes the aircraft to retrim at higher or lower lift coefficients and angles of attack with a corresponding change of airspeed. Lateral-directional control, however, is more complex. Obviously, to create a roll moment, for example to the right to bank the wing and perform a right-hand turn, the CG is shifted to the right of the aircraft's centerline. From the standpoint of stability and control dynamics, the motion and attitude of an aircraft are referenced to the CG. So with respect to the CG, there is now more of the aircraft on the left side of the CG, so the aerodynamic centers of lift and drag are now to the left of the CG. The lift therefore raises the left wing, rolling the aircraft to the right, while the drag yaws the nose to the left. This adverse yaw will induce a rolling moment counter to the commanded roll that will have a magnitude dependent on the dihedral effect of the design. If the aircraft is moderately laterally stable, the yaw-induced rolling moment will be opposite to the commanded roll, but lower in magnitude, thus reducing the control power of the CG shift control system. If the aircraft is neutrally laterally stable, there will be virtually no countering rolling moment created by the adverse yaw. If the aircraft is extremely laterally stable, the adverse yaw may induce a countering roll moment that is stronger than that of the lateral CG shift, overpowering the commanded roll and causing the aircraft to roll and turn in the opposite direction to that which was desired! Therefore, tuning the lateral stability characteristics is critical to achieving sufficient lateral control via CG shift.

The adverse yaw created by lateral CG shift is also created in conventional airplanes by wing warping and ailerons. It is typically cancelled by the pilot or flight control system by applying the rudder to neutralize the yaw angle. Unfortunately, installing a rudder or any other control surface such as spoilers or drag flaps to provide yaw control somewhat defeats the premise that CG shift control eliminates the need for aerodynamic control surfaces. So, the designer's choice is either to allow the aircraft to turn in an uncoordinated sideslip, which may be acceptable for small unmanned UASs, or find another method of lateral control without aerodynamic control surfaces. In the case of a flying wing, the simplest solution is to live with sideslipping turns, or it is necessary to use one or more internal masses that can be rotated to provide a torque about the yaw axis. For an aircraft with a fuselage, the simplest method is to install a fixed vertical tail to create directional stability that minimizes the adverse yaw angle.

As a result of extensive consideration and flight simulation, another non-control surface of lateral control became obvious for designs having a fuselage and vertical tail. For such a design in which the wing is slid fore and aft for CG shift pitch control, it is mechanically simple to also rotate the wing, skewing it relative to the fuselage. The resulting lateral asymmetry causes the aircraft to yaw into the commanded turn, and if the design has sufficient dihedral effect, this yaw induces a roll in the desired direction. Although not technically a CG shift, commanded wing skewing still provides lateral control without the need for aerodynamic control surfaces. NRL researchers developed a series of CG shift research aircraft to test each possible method, assess its feasibility, and compare each with conventional aerodynamic flight control systems.

FIRST SKEW WING CG SHIFT, C. 1992

Fig. 4.30 First Skew Wing CG Shift demonstrator with skew wing lateral control.

The first CG shift research aircraft utilized a simple fuselage with a wing that could be commanded to translate fore and aft for pitch control and skewable for lateral control (Fig. 4.30). Simulation indicated that a horizontail tail was not strictly needed for a CG shift control system, but it would obviously provide significant pitch damping. Since it was desirable to flight-test this first design as a simple RC model without an expensive or sophisticated automatic flight control system (AFCS), this design was equipped with both horizontal and vertical stabilizers in the form of a fixed V-tail.

The wing sliding and pivot mechanisms were actuated with hobby grade servos. Flight tests showed the wing manipulation control system was not very effective. The control system had play, high inertia, and low-power servos. The reaction time was slower than desirable for a useful aircraft, and authoritative control was not achieved. Custom actuators could have been developed to remedy the problem, but this was not attempted, since despite a few tantalizing results, enough design flaws were revealed to warrant a second, improved airframe. Also, the first aircraft had insufficent dihedral effect, so there was poor lateral-directional coupling and very soft lateral control. Figure 4.31 shows a three-view drawing of this research vehicle, and Table 4.10 lists its technical characteristics.

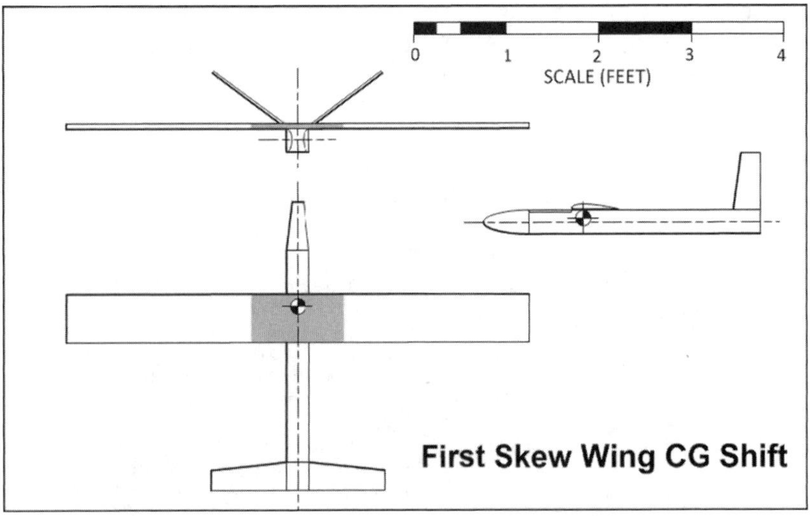

Fig. 4.31 First Skew Wing CG Shift demonstrator three-view drawing.

Table 4.10 First Skew Wing CG Shift Technical Characteristics

Parameter	Value
Wing span	5.0 ft
Length	3.0 ft
Takeoff weight	2.2 lb
Empty weight	2.2 lb
Loiter speed	20 KTAS (calc)
Dash speed	30 KTAS (calc)
Max L/D	10 (calc)

Second Skew Wing CG Shift, c. 1992

Fig. 4.32 Second Skew Wing CG Shift demonstrator.

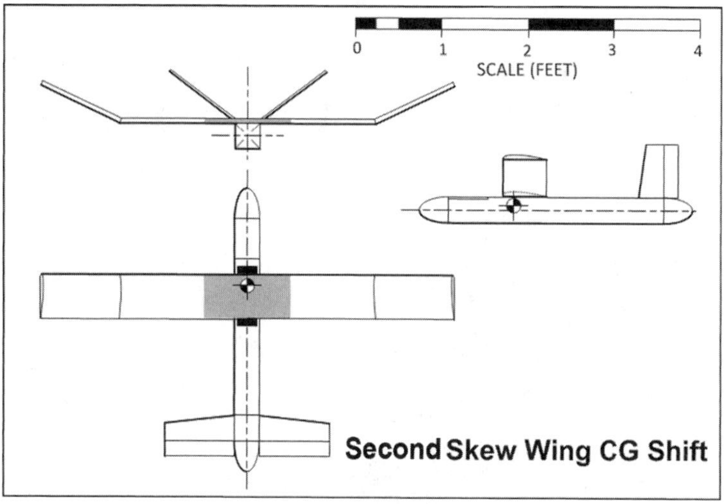

Fig. 4.33 Second Skew Wing CG Shift three-view drawing.

The second air-launch CG Shift demonstrator aircraft was designed with a polyhedral wing to provide a strong dihedral effect, and traditional tail ruddervator surfaces for backup controls to enhance the survivability with its nonrobust control system (Fig. 4.32). As in the First Skew Wing, this aircraft shifted the wing fore and aft to control pitch and pivoted the wing about a vertical axis for lateral-directional control. The control system had less play than the first design, but it still had high inertia, low-power RC hobby servos, and a relatively short operating life. Custom servo actuators were still beyond the financial resources of this small project. Nevertheless, this aircraft made several successful flights without the use of the backup ruddervators and demonstrated the feasibility of small UAS flight control without the use of aerodynamic controls. Figure 4.33 shows a three-view drawing of this research vehicle, and Table 4.11 lists its technical characteristics.

Table 4.11 Second Skew Wing CG Shift Technical Characteristics

Parameter	Value
Wing span	4.8 ft
Length	3.2 ft
Takeoff weight	2.75 lb
Empty weight	2.75 lb
Loiter speed	22 KTAS (calc)
Dash speed	32 KTAS (calc)
Max L/D	9.5 (calc)

HIGH ASPECT RATIO (HAR) CG SHIFT, c. 1992

Fig. 4.34 HAR CG Shift.

Success with the Skew Wing CG Shift research aircraft, in conjuction with 6-DOF computational flight dynamics modeling, encouraged NRL researchers to investigate a CG shift design without a horizontal tail. In addition, simulations indicated that a high aspect ratio unswept (plank-type) flying wing was feasible. Whereas high aspect ratio unswept flying wings generally suffer from having a very short pitch control moment arm, a CG shift design does not have this limitation. By being able to increase the aspect ratio and reduce the span loading compared with control-surface-controlled plank flying wings, the wing-induced drag would be significantly reduced and performance increased.

To maintain some level of conservatism, the pitch-stable Liebeck L2573a flying wing airfoil was used. Based on favorable experience with the FLYRT aircraft, Van Dam swept wing tips were used to further reduce induced drag. The wing was translated fore and aft for pitch control and skewed for lateral control. The vertical tail was a fixed directional stabilizer, and there were no aerodynamic backup control surfaces. The High Aspect Ratio (HAR) CG Shift was constructed like a conventional RC model of balsa wood and aircraft plywood skinned in MonoKote model airplane covering (Fig. 4.34).

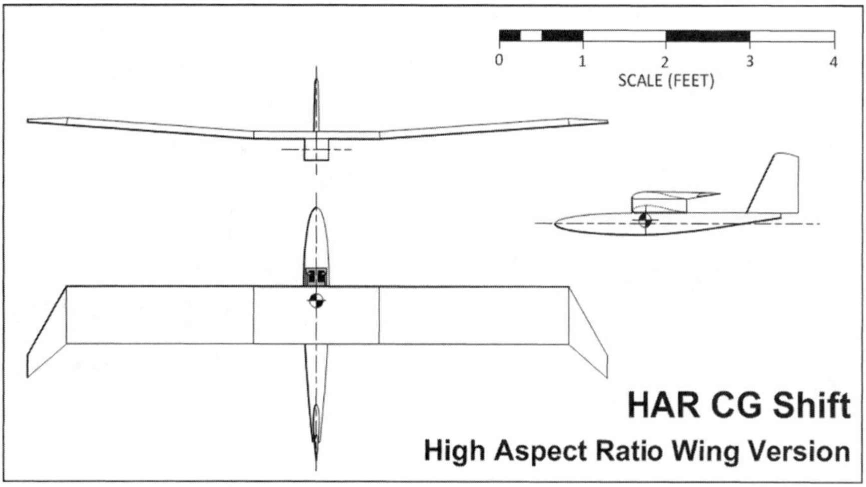

Fig. 4.35 HAR CG Shift three-view drawing.

Flight tests were successful but limited due to rapid wear in the mechanics of the control system. The HAR CG Shift proved that a tailless, high aspect ratio, unswept flying wing was certainly feasible. It was also one of the more attractive NRL research aircraft. Figure 4.35 shows a three-view drawing of this research vehicle, and Table 4.12 lists its technical characteristics.

Table 4.12 HAR CG Shift Technical Characteristics

Parameter	Value
Wing span	6.9 ft
Length	3.0 ft
Takeoff weight	2.0 lb
Empty weight	2.0 lb
Loiter speed	17 KTAS (est)
Max L/D	19 (calc)

CG SHIFT DT, 1992

Fig. 4.36 CG Shift DT prior to first flight.

The research goal for the CG Shift DT design was to investigate both longitudinal and lateral control with a fully internal CG shift control system for applications on reduced radar cross section configurations (Fig. 4.36). This demonstrator built upon the experience with the Shadow DT project and utilized displaced tails as directional and longitudinal stabilizers. An air-launched glider was produced to evaluate the concept feasibility. The aircraft spanned approximately 4 feet. The wing and tail had fiberglass skins over foam cores.

The configuration was a blended wing body with inward canted displaced tails at the tips. The inboard wing panel had high taper, and the outboard panel had constant chord. The leading edge sweep was constant for the entire span. The wing had slight geometric dihedral.

The battery was placed on a rail system that could move side to side and fore and aft via two tracks. The sliding was actuated by rotary servos with pushrods. The longitudinal motion controlled pitch, and the sideways motion controlled roll and yaw. The aircraft had a small static margin. The reduced longitudinal stability helped minimize the travel of the battery.

The CG Shift DT made one interesting flight. It was air dropped from an NDM host aircraft. The control system did not work effectively. The aircraft was excessively stable both laterally and longitudinally, and it unintentionally had just the right amount of dihedral effect for the adverse yaw-induced roll to perfectly counter the CG-shift-commanded roll. It simply would not turn or roll whatsoever when commanded. The CG Shift DT drifted downwind, off the test area, and was lost. However, the experience gained from developing this design was useful for the development of the CG Shift Bird aircraft that followed. Figure 4.37 shows a three-view drawing of this research vehicle, and Table 4.13 lists its technical characteristics.

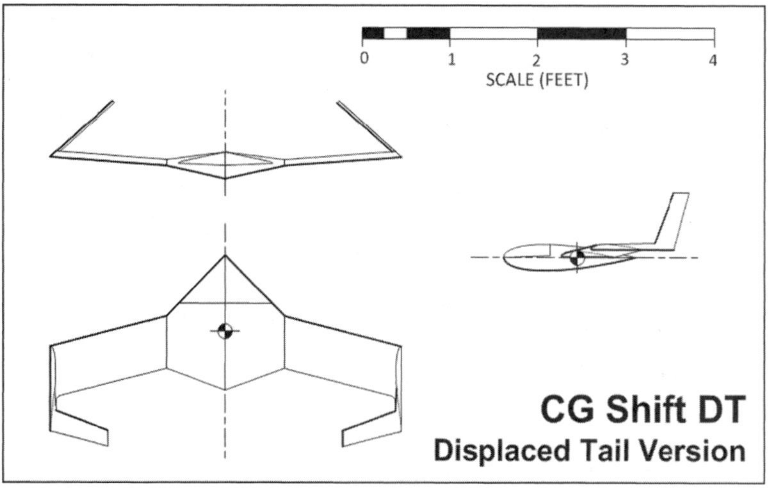

Fig. 4.37 CG Shift DT three-view drawing.

Table 4.13 CG Shift DT Technical Characteristics

Parameter	Value
Wing span	4.0 ft
Length	2.1 ft
Takeoff weight	3.6 lb
Empty weight	3.6 lb
Max L/D	12 (calc)

CG Shift Bird, 1993–1995

Fig. 4.38 CG Shift Bird demonstrator.

The purpose of the CG Shift Bird project was to develop a bird-like small UAS for reconnaissance missions. The aircraft had to have the visual appearance of a bird and also mimic bird flight behavior (Fig. 4.38). Earlier attempts at bird-like unmanned aircraft were not convincing for ground observers.

The CG shift control system enabled flight maneuvers that have similar dynamics to bird flight. The Pendulus and CG Shift DT aircraft had a wobbling motion that appeared similar to bird motion. Large raptors such as vultures tend to rock. These birds are spirally stable but Dutch roll unstable. A bird-like UAS could fly circuitous flight patterns to complete the deception. Large raptors fly for long periods of time without flapping their wings, so an ornithopter (flapping wing propulsion) was not necessary. Future versions of the aircraft would likely use electric propulsion.

The CG Shift Bird aircraft had the appearance of a large crow. The wing had multiple tapers, dihedral, and sweep angles along the span to resemble a bird planform (Fig. 4.39). The wing tips were abruptly swept aft, which helped with longitudinal stability with a low drag penalty. The wing tip design was suggested by C.P. Van Dam when he worked at the NASA Langley Research Center. He looked to nature for inspiration toward improving aircraft aerodynamic performance. The reflexed camber Liebeck (L 2573a) flying wing airfoil was used on the wing. There were no horizontal stabilizers, making this aircraft a true flying wing configuration.

The fuselage was pointy like a beak at the nose, and transitioned to a vertical line at the rear. The side view of the fuselage looked like a wedge. The large side area at the back of the fuselage served the function of a vertical stabilizer. It is difficult to determine if a triangular surface is horizontal or vertical at a distance, so the vertical fin area looked like a bird tail. Some birds roll their tails for maneuvering as well.

As on the previous Skew Wing CG Shift research aircraft, the wing pivoted about the fuselage to generate a yawing moment, which produced a sideslip. The roll-yaw coupling of the dihedral effect caused the aircraft to

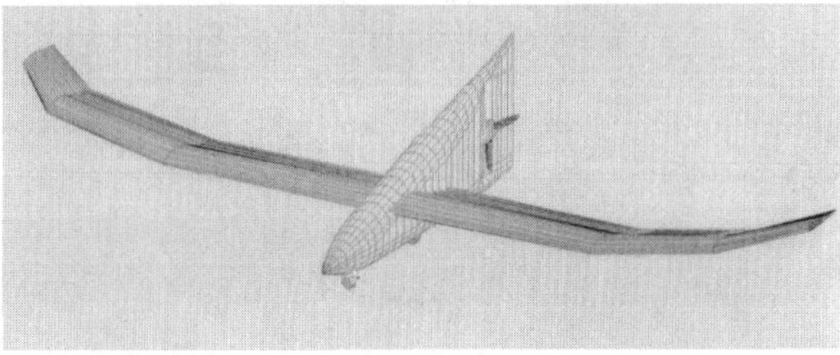

Fig. 4.39 Electric-powered mission aircraft concept.

roll. The maximum pivot angle was approximately 10 to 15 degrees. To bank to the left, the wing rotated counterclockwise as viewed from the top.

The pitch was controlled by translating a weight fore-aft on rails inside the fuselage. On the air-launched research gliders, a steel block was used for the weight, while an electrically powered mission aircraft would use its power plant battery as the weight.

The CG Shift Bird aircraft underwent extensive 6-DOF simulation analysis. The simulator had to be modified to account for the longitudinal CG shift control system. The conclusion from the simulations was that the aircraft could be made to work if there were accelerations. The aircraft had to maintain curvilinear flight for suitable longitudinal control. Also, the aircraft could not go into vertical flight motions (nose up or nose down), because the fore and aft CG shift did not affect control in these orientations.

Two half-scale, unpowered test aircraft were built to support the proposal for obtaining sponsorship of a powered mission demonstration system. The first CG Shift Bird was taken aloft on top of the NDM II host aircraft for a test flight. Unfortunately, it did not properly release from NDM II. It slid along the top of the NDM II fuselage and knocked off the host's vertical tail, causing the NDM II to crash. Foch was piloting the CG Shift Bird and heard considerable commotion but had to continue flying his aircraft. The video crew switched their subject from the CG Shift Bird to the crashing NDM II, so the rest of the test flight was not recorded.

Apparently, and not surprisingly, the impact into the NDM II vertical tail also damaged the CG Shift Bird. All pitch control was lost but it had roll-yaw authority. The sliding weight was fixed at the nose, and the aircraft made an arcing dive. The aircraft continued to accelerate without reaching trim speed when it impacted the runway and was destroyed. The released flight lasted 10 to 15 seconds.

A second CG Shift Bird was built to resume the test program (Fig. 4.40). This demonstrator flew off of the new NDM III host aircraft. The release was flawless. However, the aircraft again pitched over and impacted the ground. The descent was more gradual this time, so the crashed airframe was more intact but unable to fly again. An intern (Gundlach) successfully retrieved the downed aircraft, with some excitement, from a field containing an agitated bull! Inspection of the wreckage pointed to a failed actuator as the likely culprit.

The CG Shift Bird was a sound design and could have been developed into a mission-capable small UAS given sufficient time and resources. Unfortunately, sponsorship for a full-size mission demonstrator did not materialize. Figure 4.41 shows a three-view drawing of this research vehicle, and Table 4.14 lists its technical characteristics.

Fig. 4.40 Second CG Shift Bird atop the NDM III.

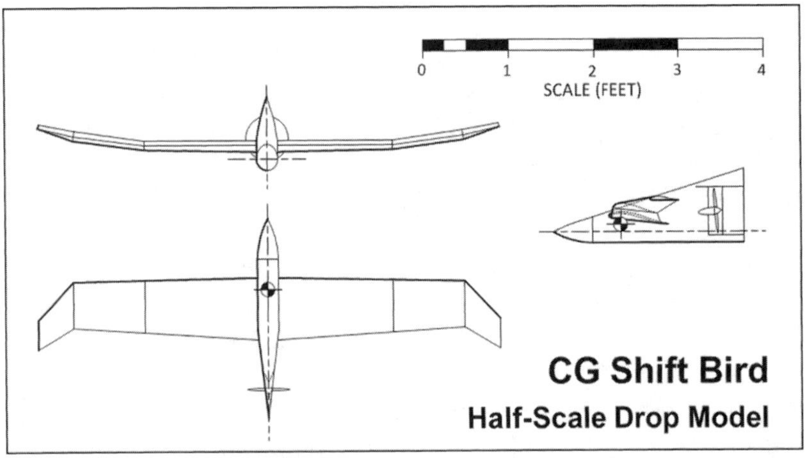

Fig. 4.41 CG Shift Bird, half-scale air-drop model three-view drawing.

Table 4.14 Subscale CG Shift Bird Technical Characteristics

Parameter	Value
Wing span	5.4 ft
Length	2.25 ft
Takeoff weight	4.5 lb
Empty weight	4.5 lb
Max L/D	16

RAMWING, 1993

Fig. 4.42 Ramwing in flight.

Ramwing was originally developed by United Technologies Corp. as an unfolding inflatable wing concept to support solid rocket booster recovery (Fig. 4.42). The wing featured a solid leading edge and separate membranes for the upper and lower surfaces. The airfoil shape that was formed by two surfaces was different in shape than a single-surface Rogallo wing. The wing inflated by a ram-air vent at the wing root apex stagnation point, and the internal air pressure helped maintain the airfoil geometry. The advantage of the concept was that the internal unfolding structure could support a much higher aspect ratio with less sweep than an equivalent Rogallo wing, giving a higher L/D. The rigid leading edge structural member ensured that the stagnation pressure on the rest of the span did not cause the wing to collapse.

Initially NRL was not interested in pursuing the Ramwing concept. The booster recovery mission called for a large, high Reynolds number device well outside the decoy flight regime on which NRL research was focusing. However, United Technologies built and tested an 8-foot-span wind tunnel model. The results looked promising, and the wind tunnel model was an appropriate size for an Active Electronic Decoy (AED)–type vehicle (Chapter 2). NRL decided to build a flight research vehicle to directly use the Ramwing wind tunnel model.

A new fuselage was developed for the program. It had a tractor propeller, tricycle landing gear, and a conventional tail. The fuselage transitioned from a large constant rectangular cross section at the front to a smaller constant diamond section at the tail, per AED requirements to launch from a shipboard

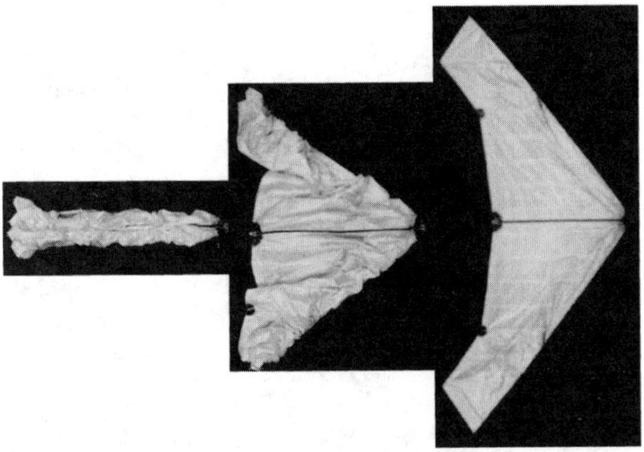

Fig. 4.43 Ramwing deployment.

MK 36 Decoy Launching System. Triangular facets formed a low-drag transition between the two cross sections. The original tail design was cruciform to represent an unfolding tail arrangement appropriate for a rocket-launched decoy (Fig. 4.43), but the rolling takeoff for the flight research airframe led to elimination of the lower vertical surface and augmented size for the upper vertical stabilizer. Rudder and elevator control surfaces provided control. The low aspect ratio swept wing provided a significant yaw-roll coupling dihedral effect.

The wing was mounted above the fuselage with struts in a parasol arrangement. The struts formed a parallelogram that enabled aft wing folding.

An O.S. BGX engine was used. This was the largest available two-cycle engine made in the appropriate power class. The engine was placed in the front of the aircraft in a tractor prop configuration.

The aircraft made a single flight at Naval Surface Warfare Center Dahlgren airfield. The wing inflated as expected as the speed built up. The aircraft was difficult to unstick, but it did eventually get airborne. There was not enough power for an authoritative climb rate. The demonstrator did not have enough altitude for a turn by the time it neared the end of the runway. The pilot had to land abruptly before overshooting the airfield, and the aircraft was not damaged. The turn characteristics were not evaluated since the aircraft flew only in a straight line down the runway.

NRL concluded that there were no significant advantages of the Ramwing over alternative folding-wing arrangements. The drag experienced in flight was higher than predicted, making the aircraft a marginal flyer. A larger engine could have been applied, but the program was internally supported with minimal resources. The demonstrator did not

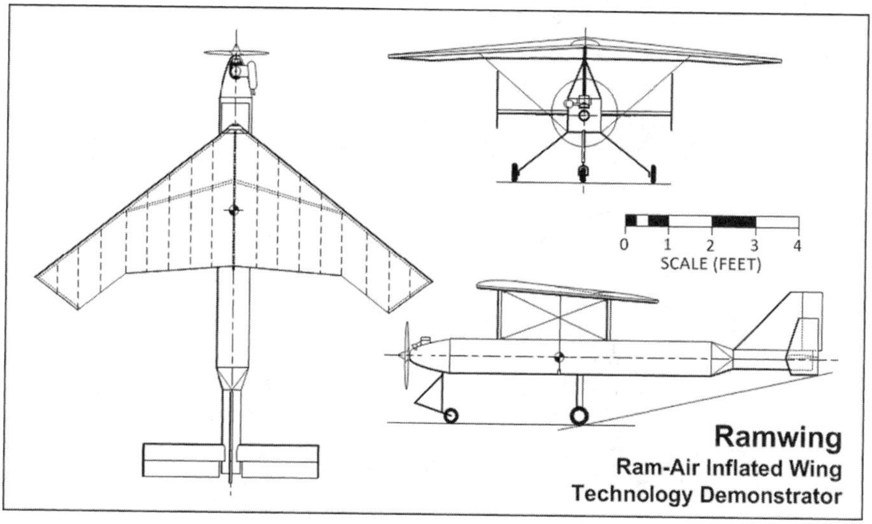

Fig. 4.44 Ramwing three-view drawing.

show enough promise to warrant further development. However, the Ramwing fuselage, engine, landing gear, and tail surfaces were reused for the NDM II. Figure 4.44 shows a three-view drawing of Ramwing, and Table 4.15 lists its technical characteristics.

Table 4.15 Ramwing Technical Characteristics

Parameter	Value
Wing span	9 ft
Wing area	15.9 ft^2
Length	9.75 ft
Takeoff weight	32.0 lb
Cruise speed	24–33 KTAS
Max L/D	4.9 (estimated as flown) 5.75 (operational decoy configuration)
Launch method	Conventional
Recovery method	Conventional
Propulsion type	O.S. 108 BGX, 1.08-cu.-in., 2-stroke engine
Power	2.9 hp

VENTURE, 1994–1995

Fig. 4.45 Venture.

The Venture program (Fig. 4.45) was a continuation of the LAURA Variable Span demonstrator effort. The LAURA Variable Span aircraft was flown with the wings folded and with the wings deployed, but the in-flight wing deployment was not attempted because resources ran out at the end of the program. The benefit of the wing deployment was to nearly double the wing area and aspect ratio in flight to transition from a high-speed cruise configuration to an efficient low-speed loiter configuration.

For Venture, the LAURA Variable Span wing and horizontal stabilizer were directly reused. Advanced Aero-Mechanisms Corp. was responsible for refurbishing the wings and horizontal stabilizer and upgrading the wing deployment software. NRL built a new fuselage and vertical tail. To simplify integration and to improve engine cooling, Venture had a tractor engine configuration. There was no requirement for a nose-mounted payload.

The Venture wings were equipped to unfold during flight, transforming from a cruise to a loiter configuration. The unlatching and refolding functionality to return to cruise configuration was not enabled, as the refolding would have required an electronically controlled latch. The missions envisioned at the time required high-speed ingress and long endurance in the mission area, after which the vehicle would be expended, so there was no desire to develop the electronic latch. However, more recent missions such as electro-optical/infrared (EO/IR) surveillance or chemical/biological agent detection would benefit from a high-speed egress capability.

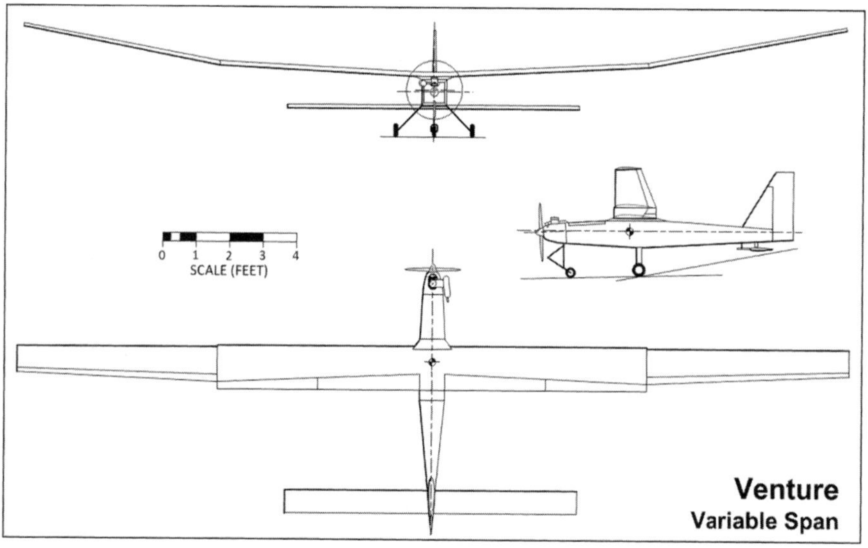

Fig. 4.46 Venture three-view drawing.

Like LAURA, the Venture program went through an incremental test program. The avionics systems and autopilot programming were first modeled in a 6-DOF simulator. Next, wing deployments on the complete Venture aircraft were demonstrated on the captive carry rig. Engine start by hand was difficult on the rig, where the aircraft was 15 feet above the ground. After the captive carry tests, the aircraft underwent a flight test program.

The flight tests successfully demonstrated the in-flight wing unfolding. The aircraft was easily trimmed between flight modes. The pilot workload was low due to the gentle, general aviation–like handling qualities. Figure 4.46 shows a three-view drawing of this research vehicle, and Table 4.16 lists its technical characteristics.

The fuel consumption was carefully measured during the flight test program. While the test flights had short duration, the fuel consumption was consistent with a very long range and long endurance capability. The tests revealed that the normally aspirated model aircraft carburetor spattered more fuel out of its throat than was sucked in and combusted. Venture, like the LAURA, was capable of flying for over 60 hours on one gallon of fuel. As such, Venture would be capable of transatlantic flight if equipped with an engine using a fuel injection system. This ambition was later demonstrated with the ONR-sponsored transatlantic flight of the Insitu Aerosonde.

Table 4.16 Venture Technical Characteristics

Parameter	Value
Wing span	12.9 ft (retracted) 24.6 ft (deployed)
Wing area	38.2 ft^2 (reference)
Length	7.67 ft
Takeoff weight	54.5 lb
Empty weight	52.5 lb
Fuel weight	2.0 lb
Loiter speed	21 KTAS
Dash speed	44 KTAS
Stall speed	17 KTAS
Endurance	60 hr on 1 gal (est)
Max L/D	16.5 (calc) folded 23.0 (calc) unfolded
Airfoils	Wing: modified Eppler 193 Tail: NACA 0010
Launch method	Conventional
Recovery method	Conventional
Propulsion type	O.S. BGX-1, 2.13-cu.-in., 2-stroke engine
Power	4.0 hp

SAIL-A-PLANE, 2005–2006

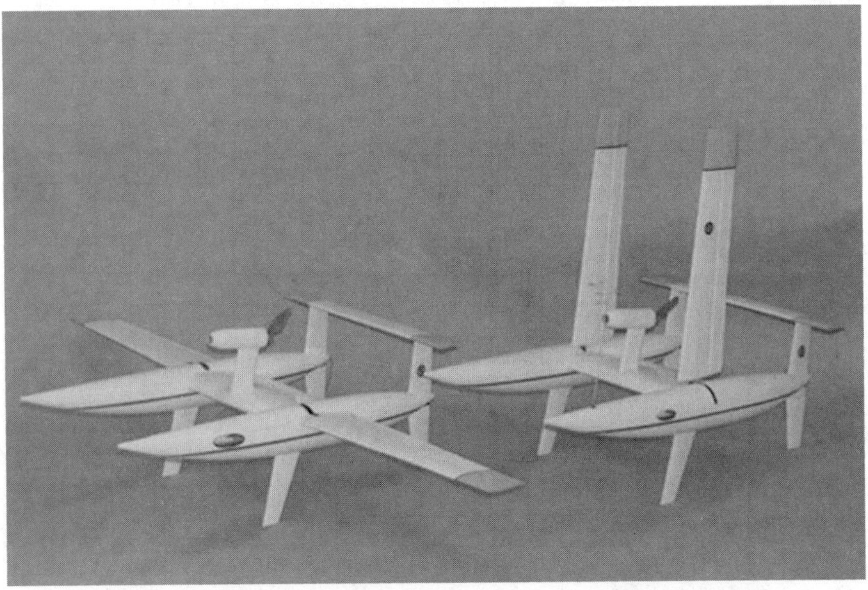

Fig. 4.47 Sail-A-Plane in flight mode (left) and in sailing mode (right).

Through the SAMPLE micro air vehicle program (Chapter 6) and other studies, NRL concluded that the optimal survivability and operational utility for autonomous systems come about when the systems are capable of two modes of locomotion. For example, a versatile vehicle can both fly in the air and drive on the ground, or be a flying submarine. NRL considered missions for larger dual-mode mobility vehicles. Several NRL engineers were both pilots and sailors, and they realized that many of the design requirements and attributes of sailplanes and high-performance sailboats are similar. The team began work on the Sail-A-Plane, which could both fly and move on the surface of water (Fig. 4.47). This system could be equipped to perform maritime surveillance, antisubmarine warfare (ASW), EW, oceanographic research, or meteorological missions.

Many high-performance catamarans use rigid sails with multielement airfoils. These sails are very similar to wings on a sailplane, except the lift vector is orthogonal to that of a wing. A wing that could be oriented vertically for sailing or horizontally for flight would be synergistic.

The first concept for Sail-A-Plane was to land a sailplane on the water and reconfigure it as a sailboat by flooding one wing (Fig. 4.48). The flooded wing would sink and act as a keel, while the upward-facing wing was a sail. The primary drawback of this design was that the submerged wing was too large to serve as a keel. Asymmetric vehicles that provided the correct keel and sail proportions were considered but would have been poor aircraft.

The next design solution was a catamaran. Two fuselages served as hulls in the water. This arrangement eliminated the need to flood parts of the vehicle. A center wing section connected the two hulls. The outer wing panels folded

Fig. 4.48 Initial Sail-A-Plane configuration with wing flooding flight configuration (left) and sailing configuration (right).

upward to serve as sails. There were two keels and two dagger boards below the water line. For flying, there were two vertical stabilizers, one on each fuselage. A horizontal stabilizer connected the tops of the vertical stabilizers, which also helped with structural rigidity.

At first, NRL considered modifying a high-performance commercial racing catamaran for flight. Many had multielement cambered rigid sails that could serve as wings. The bending moments for sailing and flying were similar as well. However, the project budget was insufficient to take this approach.

Instead, NRL designed and built subscale models to save money. Two versions were built. One was a catamaran sailboat with vertical sails. The other was a flying aircraft with horizontally oriented wings. Constructing two separate vehicles of different configurations was easier than building a single vehicle that could transform. The wing rotating mechanism could be developed later when more funding was available. Both models had molded composite structures and radio control systems. Both also had an electric motor and pusher propeller in a pylon mounted above the center wing.

Sail-A-Plane had to contend with the classic flying boat problem of breaking the water surface suction during takeoff (Fig. 4.49). The most common flying boat approach was to place a step on the hull, but this

Fig. 4.49 Sail-A-Plane operation concept.

Fig. 4.50 Sail-A-Plane in flying configuration.

generates significant base drag. The step also reduces the catamaran's sailing performance. The designers reviewed a 1950s ONR test report of a Navy Grumman Goose amphibian that used a hydrofoil. This aircraft had a single hydrofoil connected to the fuselage by a strut. As the aircraft gained speed, the aerodynamic controls became effective. For Sail-A-Plane, each hull had a keel with a small hydrofoil at the bottom. The rudder–dagger board surfaces also had hydrofoils. This was effectively a four-point hydrodynamic lifting system. The hulls would rise out of the water with the hydrofoil lift, breaking the fuselage suction. The Sail-A-Plane would continue the takeoff run and lift off the water. The hydrofoils produced little drag in flight due to their relatively small surface area.

The operational water landing would have challenges as well. The autopilot would need to determine wind speed, wind direction, and swell direction. The vehicle would need to land parallel with the swells. The operational system would likely need lidar to map and track the waves.

The operational Sail-A-Plane would be covered with solar panels (Fig. 4.50). The batteries would recharge while sailing (Fig. 4.51). Other dual-mode vehicle studies showed that the best systems have one mode that consumes one tenth the energy of the other mode. In the case of Sail-A-Plane, flying consumes about ten times as much energy as sailing. Therefore, the aircraft could fly 1 hour for every 10 hours of daylight sailing. The distances covered in the two modes would be similar. The aircraft would fly over the ocean, land on the water, sail for several hours while recharging, and then take off again.

The subscale sailing model was tested under power and under sail at the NRL indoor recreational pool and outdoors (Fig. 4.52). The sailboat behaved satisfactorily.

Next, the subscale flying model was tested. The hydrofoils were replaced with wheels to enable conventional runway takeoff and landing.

Fig. 4.51 Sail-A-Plane in sailing configuration.

The aircraft was successfully flown, but the handling characteristics were undesirable. The pylon-mounted motor system produced a large pitching moment due to the height above the CG. The horizontal stabilizer between the two hulls was not large enough for trim. The model was modified by extending the horizontal tails aft from the vertical tails on booms. The aircraft flew better with this modification. Given more time and funding, the better solution would have been aft-swept vertical tails to extend the tail arm. Funding ran out before the subscale flying model could be equipped with hydrofoils to test the water takeoff and landing.

Fig. 4.52 Sail-A-Plane water testing.

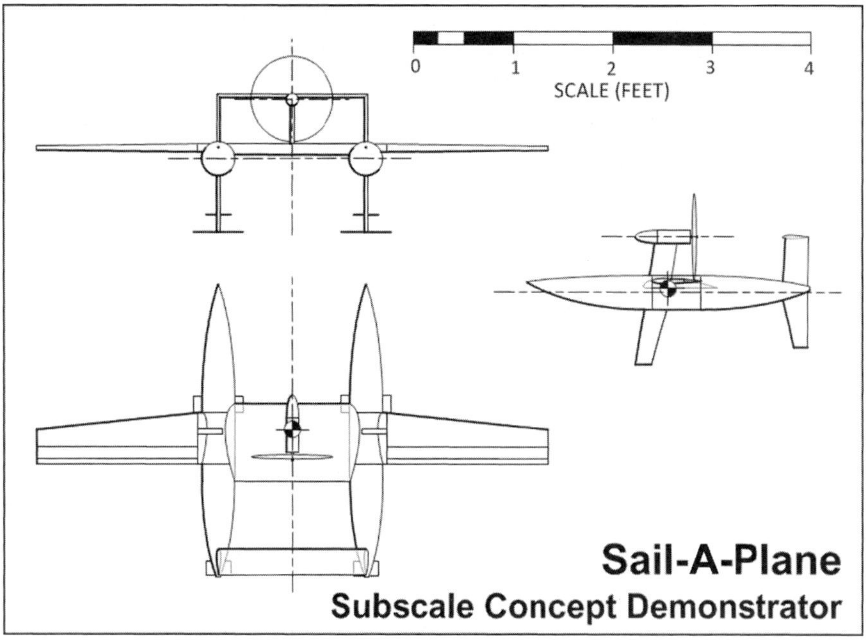

Fig. 4.53 Sail-A-Plane three-view drawing.

NRL wrote a proposal to complete the test program on the subscale model and to advance the controls technology. The follow-on program would have included autopilot development, autonomous landing algorithms, and sensors for water landing. The sponsors found the proposal to be too controversial and perhaps infeasible. The design was simply beyond what could be envisioned. Many new advances in machine vision would make the project practical today. Figure 4.53 shows a three-view drawing of this research vehicle, and Table 4.17 lists its technical characteristics.

Table 4.17 Subscale Sail-A-Plane Model Technical Characteristics

Parameter	Value
Wing span	5.2 ft
Length	2.8 ft
Takeoff weight	3.1 lb
Empty weight	3.1 lb
Max L/D	14 (calc)
Launch method	Water takeoff
Recovery method	Water landing
Propulsion type	Battery electric
Power	250 W

Chapter 5

Ground-Launched Fixed Wing Mission Aircraft

NRL research supports the development of technology demonstrators for the U.S. Marine Corps (USMC) and special operations missions, so the VRS has developed many aircraft capable of being ground launched and often, at the request of the sponsor, ground recoverable. In general, these aircraft have been designed to perform intelligence, surveillance, and reconnaissance (ISR) missions such as day/night imaging, chemical/biological agent detection and sampling, and autonomous delivery of sensors to remote locations.

Self-Navigating Drone Expendable/Recoverable (SENDER), 1994–1998

Fig. 5.1 SENDER.

The Self-Navigating Drone Expendable/Recoverable (SENDER) came about when NRL leadership realized that composite construction, electric propulsion, low Reynolds number (LRN) aerodynamics, and autopilot technology were reaching sufficient maturity to enable a small, mission-capable aircraft (Fig. 5.1). The intended missions were biological agent sampling and imagery. The SENDER program did not attempt to develop the full mission aircraft, but it sought to demonstrate key capabilities and technologies. The resulting project met or exceeded all its expectations and surprised the unmanned aircraft systems (UAS) community by having relatively high performance and mission capabilities for such a small system. Consequently, SENDER became the direct ancestor of today's small, tactical unmanned aircraft and played a significant role in the establishment of, and requirements for, the Department of Defense's (DoD's) "Small UAS" class.

The SENDER airframe and its launcher were designed to fit within a suitcase, such as a large American Tourister suitcase. This was ideal for Special Forces, because the system can be carried across borders without drawing attention. The fuselage was made in three sections. It had a circular cross section so that bayonet clips could be used to latch the fuselage sections via a simple rotation through a quarter turn, like a camera lens. The nose cone contained the motor, the motor controller, and one battery pack. The aft fuselage section contained another battery pack and the tails. The tail had a staggered V-tail arrangement, so the aft tail rotated over the forward tail for flat storage in the suitcase. This staggered arrangement both facilitated a simple folding/locking mechanism and slightly improved the aerodynamics. The central fuselage section was combined with the wing. It contained the autopilot and a payload bay. One wing panel was removable from the central section just outboard of the fuselage junction. This required only a single wing joiner, and its location was offset from the point of greatest loads. Because of its overall small size, the SENDER could have been launched out of a window from a tall building, though this capability was not demonstrated.

The aircraft had high performance for its size, and it is still impressive nearly two decades after it was conceived. It cruised at 50 knots but could dash at 100 knots in level flight. This high speed helped to penetrate wind. SENDER was optimized for range performance, whereas most electric unmanned aircraft are designed to maximize endurance. This was achieved by a high wing loading and high lift-to-drag ratio (L/D). The design one-way range with lithium sulfur dioxide ($LiSO_2$) batteries was 100 nautical miles, with the aircraft flying for 2 hours at 50 knots. Only NiCd batteries were tested on the research test flights and mission demonstrations, yielding an endurance of 20 minutes at 50 knots.

The aircraft had several aerodynamic features that contributed to a relatively high lift-to-drag value of 18. The wing used a Selig-Donnovan SD 7032 airfoil, which was popular with model sailplanes. The V-tails had a NACA 0012 airfoil section. The fuselage had a streamlined body of revolution shape, with a constant 5-inch-diameter circular cross section for the middle fuselage. The high wings were effectively blended with the fuselage at the juncture to minimize interference drag. An advantage of the staggered V-tails was that the pressure peaks did not line up, so the tail interference drag and fuselage/tail juncture drag were reduced. Small, aluminum, triangular belly skids were used in place of landing gear, which eliminated a major drag contributor. The measured C_{D0} value (drag coefficient at zero lift) of 0.032 was very good for a low Reynolds number vehicle.

SENDER had a simple control system consisting of ruddervators on the V-tails. Ruddervators act as rudders when deflected differentially and as elevators when deflected together. A micro servo was buried within each tail, and a short pushrod connected the servo arm to the control horn with very little linkage play. The high-wing, geometric wing dihedral, and slight wing sweep all contributed to the dihedral effect (negative $C_{l,\beta}$), so the yaw control induced the roll moment by lateral-directional coupling. The control surfaces had a live hinge, whereby the upper tail skins served as the hinge. The tails were built in a female mold with upper and lower halves bonded together, and then the lower skin below the hinge line was simply cut away. No fatigue was observed on this skin hinge for an aircraft that flew more than 30 sorties. Advantages of the skin hinges included simpler manufacturing, and elimination of flow leakage, which improved control effectiveness. The controls were found to be so highly effective that the SENDER was shown to be very aerobatic and capable of performing near-axial rolls, despite having no ailerons and relying on yaw-roll coupling for its lateral control! It was also hoped that the staggered V-tail arrangement might reduce Radar Cross Section (RCS) by not creating a 90-degree angle that would be a corner reflector, but this was never measured.

The airframe was constructed of composite materials. Most of the structure was made of a sandwich of thin glass layers with 2-millimeter Rohacell (polymethacrylimide) structural foam core in the middle. The spar consisted of a foam and carbon member. The wing did not have any ribs except at the wing joiner. The wing structure passed through the fuselage. Upper fuselage fairings covered the wing-fuselage junction. Fuselage bulkheads at the joiners were made of G10 fiberglass boards (fiberglass-epoxy laminate) with cut out features for the machined aluminum bayonet clips. The wings and tails were hollow to allow for internal components. For convenience and easy repair, the airframe was painted with Rustoleum spray paint.

The payload bay was located in the center fuselage section such that it was near the longitudinal center of gravity (CG). An advantage of the CG payload bay location is that droppable payloads that do not affect the CG when released can be installed. A payload hatch was positioned on the bottom fuselage to provide fuselage access. The high wing allowed large, unimpeded payload field of regard looking nadir and to the sides. The payload hatch was nonstructural, so apertures could be cut into this fairing. The small landing skids were fitted fore and aft of the payload bay to protect the aircraft during landing. The reason for the slight wing sweep was to position the CG location required for stability at the midpoint of the central fuselage section. SENDERs were flown with gimbaled camera turret, biological agent sampler, and chemical agent detection payloads.

SENDER had an electric propulsion system that closely resembles modern systems. It used an Aveox brushless motor with a planetary reduction gear box. This motor was of an “inrunner” configuration, to use modern terminology. The folding propellers were commercial off-the-shelf electric sailplane propellers. Two NiCd battery packs were located in the front and aft fuselage sections. These packs were connected with low loss Sermos connectors. The brushless motor had very low electromagnetic interference (EMI), since there was no brush arcing. The motor also had very high efficiency, which contributed to the long-range capability.

Three methods were used to launch SENDER. The first was an air launch from the NDM III host utility aircraft. SENDER rested on a release platform on top of the fuselage. Upon reaching the desired deployment altitude, the NDM III entered a shallow dive to accelerate beyond the SENDER’s trimmed release airspeed, and SENDER was released. SENDER immediately climbed away from the NDM III. Each aircraft then turned in opposite directions to ensure separation. The air launch method was generally uneventful, except on two occasions when the host aircraft’s release mechanism failed to operate properly. The first release failure on an NDM III required the host aircraft to land with the SENDER still aboard. The second failure occurred when the SENDER was being launched from a Dakota utility aircraft in support of chemical and biological agent sampling missions performed with the U.S. Army at Aberdeen Proving Ground, Maryland. Shortly after takeoff, the release mechanism structurally failed, and the SENDER was blown aft by the slipstream and locked itself to the Dakota tail surfaces. The immediate aft shift of the host aircraft CG caused the Dakota to abruptly enter a low-altitude, full-power, deep stall. The resulting crash extensively damaged both aircraft, but they were both repaired to fly again.

The second launch method used a bungee-powered ground dolly. Bungee cord was staked to the edges of the runway. The wheeled dolly interfaced to the bungee with a pulley to even out the tension and to facilitate a straight run down the runway. The aircraft rested in a saddle on the dolly. The ground crew pulled the dolly back until sufficient bungee cord tension was achieved. Then the dolly with aircraft would roll down the runway and build up speed. When the dolly and the aircraft reached the desired airspeed, the motor would be turned on and the aircraft climbed away. This method sometimes resulted in a crash when the aircraft climbed away before reaching the necessary airspeed. In these cases the torque from the propeller would snap-roll the stalled aircraft to the left.

The third SENDER launch method was from a launch rail. Two forms of rail launchers were developed. One used a bungee cord to pull the aircraft along the rail. The second rail launcher fired a solid rocket booster down a slotted hollow rail with a square cross section. The rocket had a fin extending through the rail slot that interfaced to a fuselage hardpoint. The rocket booster would fly out the far end of the rail after the aircraft launched and travel downrange. Sometimes the aircraft would jump away from the booster fin under the shock of rocket ignition, leaving the aircraft behind. The rails were segmented to allow packaging into the suitcase.

SENDER flew like a radio-controlled (RC) pylon racer with the high L/D and high wing loading. The tails were relatively small to enhance range performance, which gave limited control authority at low speeds. The wing tended to tip stall at low speed, indicating the amount of wing washout was insufficient. The aircraft also tended to Dutch roll at low speed due to the strong dihedral effect. The Dutch roll oscillation could eventually result in a tip stall as the roll amplitude increased. The relatively strong motor torque exacerbated the problem. The near-stall, low-airspeed regime had to be avoided, but the high-speed handling characteristics were excellent. SENDER was a demanding aircraft to fly manually, but flying it was well within the capabilities of a skilled RC pilot.

SENDER performed pioneering research in digital GPS-based autopilots for small unmanned aircraft. NRL developed the Lightweight Autopilot (LWAP) to support the autonomous flight. The autopilot weighed approximately 2 pounds and consumed most of the payload bay. Clearly this was an interim step. However, SENDER may have been the first small unmanned aircraft to fly with such an autopilot system. The demanding flight characteristics required an autopilot that was operationally useful for the average operator (rather than test pilots). The LWAP used an Onset Model TT8 microcontroller and a Trimble SV6-CM2 GPS receiver. NRL conducted extensive flight tests with the LWAP installed on SENDER. The system often worked well, and many autonomous flights were demonstrated. However, the poor quality of small accelerometers and air data sensors of the time led to problems with sensor drift and variability. The test pilot could take over with RC control when the autopilot had difficulty. The conclusion of the LWAP tests was that small autopilots would be feasible with further investment. Miniaturized and reliable small autopilots became available around 2000, due in part to the Defense Advanced Research Projects Agency (DARPA) Micro Air Vehicle (MAV) program.

Fig. 5.2 SENDER head-on perspective.

NRL was never tasked with producing the final mission aircraft. SENDER set the standard for small unmanned aircraft and demonstrated a practical design. It proved the utility of this vehicle class and demonstrated several mission capabilities. The system flew for the Army, Navy, and other customers. SENDER was a direct ancestor of Dragon Eye, which became the

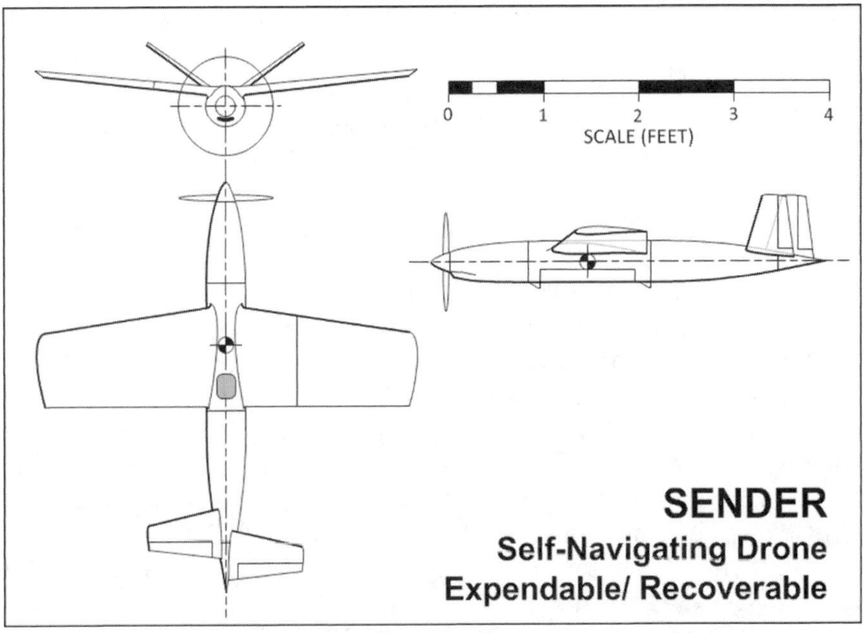

Fig. 5.3 SENDER three-view drawing.

Table 5.1 SENDER Technical Characteristics

Parameter	Value
Wing span	4.0 ft
Wing area	3.33 ft^2
Length	4.1 ft
Container size	4.0 ft × 1.25 ft × 0.75 ft
Payload volume	100 cu. in.
Takeoff weight	10.0 lb (excluding booster)
Empty weight	3.5 lb
Payload weight	2.5 lb max
Battery weight	4.0 lb
Cruise speed	50 KTAS
Dash speed	90 KTAS
Stall speed	40 KTAS
Range	100 nmi (one way); 50 nmi radius
Endurance	2 hr with lithium primary batteries
Max L/D	18
C_{D0}	0.032
Airfoils	Selig SD 7032 (wing) NACA 0012 (tail)
Launch method	Air launch, rail launch, dolly
Recovery method	Expendable or skid landing
Propulsion type	Battery electric

first U.S. military program of record Small UAS (SUAS) that was produced in quantity and used effectively in combat.

Two SENDERs were put on display by the Army at Aberdeen Proving Ground. These two systems had demonstrated the capabilities and limitations of small airborne chemical agent detectors. The aircraft had the names "Anthrax Eater" and "Ebola Gay" painted on their noses. The engineers who created the names had second thoughts as they delivered the systems, as these names were as irreverent as most engineers enjoy being, but the Army loved them. Figure 5.2 shows a head-on perspective of SENDER, Fig. 5.3 a three-view drawing, and Table 5.1 a list of its technical characteristics.

SENDER: Further Reading

Foch, R.J., "SENDER, A Low Cost Airobotic Platform," *Proceedings of the 1996 Association of Unmanned Vehicle Systems International (AUVSI) Conference*, Orlando, FL, 1996.

SENDER 2, 1998–1999

Fig. 5.4 SENDER 2.

SENDER 2 was a growth version of the original SENDER. Its payload capacity grew from 2 pounds to 5 pounds. SENDER 2 also leveraged more advanced autopilot technology. This system was informally known as "Super Sender."

SENDER 2 featured a conventional airframe configuration. The three-piece wing had dihedral in the outer panels. The wing was mounted as a high wing configuration. The tail was conventional. Two SENDER stock motors and propellers were used for propulsion to minimize development time and cost. The motors were mounted on the wing in a tractor arrangement.

Fig. 5.5 SENDER 2 follow-on concepts.

One prototype was built and successfully flown. The first flights occurred on May 21, 1998, at Webster Outlying Field in St. Inigoes, Maryland. The demonstrator used rechargeable NiCd batteries to keep costs down. One flight lasted for 2 minutes and the second flight was 12 minutes. SENDER 2 was launched from a moving vehicle, as shown in Fig. 5.4, as an expedient to avoid the development of a larger dolly, longer catapult rail, or bigger host aircraft.

Sponsors hinted that air launch requirements were desirable for future programs for SENDER 2–sized systems, so new airframe configurations were conceptualized and studied. A twin-boom design with a single folding vertical tail was considered (Fig. 5.5, left). This design had a 7-foot span. It could be transported in a standard size SENDER suitcase but it would have been tricky to unfold for an air launch. A second configuration was then developed with a single fuselage, a top-mounted pivoting wing, and a rear-mounted propeller, which could be readily unfolded in free-fall after an air launch (Fig. 5.5, right). This 8.6-foot-wingspan conceptual design was the direct ancestor to the Defense Threat Reduction Agency (DTRA)–sponsored NRL FINDER Advanced Concept Technology Demonstration program (Chapter 3). This design's overall configuration and launch sequence were adopted for FINDER. The SENDER 2 did not proceed beyond the demonstrator and follow-on conceptual design, being supplanted by the FINDER program and the Extender project. A three-view drawing of SENDER 2 can be found in Fig. 5.6 and a list of its technical characteristics in Table 5.2.

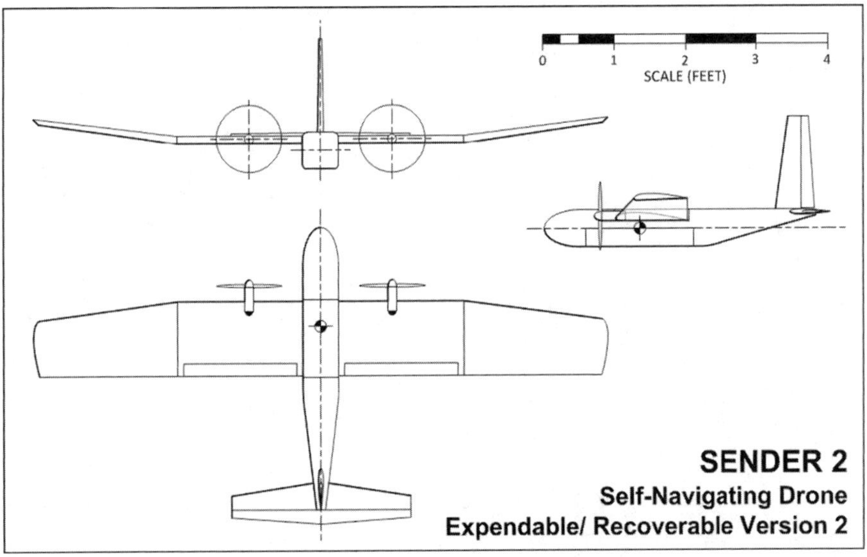

Fig. 5.6 SENDER 2 three-view drawing.

Table 5.2 SENDER 2 Technical Characteristics

Parameter	Value
Wing span	8.0 ft
Wing area	7.5 ft^2
Length	4.0 ft
Takeoff weight	12.0 lb
Empty weight	6.0 lb
Payload weight	3.0 lb
Battery weight	3.0 lb
Loiter speed	30 KTAS
Dash speed	40 KTAS
Stall speed	22 KTAS
Endurance	0.2 hr
Airfoils	RF 1165FB (wing) NACA 0009 (tails)
Launch method	Car launch or air launch
Recovery method	Belly landing or expendable
Propulsion type	Battery electric
Power	500 W

CRAWLER DELIVERY AIRCRAFT, C. 1995

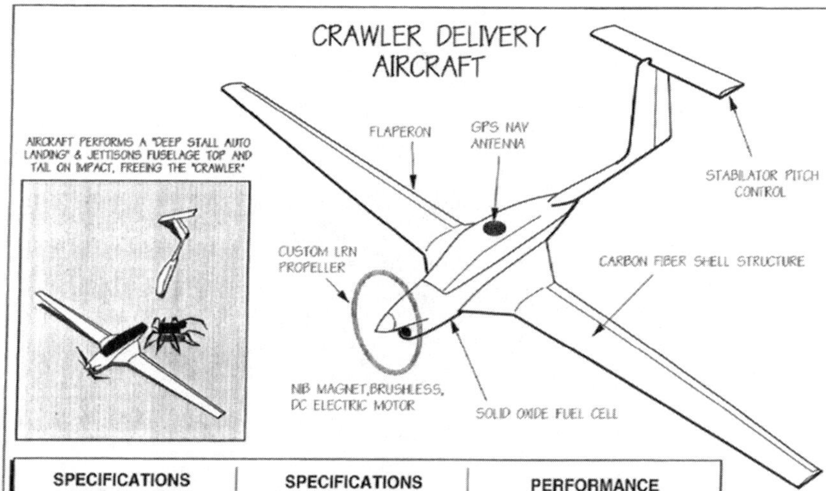

SPECIFICATIONS 24.2 LB PAYLOAD		SPECIFICATIONS 32.2 LB PAYLOAD		PERFORMANCE BOTH VERSIONS	
WINGSPAN	12.8 FT	WINGSPAN	14.1 FT	MAX RANGE	100 NM
LENGTH	7.0 FT	LENGTH	7.5 FT	MAX ENDURANCE	2.2 HR
WING AREA	10.85 SQ FT	WING AREA	12.9 SQ FT	CRUISE AIRSPEED	50 KT
GROSS WHT	55.0 LB	GROSS WHT	66.0 LB	MAX AIRSPEED	80 KT
PAYLOAD TYPE	ROBOT CRAWLER	PAYLOAD TYPE	ROBOT CRAWLER	STALL AIRSPEED	35 KT
				AUTO LAND	DEEP STALL

Fig. 5.7 Crawler Delivery Aircraft design and major features.

The Crawler Delivery Aircraft was a concept for a system that could deliver a crawler unmanned ground vehicle (Fig. 5.7). The crawler had eight legs, which would provide robust locomotion, and resembled a spider. NRL would develop both the unmanned aircraft and the crawler. The expendable aircraft fuselage would break open aft of the wing to enable the crawler to exit. The electric power source was to be a JP-5 fueled solid oxide fuel cell being developed by industry under a Navy contract; this would provide about 2.5 times the endurance of the best available batteries. This design and mission concept were presented to the Special Forces community, but funding was not available for new system development at that time.

SWALLOW, 1996–1998

Fig. 5.8 Swallow.

Swallow was developed for DARPA to carry a real-time biological agent detection payload (Fig. 5.8). It was a tactical-class unmanned aircraft that served as a flying miniature bio lab. Unlike previous efforts, which relied on the unmanned aircraft to bring back a sample that was later analyzed in a laboratory, Swallow conducted in-flight detection and relayed the information to an operator on the ground. Like its namesake bird, the Swallow ate "bugs."

This project was originally called the Airborne Biological/Chemical Detection (ABCD) unmanned aircraft. It was intended to carry a 25-pound payload (Fig. 5.9), though the payload weight was greatly reduced during the Swallow program. The ABCD configuration had two motors, as only smaller brushless motors were available at the time (Fig. 5.10).

NRL's Center for Bio/Molecular Science and Engineering developed the miniature, real-time biological detection payload. This group had a tabletop bio lab that included sizable glassware. The lab operated by pumping the collected sample agent in a fluid solution over antigen-coated fiber-optic cables. When the antigen died, the fiber-optic fluorescence changed. Within minutes it was possible to determine the agent. The system had a bank of antigen-coated fiber-optic cables to detect multiple biological agents of interest.

The major challenges were to miniaturize and to ruggedize the tabletop experiment. The whole system had to fit within the nose of the unmanned aircraft. NRL's Offboard Countermeasures Branch was accustomed to miniaturizing electronic warfare (EW) payloads in order to place them on aircraft constrained by size, weight, and power. The same methods were applied to this experiment. NRL worked with industry to create a 10-pound payload.

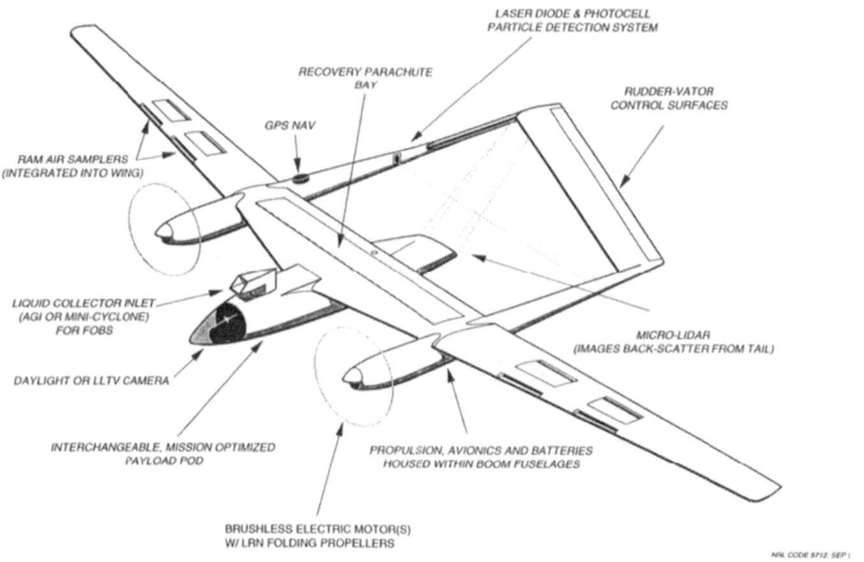

Fig. 5.9 Follow-on ABCD concept.

A key innovation was a method of putting the biological agent into the liquid. The biological agent had to be kept alive for the method to work properly, so pumps and other techniques would not be appropriate. The approach taken was to create a water cyclone. The air entered the cyclone chamber via a ram-air probe that stuck out of the nose of the aircraft. The water spun like a whirlpool on a cylindrical surface. The agent ingested through the probe embedded in the fluid wall. The air inlet and water cyclone were then turned off, and the water drained into the bank of fiber-optic cables.

The payload functionality was first evaluated in a wind tunnel. The payload was then flown on a large Telemaster utility aircraft. At the time of the flight test, there were still glass components in the system. The

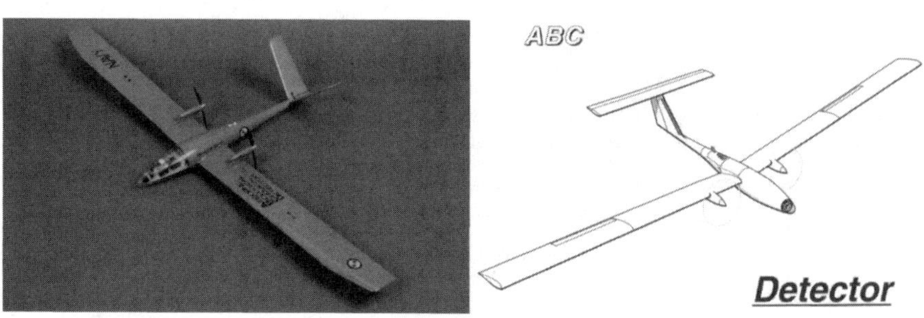

Fig. 5.10 ABCD model (left) and original concept drawing (right).

payload functioned well, but a hard landing caused some of the glass components to break.

The first operationally representative payload flights, now well hardened for field use, were conducted at Dugway Proving Ground in Utah, using the Dakota I utility aircraft. The aircraft took samples of biological agent simulants that were released over a calibrated test range at night. The research team wore biohazard gear and aviator's night vision goggles.

The Swallow aircraft was developed in parallel with the bio agent detector payload. The aircraft was developed very rapidly to address an urgent need. Swallow was intended to fly at short range and endurance, but be rapidly deployed. The short endurance was compatible with battery electric propulsion, which greatly simplified the system relative to gas engines. The aircraft would be plugged together and take off from a road or runway. Alternatively, the aircraft could be rail launched and parachute recovered. Due to the lack of available autopilots and short program duration, the system was flown via radio control with the pilot looking through a camera.

The aircraft had a conventional configuration with distinctive features. It had an appearance reminiscent of a Windex sailplane. The wings were positioned on the fuselage with a high wing arrangement. The wings had slight dihedral. The horizontal and vertical tails were located at the aft of the single fuselage. The payload was housed in the fuselage nose with a large nose hatch for access. The batteries were located under the wing in the fuselage. A compartment aft of the wing contained the parachute and the avionics. The parachute hatch flipped back toward the tail. A fixed tricycle landing gear was used for the early flight testing.

The aircraft fit within a storage container measuring 5.75 feet by 2 feet by 1 foot. The fuselage was kept short to keep the box dimensions small, so the tails were close coupled. The wing came apart in three pieces, such that the center section had a continuous spar to achieve a lightweight structure. The center section was bolted to the fuselage, and the outer wing panels plugged in with circular wing rod tubes and alignment dowels. The horizontal stabilizer and rudder halves unplugged from the vertical tails.

The tail had a highly unconventional cruciform arrangement with the propeller arc passing through the horizontal and vertical surfaces (Fig. 5.11). This tail arrangement was adopted to enable good control authority at low speeds and to protect the propeller from the parachute during deployment. The slipstream blowing over the control surfaces allowed the fuselage to be shorter. The horizontal surface consisted of a fixed front surface (horizontal stabilizer) and a moving rear surface (elevator). Each of these surfaces had an airfoil cross section. The separation of the two surfaces created a slot in which the propeller passed. Similarly, the vertical stabilizer and rudder were

Fig. 5.11 Swallow during testing.

separated. The motor mount was positioned at the intersection of the horizontal and vertical surfaces.

There were concerns about vibration and noise as the propeller blades passed through the wake of the stabilizers. The chosen remediation was to ensure that propeller blades did not pass through the wake simultaneously so that there was not an additive noise and vibration magnitude increase. A five-blade propeller was used so that no two blades passed through a wake simultaneously. With a two-blade or four-blade propeller, the blades would pass the structure in opposing pairs at lower frequencies, which would cause larger-magnitude vibration loads on the tail. With a five-bladed propeller, blades individually pass through the wakes 20 times per revolution. This increases the frequency, and higher frequencies damp out more rapidly over distance. However, the aircraft was still quite noisy for the power and it had a unique acoustic signature. The lower amplitude and higher frequency also aided vibration damping.

NRL adapted a method used for helicopter rotors for dynamic balancing of the propeller blades as they passed through the stabilizer wakes. Each of the composite blades was sandpapered and polished to the same weight. The blades were housed within a five-star aluminum hub via a lead-lag hinge interface. In theory, the blades would be self-balancing, and they could adjust their relative positions to run as smoothly as possible as they passed through the wake.

Given the development risk of this untried propeller arrangement, NRL decided to conduct wind tunnel testing of the tail section prior to building the flight vehicle. The tests helped characterize the vibration of the propeller and tail surfaces. NRL's Optical Sciences Division developed a new fiber-optic cable technology for high-rate, real-time strain measurements. It was discovered that backscatter properties changed when the cable was damaged due to microscopic fractures. The properties would partially reset when the strain was removed. This system could provide real-time strain measurements at very high frequency. The wind tunnel model's horizontal stabilizer, elevator, vertical stabilizer, and rudder were made with fiber-optic cables embedded during layup of each in its female mold, so the outer surface was not altered. No unusual structural problems were discovered during the wind tunnel test, so the development of the Swallow prototype proceeded. The fiber-optic strain gage technology was later used for helicopter rotor blade testing. The Swallow program was the first application of this technology.

The fuselage and wings were hollow molded composite structures made of fiberglass, carbon fiber, and sandwich (core) materials. The wing used the "lobster claw" construction technique employed on SENDER, with a carbon fiber box spar. SENDER and Swallow were contemporary programs. Early prototype wings were made from hot-wired foam with vacuum-bagged composite skins while the final tooling was under fabrication.

The Swallow propulsion system consisted of a motor, motor controller, and battery. The brushless motor system was produced by Aveox. The batteries for the test program were NiCd, which enabled approximately 15-minute endurance. Operational variants would use lithium primary batteries (nonrechargeable).

The flight test program was uneventful. The close-coupled tail created minor controls challenges, but these were remedied through adjusting the CG and the radio control programming.

Swallow and the payload functioned as intended, so it was time to bring the system together. The payload was integrated onto the airframe. The system was rail launched (Fig. 5.12). The parachute recovery was not demonstrated, and recovery was by conventional wheeled landing. The aft

Fig. 5.12 Swallow on a rail launcher. Note the tail extension.

Fig. 5.13 Swallow with sniper-detection payload.

fuselage was slightly lengthened on the second prototype to increase the tail arm for enhanced stability and reduced pilot workload. The payload-equipped Swallow was successfully operated at Dugway Proving Ground, and the integrated system was proven. This was the first purpose-designed unmanned aircraft in-flight demonstration of a bio-detection payload. The capability was ready to deploy, but the perceived threat wound down. DARPA was pleased with the system, and the system can be produced in quantity with current technological updates, if the need for such a system arises again.

Swallow was later used to demonstrate a counter-sniper capability. A side-looking infrared camera could detect the muzzle flash of a sniper rifle. With the aid of autopilot components that served as an inertial navigation system (INS) to provide the position solution, the estimated threat position could be calculated. A counter-sniper system could then act upon this target information. This capability was successfully demonstrated (Fig. 5.13), but it was not fielded.

Figure 5.14 shows a three-view drawing of Swallow, and Table 5.3 lists its technical characteristics.

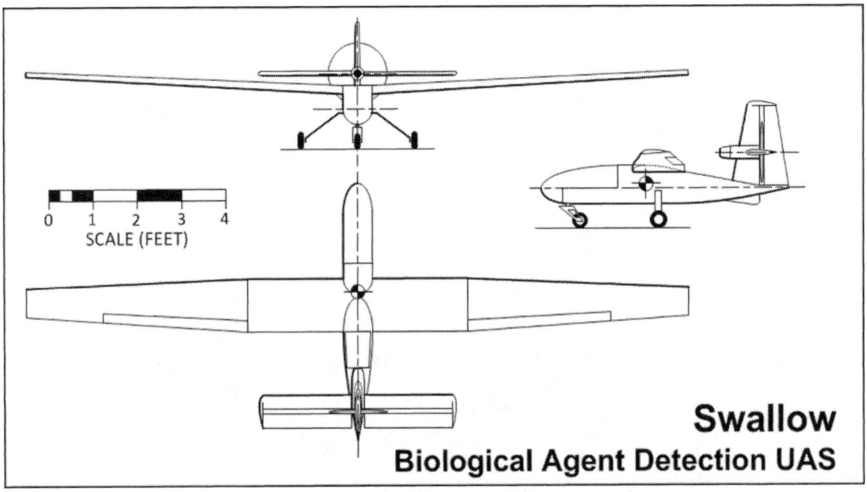

Fig. 5.14 Swallow three-view drawing.

Table 5.3 Swallow Technical Characteristics

Parameter	Value
Wing span	15 ft
Wing area	14.75 ft^2
Length	5.6 ft
Takeoff weight	62.0 lb
Empty weight	52.0 lb
Payload weight	10.0 lb max
Payload volume	900 cu. in.
Cruise speed	50 KTAS
Dash speed	90 KTAS
Stall speed	35 KTAS
Endurance	2.8 hr
Range	110 nmi
Launch method	Conventional or rail launcher
Recovery method	Conventional or parachute
Propulsion type	Battery electric
Power	1500 W

Swallow: Further Reading

Anderson, G.P., King, K.D., Cuttino, D.S., Whelan, J.P., Ligler, F.S., MacKrell, J.F., Bovais, C.S., Indyke, D.K., and Foch, R.J., "Biological Agent Detection with the Use of an Airborne Biosensor," *Field Analytical Chemistry and Technology,* Vol. 3, Issue 4–5, 1999, pp. 307–314.

Anderson, G.P., Ligler, F.S., Stenger, D.A., Davidson, P.T., Foch, R.J., and MacKrell, J.F., "Flight of the Biosensor," *1997 NRL Review*, Naval Research Laboratory, Washington, DC, 1997, pp. 55–58.

Moroz, S.A., Pierson, R.B., Ertem, M.C., Burchick, D.A., Sr., and Ippolito, T., "Airborne Deployment of and Recent Improvements to the Viper Counter Sniper System," *IRIS Passive Sensors Symposium*, March 1999.

SEARCHER, *1997*

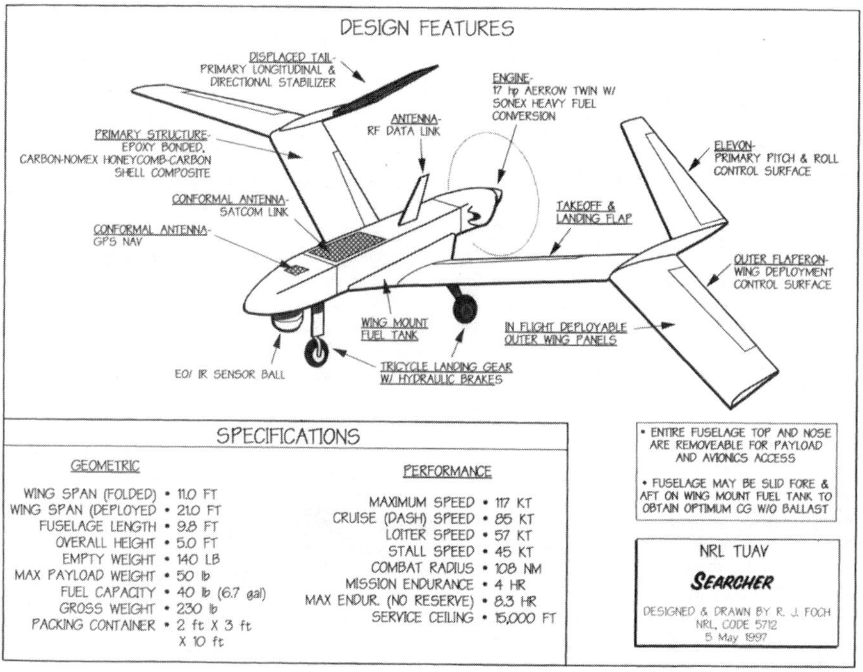

Fig. 5.15 Searcher design and major features.

Searcher was a tactical unmanned aircraft concept that used the displaced tail configuration. The outboard wing panels would deploy in flight to enable good range performance and good endurance performance in different phases of flight. This design built upon lessons learned from the LAURA program. Searcher would take off and land conventionally from a runway, and was not designed for air deployment. The Searcher design (Figs. 5.15 and 5.16) led to the configuration of the Extender air-launched unmanned aircraft project.

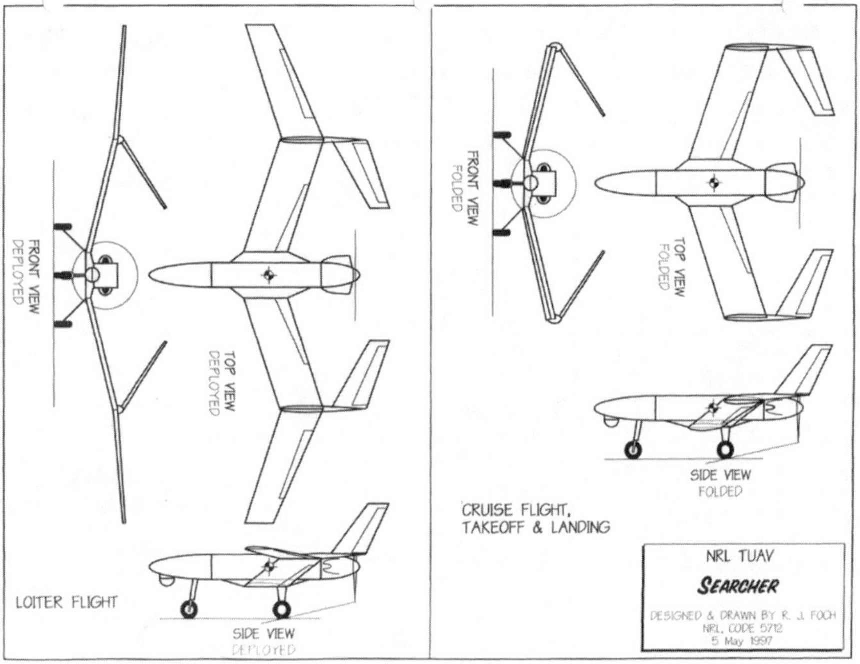

Fig. 5.16 Searcher in loiter configuration (left) and cruise configuration (right).

Ghost, 1997–1998

Fig. 5.17 Ghost.

NRL was asked to build a testbed aircraft to enable an electronic payload to operate in the air without being impeded by the airframe. In other words, the aircraft had to be as transparent as possible at the payload's radio frequencies. The aircraft magnetic and electrical properties were also important.

Fig. 5.18 Ghost structure under construction.

Ghost's payload was housed in a cylindrical payload bay in the wing ahead of the main spar. It was oriented orthogonal to the direction of flight. The airfoil was a custom flat-bottom design with high thickness-to-chord ratio. The thick airfoil was required to house the wingtip-to-wingtip cylindrical payload.

Ghost was a large aircraft for NRL, with a 1200-pound gross weight and 30-foot span (Fig. 5.17). It had a mission payload capacity of 400 pounds. The structure was designed to have a small deflection under load rather than for strength, so the airframe was considerably stronger than required for the design loads. As a result, larger payloads could have been carried before reaching structural limits. Ghost was the size of a typical two-place general aviation aircraft.

A simple airframe solution was selected. Composite materials could not be used, because many of the materials such as carbon fibers, fillers, paints, and primers are semiconductors or conductors. Instead, NRL selected a wood airframe (Fig. 5.18). Sitka spruce was used for the spar beams, longerons, and compression members. Birch plywood was used for ribs, skins, and formers. The landing gear struts were made of fiberglass. The wheel hubs were ceramic, and the tires were traditional rubber. All fittings were nonmetallic. The propeller was made of wood. The airframe was painted with clear maritime spar varnish, which was nonconducting. This gave the aircraft a natural wood finish appearance. The wing tips were colored orange for visibility with nonconducting, colored tape rather than paint.

Not all components could be nonconducting, such as the avionics, engine, and actuators. These components were placed inside a metal mesh to create a

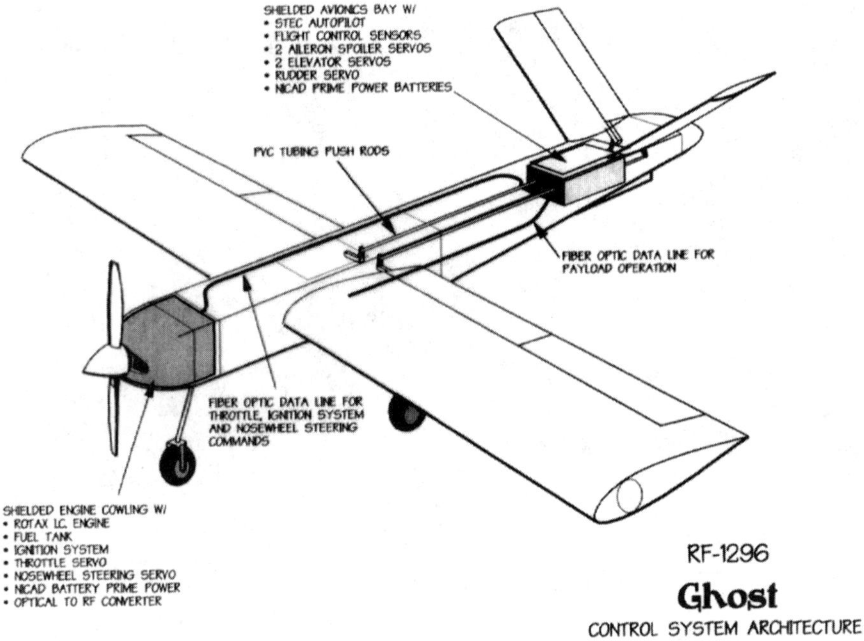

Fig. 5.19 Ghost internal systems arrangement and features.

Faraday cage. The signal connections between components were handled through fiber-optic cables (Fig. 5.19). The Rotax 912 engine and fuel were housed in a forward Faraday cage. The RF characteristics of the Faraday cage geometry were well understood, and calculations could account for this effect on the payload performance.

All servos were housed within a shielded avionics box. A more traditional unmanned aircraft approach would be to have actuators located at the control surfaces. Ghost's servos actuated Kevlar cables, which in turn moved through fiberglass pulleys. The pulleys attached to the airframe with fiberglass and nylon bolts. The control surface control horns were constructed of fiberglass. The mechanical pulley system enabled differential ailerons through mechanical means to minimize adverse yaw.

The wings came in two halves for transportation. To eliminate conductive metal fittings, the wing was joined with a Chinese scarf joint. The wing spar halves fit together in a socket. The fit was so precise that the wing was retained with only two nylon bolts. The loads were transferred through wood-to-wood contact via the precise fit.

Wood airframes can be labor intensive, so innovative approaches were necessary to keep costs low. A water jet cutter was used to cut all ribs and formers. Laser cutting was not yet commercially available. Even if laser

Fig. 5.20 Ghost in flight.

cutting had been available, the carbon coating on the cut surfaces may have posed conductivity problems. Flow Research Inc. water jet cut the parts in an automated process that saved hundreds of work hours for NRL.

Ghost was flown at NASA's Wallops Island Test Range for initial test flights (Fig. 5.20). The aircraft was designed to have general aviation–like handling qualities, so its good flight characteristics were expected. The aircraft required a remote pilot for launch and recovery operations, and it flew semiautonomously via the autopilot once aloft.

Ghost then went on to perform successful payload demonstration flights at another facility. Unfortunately, the aircraft crashed on its 13th flight due to an avionics failure. However, the program objectives were fully met. Figure 5.21

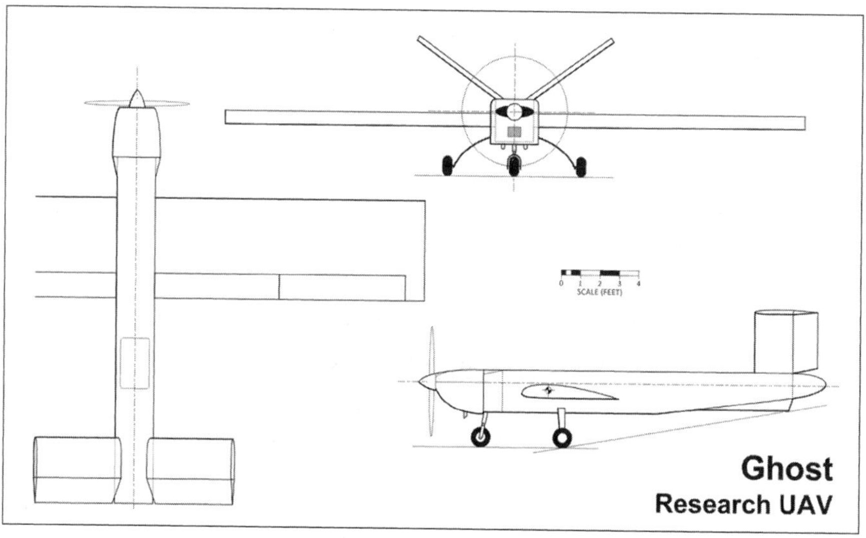

Fig. 5.21 Ghost three-view drawing.

Table 5.4 Ghost Technical Characteristics

Parameter	Value
Wing span	30.0 ft
Wing area	150 ft^2
Length	22.0 ft
Takeoff weight	1200.0 lb
Empty weight	800.0 lb
Payload weight	400.0 lb
Fuel weight	48.0 lb
Cruise speed	70 KTAS
Dash speed	105 KTAS
Stall speed	40 KTAS
Airfoils	Custom flat-bottom airfoil based on thickened RF 1165FB
Launch method	Conventional
Recovery method	Conventional
Propulsion type	Rotax 912, 2-stroke reciprocating engine
Power	81 hp

shows a three-view drawing of this ground-launched aircraft, and Table 5.4 lists its technical characteristics.

PARA EYES, 1997–1998

Fig. 5.22 Para Eyes.

Para Eyes was a simple, electric-powered parafoil unmanned aircraft (Fig. 5.22). It weighed 45 pounds and could be transported in the back of a High Mobility Multipurpose Wheeled Vehicle (HMMWV). Para Eyes was designed to be operated by nonaviation-trained U.S. Army personnel under field conditions. The mission was to provide enhanced day and night situational awareness in the local vicinity via close-range aerial surveillance. One demonstration system was sponsored by the Army.

A parafoil aircraft configuration was selected for ease of operation and packaging efficiency. If the vehicle lost power, it would simply land in a low-speed, low-descent-rate glide. The parafoil was stall-proof, enabling safe operations.

Para Eyes did not have traditional aerodynamic controls. Most parafoils have lines that can be pulled differentially for roll or together for pitch and flare control. Para Eyes had two motors connected by a cross bar. The motors could operate differentially for yaw control. The total propulsion power controlled rate of climb or descent. Additionally, climb rate control was augmented by the ability for the pilot to rotate the motor arms in pitch.

Per requirements, Para Eyes had no autopilot and was flown by an external operator in the loop. It had no systems to provide autonomy, except for being inherently stable by design and having slow control response. These characteristics made operation simple, so flying the Para Eyes required minimal training and it could easily be operated by nonaviation personnel.

Launching Para Eyes was also simple. First, the vehicle was removed from the storage container. The parafoil was laid out on the ground behind the aircraft. Power was applied, and the propeller wash inflated the canopy. The thrust also generated forward velocity. The aircraft lifted off the ground and climbed away, as shown in Fig. 5.23. The aircraft was then manually flown to the areas of interest within Radio Frequency (RF) link range of the operator as the payload imagery was observed. Landing was even simpler than takeoff. To land, Para Eyes was flown over and several feet short of the desired landing spot. Then the throttle was closed, and the aircraft simply paraglided to the ground. As such, the Para Eyes was easy to operate, being completely stable about all three axes and stall-proof. These features came at the price of very poor aerodynamic performance and high power consumption. However, for meeting the project's requirements, trading aerodynamics for near foolproof operation was the correct choice.

NRL conducted numerous demonstrations for the Army. They were pleased with the Para Eyes system and accepted delivery of the prototype, but there was no transition to an operational system. Figure 5.24 shows a three-view drawing of this ground-launched aircraft, and Table 5.5 lists its technical characteristics.

Fig. 5.23 Para Eyes preparing for launch.

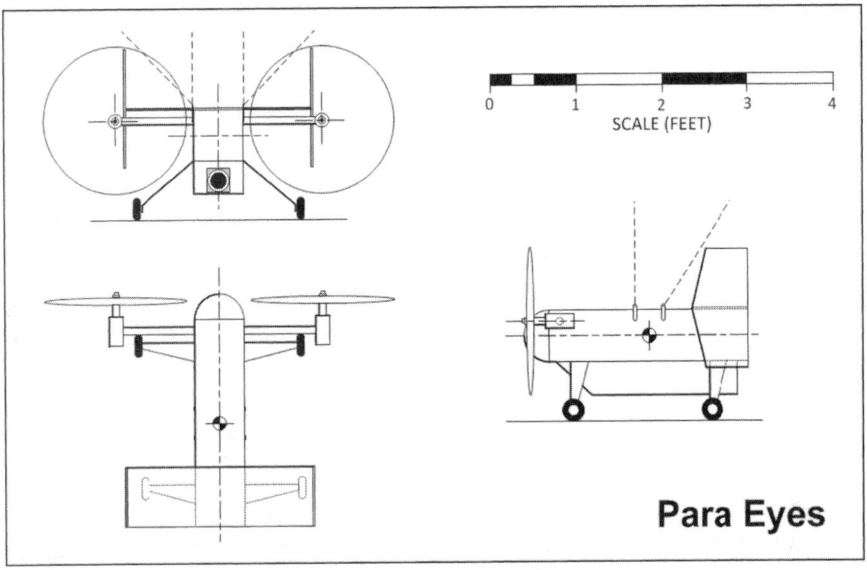

Fig. 5.24 Para Eyes three-view drawing.

Table 5.5 Para Eyes Technical Characteristics

Parameter	Value
Wing span	10.0 ft
Wing area	57 ft^2
Length	2.6 ft (fuselage)
Takeoff weight	45.0 lb
Empty weight	26.0 lb
Payload weight	4.0 lb
Battery weight	15.0 lb
Loiter speed	17 KTAS (estimated)
Endurance	0.2 hr
Max L/D	2 (approx)
Launch method	Conventional
Recovery method	Conventional
Propulsion type	Battery electric
Power	1500 W

MOTH, 1999

Fig. 5.25 Moth undergoing flight testing (taken from video).

NRL was tasked with rapidly fielding an expendable unmanned aircraft during the Kosovo campaign. The Navy had difficulty obtaining imagery of a target in its theater of operations. DoD's Pioneer unmanned aerial vehicles were repeatedly shot down over this target, so their continued use became cost prohibitive. The Navy only required periodic low-resolution imagery of the target, so a small, affordably expendable system would be ideal (Fig. 5.25).

NRL responded by designing and building a new UAS in only 9 days. The aircraft was originally called the "Kosovo Kluge" by the development team, but the official name was changed for obvious reasons. Each team member worked two of three shifts, enabling work to continue around the clock to have the system ready for deployment as quickly as possible. The Vehicle Research Section (VRS) shop kept a large stock of building materials and construction supplies on hand to help expedite prototype development, so the Moth was designed to utilize these, and no procurements were necessary. There was very little program documentation due to the quick timeline. Today, the only traces of the aircraft are rib templates for the foam core wings and a few hand sketches. No detailed drawings were produced and only video was taken.

The initial layout had a conventional configuration with a single fuselage, a mid wing, and a V-tail. The side-looking camera was mounted inside the fuselage aft of the wing. The aircraft initially had a tractor engine configuration and tricycle landing. The design and fabrication progressed for 4 days, after which a major design change was needed, and this required a complete redesign!

The delivered camera was larger than the one originally specified, so it became necessary to house the camera in the nose and to add a pan stage pointing control. The nose was given a circular window that swept half a cylinder. The wing was kept the same, but the rest of the design changed. The fuselage was now a short pod to enable a pusher engine installation. Two booms extended from the wings with inward-canted, displaced tails. The aircraft was redesigned in one work shift, and it flew after 5 days had elapsed from the start of the new design. It was powered by an O.S. 70 four-stroke model aircraft engine.

The aircraft had poor handling characteristics. The wing had no dihedral, and the displaced tails provided an anhedral (negative dihedral) effect. The aircraft was highly spirally unstable and had marginal directional stability, an unusual combination. Immediately after takeoff, the test pilot asked, "Where would you like me to crash it?" However, demonstrating considerable skill and flying experience, he was able to return to the runway and execute a successful landing.

The booms were immediately sawed through ahead of the tails. They were reinstalled with rapid-setting epoxy glue, swapping right for left and vice versa. They now canted outward to improve the lateral stability, and small vertical ventral fins were added below the booms under the tail surfaces to improve the directional stability. Within 1 hour, the prototype was ready for its second test flight. The hasty field modifications resulted in modest improvements to the handling qualities. Flight tests indicated that the nose geometry caused problems with the wing root flow. Also, close-coupled dynamic pitch problems were noted, and it was felt that the tail booms were too short. Despite these limitations, the Moth could be flown by a skilled RC pilot.

A few days after the prototype's flight tests, a second aircraft was nearly completed. NRL was planning to operate both aircraft in theater to fly the required mission. Unfortunately, the deployment was called off when the military situation changed. So, this herculean effort that resulted

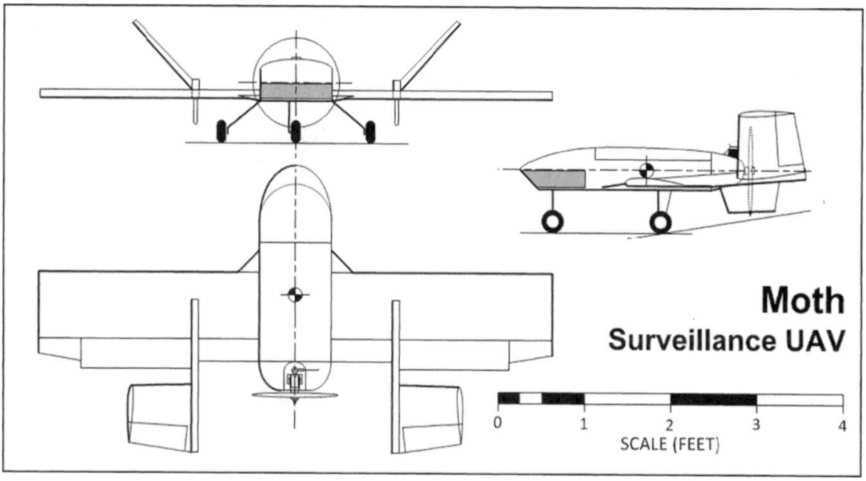

Fig. 5.26 Moth three-view drawing.

in the design, fabrication, and flight-testing of a new unmanned aircraft design in less than 2 weeks' elapsed time was all for naught. On the positive side, the development team members demonstrated an extremely high level of skill, integration, and dedication. On the down side, they were thoroughly exhausted and required a vacation to rest and recuperate. Figure 5.26 shows a three-view drawing of this ground-launched aircraft, and Table 5.6 lists its technical characteristics.

Table 5.6 Moth Technical Characteristics

Parameter	Value
Wing span	6.0 ft
Wing area	5.8 ft^2
Length	3.1 ft
Takeoff weight	8.0 lb
Empty weight	6.0 lb
Payload weight	2.0 lb
Loiter speed	50 KTAS (estimated)
Dash speed	65 KTAS (estimated)
Stall speed	35 KTAS (estimated)
Endurance	1.0 hr
Max L/D	5 (estimated)
Launch method	Conventional
Recovery method	Conventional
Propulsion type	O.S. 70FS 4-cycle model aircraft engine
Power	1.2 bhp

NEXUS, 1999–2000

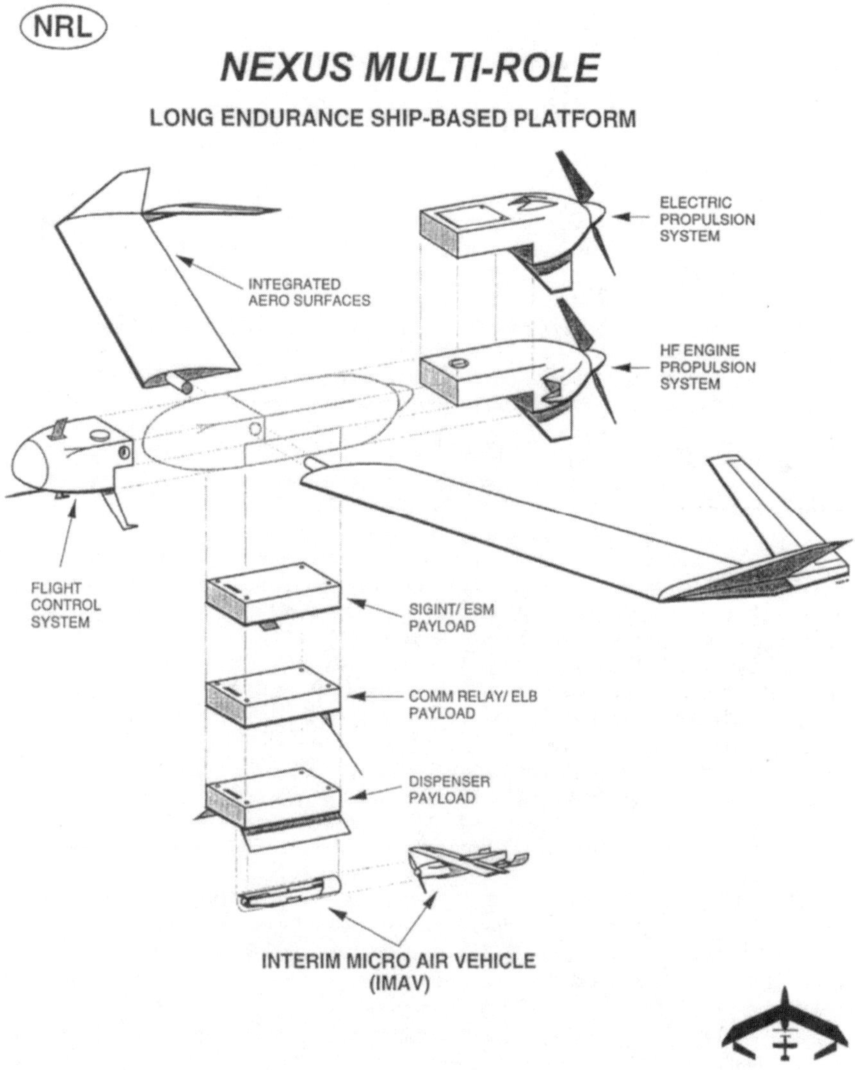

Fig. 5.27 Nexus concept.

In 1999, the Secretary of the Navy (SECNAV), the Honorable Mr. Richard Danzig, visited NRL to review the Lab's unmanned aircraft projects. He was favorably impressed with the autonomy, deployment, and propulsion technologies. He inquired whether these technologies could be applied to traditional unmanned aircraft systems instead of just EW decoys. Up to this point, Naval Air Systems Command (NAVAIR) and other UAS development organizations had not tasked NRL to look beyond EW systems. SECNAV asked NRL to write a white paper that described the Lab's technologies and their application for future UAS development.

A preliminary design, called Nexus, was developed to illustrate the application of the NRL unmanned vehicle technologies for UAS applications (Fig. 5.27). Nexus was envisioned to be an affordably expendable, multimission, small unmanned aircraft for over-the-hill reconnaissance, signals intelligence (SIGINT), and communications relay missions. The flight had to be highly autonomous with no RC operation or piloting skills, so no aviation training would be necessary. Contemporary tactical unmanned aircraft did not yet employ autonomous technology.

Another focus of the white paper was how to rapidly procure and field a system. Small unmanned aircraft technology was progressing rapidly at the time. The traditional procurement cycle would take so long to field a system that the system would be obsolete by the time it became operational. Foch proposed building the system to best commercial practices, much like desktop computers were built. This would streamline system acquisition when compared with Mil-Spec systems (systems built to military specifications) with their voluminous requirements. It should be noted that in the past decade, several unmanned aircraft have been fielded under contractor-owned, contractor-operated (COCO) ISR services contracts, with the full program of record acquisition cycle circumvented for expediency. However, the Nexus proposal predated the widespread COCO ISR services contract practice.

NRL proposed two aircraft: Nexus and a smaller aircraft to be deployed from Nexus. The main Nexus aircraft was a modular, plug-and-play system. It could accommodate different propulsion modules, payload modules, and wing panels. The modular nature of the aircraft would facilitate spiral upgrades for a production system. The avionics, payloads, engines, and other systems could be swapped out when improvements were available.

The Nexus vehicle had a novel modular design. The nose module housed the avionics. The propulsion module attached to the back of the nose module. The interchangeable payload bay fit underneath both the nose module and the propulsion module. The three fuselage modules were interlocking boxes. The wings plugged into the nose module. A displaced tail boom attached to each wing, so the tails did not need to interface with any other fuselage module. The wing sweep allowed attachment to the nose module while placing the required CG over the centroid of the payload module. An upgrade for span increase or decrease would require a sweep modification. The Nexus's displaced tail configuration inherited strongly from the Extender design. Unlike Extender, Nexus did not need to air deploy. The displaced tail configuration enabled wing-tail system modifications independently of the fuselage system. The propulsion module could include fuel and a reciprocating engine or batteries and an electric motor. The entire payload module with integrated mission payloads could be swapped as a single module. The payload modules and propulsion modules could be changed without impacting the wings or any other module.

Nexus today would fit DoD's new category of small tactical unmanned aircraft system (STUAS), or Tier II, though this designation did not exist in 1999–2000. Nexus had a gross weight of 50 pounds and a 10-pound payload capacity. The deployed wing span was 14 feet, and the vehicle could stow in a 1.5-foot by 1.5-foot by 6-foot package. The initial electric-powered version could fly for 2 hours with lithium batteries. The heavy fuel reciprocating engine variant was designed for 8 to 12 hours' endurance. Nexus would operate at an airspeed of 45 to 50 knots and had a service ceiling of 10,000 to 15,000 feet.

The second aircraft was a small, air-launched unmanned aircraft that would be dropped from the Nexus payload bay (Fig. 5.28). It fit within a tube measuring 4.875 inches in diameter and 30 inches in length. The

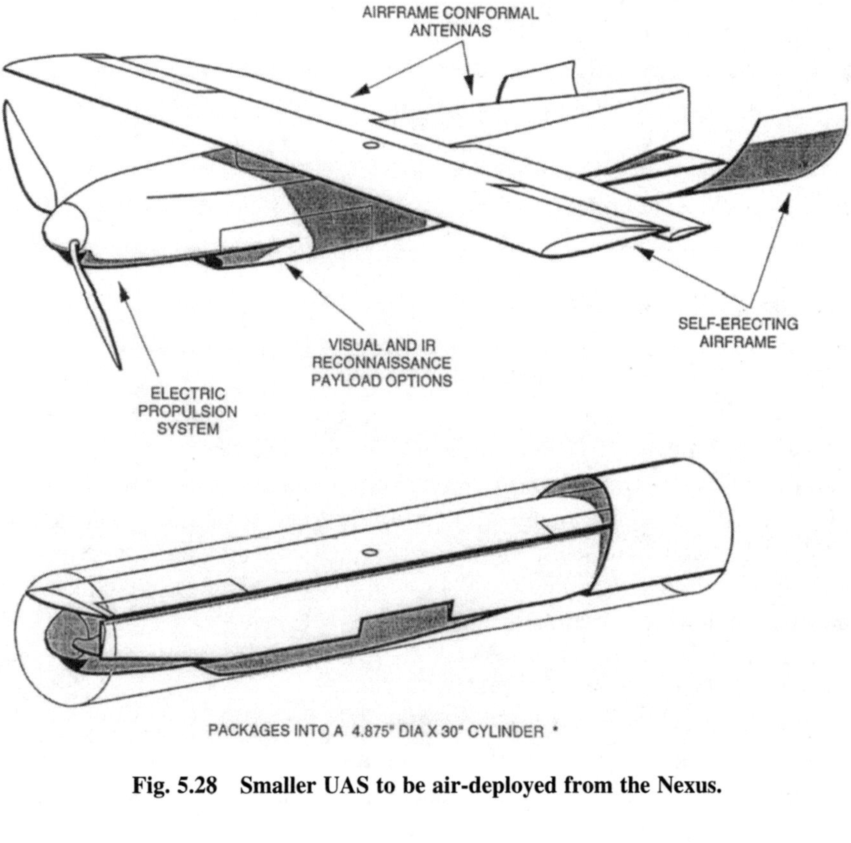

Fig. 5.28 Smaller UAS to be air-deployed from the Nexus.

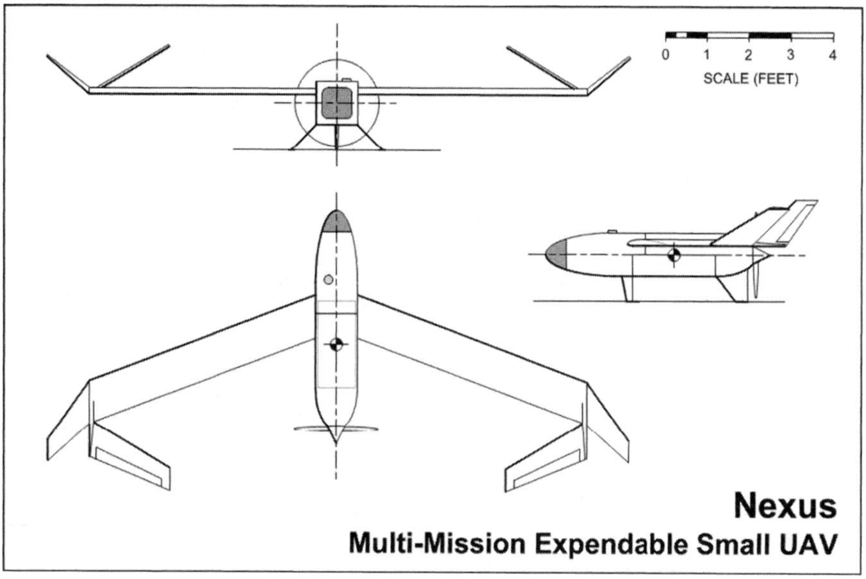

Fig. 5.29 Nexus preliminary design three-view drawing.

deployed wingspan was 2.4 feet. The gross weight was 4 pounds, and it could carry a 1-pound payload. The electric propulsion would enable 30 minutes' endurance with an airspeed of 40 to 45 knots. It could also be ground launched from a catapult. The combat radius and communications range were 10 nautical miles. Figure 5.29 shows a three-view drawing of this ground-launched aircraft, and Table 5.7 lists its technical characteristics.

Table 5.7 Nexus Technical Characteristics

Parameter	Value
Wing span	14.0 ft
Wing area	13.5 ft^2
Length	6.0 ft
Takeoff weight	50.0 lb
Payload weight	10.0 lb
Battery or fuel weight	15.0 lb
Loiter speed	45–50 KTAS
Endurance	8–12 hr (heavy fuel engine)
	1.3–2.4 hr (electric)
Launch method	Ground launch (method not specified)
Recovery method	Recoverable or expendable
Propulsion type	Heavy fuel reciprocating engine or electric

NRL delivered the white paper to SECNAV. It described a 1- to 3-year system delivery rather than a traditional 5- to 10-year program. This new development paradigm was not viewed favorably by the conservative acquisition community. The modular system enabled a rapidly delivered initial capability and then spiral upgrades or modifications, rather than a single system that could satisfy myriad requirements at delivery. SECNAV challenged the Navy to find a way to streamline unmanned aircraft acquisition. The objectives of the proposed program were to leverage NRL technology and to demonstrate a streamlined acquisition approach to develop a small UAS.

NRL was tasked to find a user and a resource sponsor that would field and produce a demonstration system. NRL briefed the Nexus to the Vice Chief of Naval Research, who was also the Commanding Officer of the Marine Corps Warfighting Laboratory (MCWL). He was not interested in the large Nexus but highly enthusiastic about the small aircraft to be dropped from the Nexus. The NRL estimate to deliver the small second aircraft was $4 million, which was considered a small quantity for Fleet investment in a new technology and a new acquisition paradigm. The USMC had a requirement and funding for the Interim Small Unit Remote Scouting System (ISURSS). The purpose of this system would be to perform local surveillance in a USMC small unit or platoon. He suggested that modifying the Nexus small UAS design to be ground launched and easy to operate by nonaviation Marines would make it a contender for the ISURSS. He also stated that this system was to be named "Dragon Eye." NRL quickly took advantage of this opportunity and delivered a detailed proposal to MCWL. This resulted in the Dragon Eye program, which became the first program of record in the Small UAS category.

Although the Nexus designs were never developed beyond the preliminary design phase, their concepts influenced several subsequent NRL projects including the Sea Wind and modular aircraft studies. Most significantly, the Nexus design strongly influenced one of the authors (Gundlach) in his design of the Integrator unmanned aircraft while he was employed at Insitu Inc. The Insitu Integrator went on to win the Navy/Marine Corps STUAS program of record competition and is now designated the RQ-21A. Gundlach spent several years considering variations of the Nexus fuselage modular concept and wing design before settling on the initial Integrator configuration. Integrator has a nose payload module, a CG payload bay, a central fuselage module with wing attachment, and a propulsion module. In essence, it is a Nexus fuselage with an additional nose payload bay in front. The initial Integrator design had a displaced tail consisting of a vertical and horizontal tail on each boom. This design later evolved to a connected twin-boom tail arrangement.

DRAGON EYE, 2000–2002

Fig. 5.30 RQ-14A Dragon Eye.

Dragon Eye was a small unmanned aircraft (Fig. 5.30) developed for the USMC as a contender for being placed into production under the ISURSS. Dragon Eye was successful and became the first NRL-developed UAS to achieve program of record status. It was designated the RQ-14A, and more than 3000 vehicles have been produced. They accumulated more than 10,000 combined flight hours serving in combat operations during Operation Iraqi Freedom and in the skies of Afghanistan.

NRL worked in conjunction with MCWL to develop the Dragon Eye system. MCWL generated the system requirements, supported testing, conducted training, and led the mission planner development. The Marine Corps System Command (MARCOR SYSCOM) was the acquisition sponsor. Dragon Eye became the system funded under the USMC ISURSS program of record.

The aircraft configuration was driven by the requirement to fit within an all-purpose lightweight individual carrying equipment (ALICE) military backpack. The equivalent internal backpack space was a box shape measuring 15 inches by 15 inches by 7 inches. The initial Dragon Eye configuration was similar to the Micro Tactical Expendable (MITE) Micro Air Vehicle (MAV) with swept wing tip extensions (Chapter 6). A prototype aircraft was built and tested in 5 weeks and was first flown on July 28, 2000. It was successfully demonstrated for MCWL personnel a few days later.

There was a problem, however, because the unusual configuration stood out, drawing excessive attention from ground observers. The USMC asked if NRL could develop an aircraft that was more difficult to recognize as unusual. NRL drew upon experience from earlier research on the CG Shift bird-like designs. The earlier programs revealed that using a bird-like wing

Fig. 5.31 AeroVironment-produced Dragon Eye System.

aspect ratio with triangular vertical tail made the aircraft difficult to distinguish from a bird. The vertical triangular tail looked like a bird's triangular horizontal tail at a distance. With low-visibility gray paint and typical haze conditions, the aircraft would be difficult to distinguish at a distance by typical ground observers (Fig. 5.31).

The final Dragon Eye configuration had a plank flying wing, a single fuselage, and a vertical tail. The wing had an aspect ratio of 4. The airfoil was the Liebeck 2973A, the same type used on the AeroVironment Pathfinder flying wing. At Dragon Eye wing chord Reynolds numbers, this airfoil requires a boundary layer trip to reach its design parameters. Overall, the general configuration, wing proportions, and wing airfoil with its boundary layer trip were identical to those of the Flying Wing Research Testbed (FLYRT). Data from the FLYRT program was therefore applied to the Dragon Eye design and eliminated the need for wind tunnel testing. Two electric motors were placed on the wing, with tractor propellers. Flight control surfaces consisted of wing elevons, and there was no rudder control. Dragon Eye rolled to turn. For storage, the wing came in three segments, and the fuselage nose and tail sections were removable.

Components were distributed among airframe modules (Fig. 5.32). The center wing-fuselage module housed the autopilot, motors, motor controllers, elevon servos, and battery pack. The communication system was built into the vertical tail. A plug-and-play payload nose section could carry a variety of payloads. These could include forward-looking cameras, side-looking cameras, chemical detection payloads, or biological samplers.

Assembling the aircraft in the field was simple. The operator placed the battery in the main module, snapped the fuselage nose and tail sections into

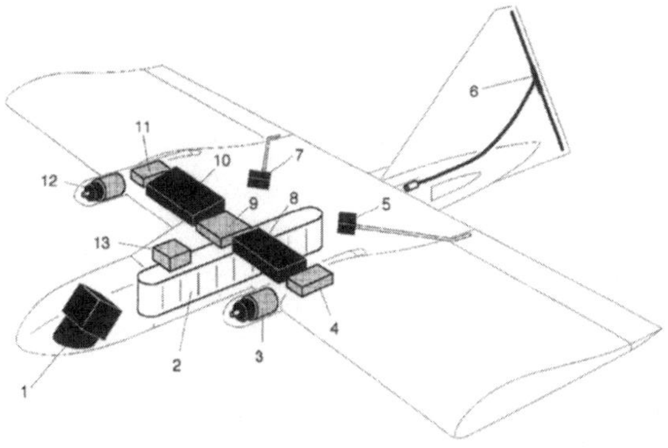

1) EO camera payload
2) Prime power battery pack
3) Brushless DC motor & gearbox
4) Sensorless motor controller
5) Left elevon servo actuator
6) Dipole antenna
7) Right elevon servo actuator
8) Communication link system
9) DC-DC voltage converter
10) Autopilot & GPS navigator
11) Sensorless motor controller
12) Brushless DC motor & gearbox
13) FCS air data sensors

Fig. 5.32 Dragon Eye internal layout.

place, and attached the outer wing sections. The nose and tail sections were held on by small spring steel clips.

The airframe modules were designed to separate during a rough landing to help absorb impact energy. Any damaged modules could be replaced in the field, as all modules were interchangeable between airframes. Interchangeability was achieved with generous tolerances at interface points. Over time, the joints would wear out from being popped loose during almost every landing. The design life was only 20 flights, but in practice, some vehicles lasted for 80 to 100 flights. These long-lived airframes were being held together by tape by the end of their service. The extended use in the field was a matter of operational necessity, but the modules were not supposed to be maintained in the field. Rather, damaged modules would be shipped back to the manufacturer for repair. Motors and motor controllers were reused when the airframes wore out.

One operator decided to try to increase the airspeed of his Dragon Eye by scraping away the molded composite boundary layer trip. Perhaps he thought the spanwise bump was a manufacturing defect. Without this trip, at Dragon Eye's Reynolds numbers, the Liebeck 2973A airfoil has a large laminar separation bubble on its upper surface. This modified, tripless Dragon Eye flew slower, had lower climb performance, and had shorter flight endurance due to the drag increase.

Dragon Eye had two launch methods. The main method was hand launch using an overhead hand toss technique. The relatively high wing loading and sensitivity to angular rates required some skill and strength to enable a successful launch. The second and primary launch method used a bungee launch system (Fig. 5.33). A bungee cord with a string attached would be used to impart energy to the aircraft. One end would be staked to the ground, attached to a heavy object, or even held by another Marine. The other end

Fig. 5.33 Dragon Eye bungee launch.

would attach to a hook on the underside of the fuselage. The Dragon Eye would be walked back to extend the bungee to a predetermined tension and then be released.

A system deployed to a USMC Small Unit consisted of two aircraft, battery packs, a variety of payload noses, and a ground control station (GCS). All system components fit within ALICE backpacks. Two Marines would each carry one aircraft and the payload noses. Another would carry the GCS, and a fourth Marine would carry the battery packs. The GCS was a CF-34 laptop computer that connected to a separate communications box. The GCS weighed 9.3 pounds, excluding the battery. The operator could use either video goggles or the laptop for the display.

The propulsion system consisted of geared brushless motors, motor controllers, and a $LiSO_2$ battery. The $LiSO_2$ primary battery (nonrechargeable) was a military battery in use by the USMC.

The flight tests progressed from RC flights to autopilot flights. In 2000, there were few autopilots available at the appropriate scale. The SENDER LWAP autopilot was too heavy and unreliable. NRL selected a Lockheed Martin autopilot based on technology developed for the DARPA MAV program. NRL worked with this vendor to close the control loops and to develop the autonomous flight capabilities. Through the autopilot development testing, NRL learned that the autopilot could handle much lower static margins than would be appropriate for RC control. Flying the Dragon Eye with a lower static margin improved the flight performance.

The most popular camera payload was a fixed, side-looking camera. Therefore, it was not possible to fly the aircraft through the camera. Instead, the GCS allowed the operator to select a point on the ground to image. The system would find a circular flight path that would keep the point on the ground within the camera field of view. The system also enabled a "U-Steer" mode, in which the operator could navigate the aircraft using the outer loops. Non-aviation personnel could learn all the parameters of the mission planning system with minimal training. The semiautonomous flight operations improved reliability.

The USMC flight performance requirements were not overly demanding. NRL did not need to push aerodynamics or propulsion design in order to satisfy the requirements. The airframe could be made robust to take abuse in the field at the expense of higher weight. The airframe minimum gage was thick compared to that of other small unmanned aircraft, and the airframe could be stepped on by a Marine. The structure was much stronger than would be required to withstand flight loads. This made obtaining a NAVAIR flight clearance easy. The V-n (gust load) diagram showed that it was not possible to overstress the airframe in a power-on dive with full control surface deflections. Further airframe refinement could have reduced weight and improved performance, but it was felt that the system was adequate as developed.

The NRL-fabricated prototype airframes had single-layer Kevlar skins over a foam core. Production versions had hollow molded structures.

NRL completed the advanced flight testing, and then contracted to industry to produce the system. NRL is set up to develop technology and prototype systems but is not interested in production. NRL provided the design information and test data. Manufacturers were asked to produce the initial batch to the NRL design with upgrades being allowed (subject to Navy/USMC approval) in subsequent production batches. The flexibility allowed the producers to modify the design for improved performance, reliability, and manufacturability. The competition yielded 10 bids from industry. Two proposals that met the requirements at a reasonable price were selected. The top award was made to AeroVironment in November 2003, and this company would provide the systems exclusively to the USMC. The second award went to BAI Aerosystems, who could provide systems to other customers such as the Navy and other operators. AeroVironment produced about 2000 aircraft, and BAI produced another 1,000. BAI later developed an improved variant with extended span and increased aspect ratio. The NRL, AeroVironment, and BAI variants are all unique, and their components are not interchangeable with each other.

The program went from concept to transition to USMC use in 37 months, and to high-rate production in 48 months. Its development verified the validity of the new acquisition approach for simple military systems proposed by Foch to SECNAV. This system then served in combat operations for several years with the Marine Corps, until it was replaced by the AeroVironment RQ-17 Raven. A Dragon Eye system is on display at the Smithsonian National Air and Space Museum in Washington, DC. Figure 5.34 shows a three-view drawing of this ground-launched aircraft, and Table 5.8 lists its technical characteristics.

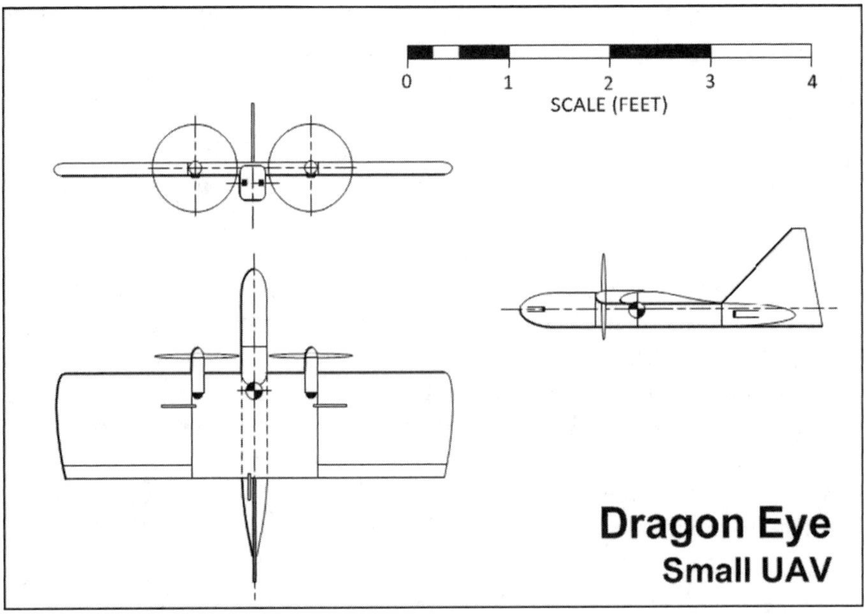

Fig. 5.34 Dragon Eye three-view drawing.

Table 5.8 Dragon Eye Technical Characteristics

Parameter	Value
Wing span	3.75 ft
Wing area	3.8 ft^2
Length	3.0 ft
Takeoff weight	5.5 lb (4.5 design, 6.0 max)
Payload weight	1.0 to 2.0 lb
Battery weight	1.75 lb
Loiter speed	35 KTAS
Dash speed	40 KTAS
Stall speed	20 KTAS (estimated)
Endurance	0.75 to 1 hr
Airfoils	Liebeck 2973A (wing)
	Flat airfoil (vertical tail)
Launch method	Hand launch or bungee launch
Recovery method	Belly landing
Propulsion type	Battery electric
Power	300 W

DRAGON EYE: FURTHER READING

Bovais, C., MacKrell, J., Foch, R., and Carruthers, S., "Dragon Eye™ UAV: Concept to Production," *Bristol UAV Conference*, Bristol, U.K., 2003.

Foch, R.J., and Dahlburg, J.P., "Dragon Eye: Airborne Sensor System for USMC Small Units," *2001 NRL Review*, Naval Research Laboratory, Washington, DC, 2001, pp. 183–184.

SEA AIRBORNE LEAD LINE (SEA ALL), 2001–2002

Fig. 5.35 Sea ALL airframes.

During the Dragon Eye development, the Navy, via a direct request from the Fleet in theater, asked NRL to develop a small unmanned aircraft system to provide surveillance for small ships deployed in the Persian Gulf. The request came around the time when a small craft loaded with explosives detonated next to USS *Cole*, causing casualties and major damage. Small craft are difficult to defend against because they are hard to see, fast, and difficult to destroy. The Dragon Eye system was the most appropriate system at the time, but its development was still under way.

The requirement was to produce a small unmanned aircraft system that could be easily launched and recovered from a ship (Fig. 5.35). The system had to be capable of operations at sea and while the ship was in port. The aircraft had to produce close-range video imagery in near real time. The purpose of Sea Airborne Lead Line (Sea ALL) was to serve as a sentry while ships were in port. It could perform harbor patrol, maritime interception, surveillance, and search and rescue missions. The Chief of Naval Research tasked NRL to develop and field the system.

NRL determined that the Dragon Eye system could be adapted to the mission. This impacted the main Dragon Eye development program slightly.

The intended operators were U.S. Navy enlisted personnel, NATO forces, and indigenous forces at the port facilities. The system consisted of four airframes, ground control equipment, a net recovery device, and spares. The aircraft had a 10-kilometer range and up to 1-hour endurance.

The full autopilot capability was not yet available, so Sea ALL had to employ man-in-the-loop control. The pilot would fly through the forward-facing camera by looking at the GCS monitor. This approach was common for small unmanned aircraft developed in the 1980s and 1990s, such as the AeroVironment Pointer. Telemetry from the aircraft was displayed on the monitor. The aircraft was equipped with a GPS that provided position. The display also showed a bearing back to the ship as a navigation aid. This feature permitted flight beyond visual range without risking the loss of the aircraft. The operator would simply align Sea ALL with the appropriate heading and wait for the ship to reappear. The controls were direct stick-to-surface and throttle without any autopilot inner loop control logic. However, the aircraft was easy enough to fly through the camera that operators with no prior RC experience could fly it without much difficulty. New RC pilots find it challenging to control turning when the aircraft is facing them when the visual behavior of the aircraft is part of the feedback loop. However, flying through the camera eliminates this problem. A simple flight simulator helped with operator training.

Dragon Eye was not capable of landing on a ship, so the Sea ALL system required development of a deck recovery system. The solution was a small square net measuring 12 feet on each side (Fig. 5.36). The net was located near the fantail of the ship. A fluorescent square was painted on a portion of the net to provide a target. The pilot would fly the aircraft via the camera toward the net target. Airframe components would come off during a rough landing, but without damage.

When development was complete, the Sea ALL systems were sent overseas. After brief training, the Navy crew operated the system. The Sea ALL system performed multiple missions. All shipped hardware was left behind with the Navy, and no further systems were developed. The capability eased the minds of Navy leadership that if the terrorist problems persisted, a small solution could be rapidly deployed.

A later operation aboard a NATO frigate showed that the ship's radar created radio frequency interference (RFI) problems with the Sea ALL video system. The video would degrade whenever the radar painted the aircraft. This problem was manageable and did not pose a flight safety risk. The British operators liked the system so much that the deployment was extended by three months.

NRL continued to develop net recovery system technology after the conclusion of the Sea ALL program. A later exploratory development

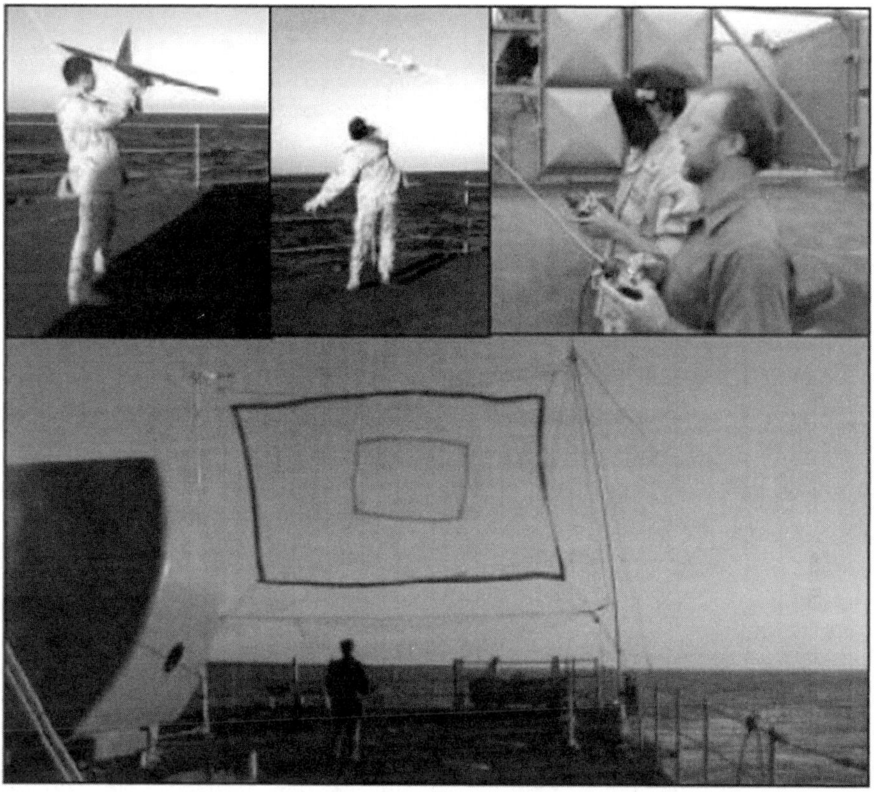

Fig. 5.36 Sea ALL launch and recovery operations.

program for an optical net recovery system helped to advance automated recovery. LED lights were distributed around a square on the net. The autopilot system was able to resolve position and orientation based on the distance between the LED square vertices. This system worked when the aircraft was close enough to the net to permit the net features to be spread across multiple imager pixels. This system was passive in that it required no RF communications. The system required either a landing camera or a gimbaled camera locked in the forward pointing position. In principle, this system could be applied to any unmanned aircraft with the appropriate software and camera system.

GOLDEN EAGLE, 2002

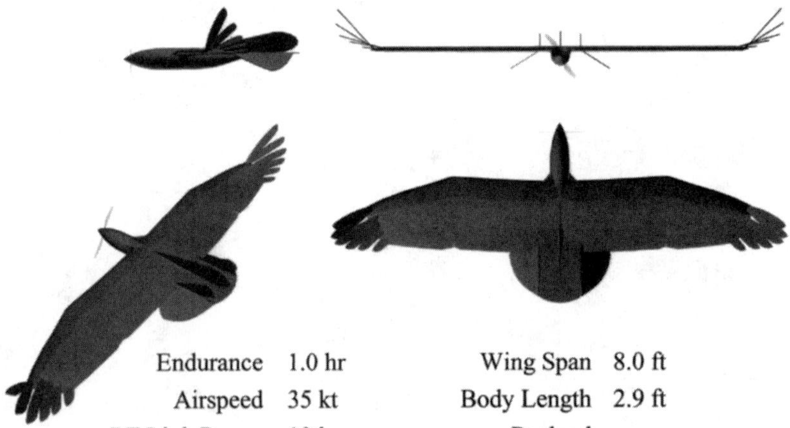

Endurance	1.0 hr	Wing Span	8.0 ft
Airspeed	35 kt	Body Length	2.9 ft
RF Link Range	10 km	Payload Capacity	1.5 lb
Altitude Range	Sea Level to 12,000 ft	Gross Weight	9.0 lb
		Propulsion	Electric w/ $LiSO_2$ Battery

Fig. 5.37 Golden Eagle views and characteristics.

The Golden Eagle was a concept for an electric-powered surveillance unmanned aircraft that could be disguised to look like an indigeneous bird (Fig. 5.37). Unlike the CG Shift Bird, the Golden Eagle used traditional control surfaces rather than CG shift controls. Control surfaces included wing ailerons, an elevator on the tail, and drag rudders near the wing tips. The wing tip surfaces that resembled feathers would help reduce induced drag while augmenting the visual appearance. Experience with Dragon Eye showed that adopting a bird-like aspect ratio and low-visibility coloring could be sufficient to effectively mimic a bird at a distance, so perhaps this feather approach was excessive.

The study took place over a two-day period, culminating in a briefing. The study did not result in a program. NRL built two small hand-launched gliders that were scale models of the Golden Eagle design, and they flew well.

Later studies evaluated slight flapping on outer wing panels through dihedral changes. The wing flapping was for visual appearance rather than ornithopter propulsion. It is not clear whether this approach would be effective at helping disguise a loitering small UAS as an indigenous bird.

SEA QUELL, 2003

Fig. 5.38 Sea Quell prototype.

The USMC Dragon Eye program requirement for flight endurance was only 1 hour, even though mature electric propulsion and airframe technologies would enable 2 to 2.5 hours of endurance while meeting other program requirements. The USMC had requested that NRL not design Dragon Eye to exceed the requirements, to reduce development time and cost. Sea Quell was an internally funded project by NRL to demonstrate the potential for longer endurance. The name Sea Quell was a play on "sequel," since Sea Quell could be the sequel to Dragon Eye.

Like Dragon Eye, Sea Quell was hand launched and landed on its belly. Similarly, the flight controls were elevons and throttle. It had a plank flying wing configuration, with a vertical stabilizer and single fuselage (Fig. 5.38). The proven Liebeck 2973A airfoil was used. Different from Dragon Eye, Sea Quell had a single tractor motor at the front of the fuselage, an arrangement that had lower drag than the former's twin-motor configuration. The penalty was that Sea Quell had no forward payload bay. Sea Quell's aspect ratio was higher for reduced induced drag. The outer wing panels were tapered to reduce the induced drag and structural weight. Like Dragon Eye, Sea Quell components could be unplugged to fit within an ALICE backpack.

As seen on the Dragon Eye and CG Shift Bird programs, a wing of bird-like aspect ratio and a triangular vertical tail could visually appear like a bird to ground observers. Sea Quell's wing had a similar aspect ratio to a sea gull, and so to make the test aircraft a little more interesting than the overall gray color of the typical UAS, a sea gull's color scheme was adopted. Coincidentally, this paint job looked much like those of earlier U.S. Navy manned aircraft.

The prototype was rapidly built using simple, model airplane construction techniques. The wing was balsa wood sheeting over a foam core. The fuselage and vertical tail were constructed of balsa wood and birch aircraft plywood.

The propulsion system consisted of a lithium polymer battery, brushless motor, and motor controller. The lithium polymer battery was rechargeable to reduce cost for flight research. A mission aircraft would have used a primary (nonrechargeable) lithium battery to improve endurance.

NRL flew the aircraft one time. The motor controller failed about 10 minutes into the flight. The motor controller was rated for the current, but there appeared to be quality issues with its manufacture. The controller air cooling may have also been inadequate. Flight data showed that the power consumption was consistent with a 2.5-hour flight, so the performance was partially verified.

The motor controller problem could have been quickly fixed, but the program was terminated. NRL was notified by the USMC to cease development. The concern was that the Sea Quell program could jeopardize Dragon Eye's transition to operations. It made no sense for NRL to threaten the acquisition program they worked so hard to develop.

Sea Quell showed the level of flight performance possible for a small, battery electric, unmanned military aircraft. Figure 5.39 shows a three-view drawing of Sea Quell, and Table 5.9 lists its technical characteristics. Though an unrelated effort, a later version of the BAI-produced Dragon Eye had extended wings. This design also showed significant performance improvement over the baseline Dragon Eye, and it had similar characteristics to the Sea Quell.

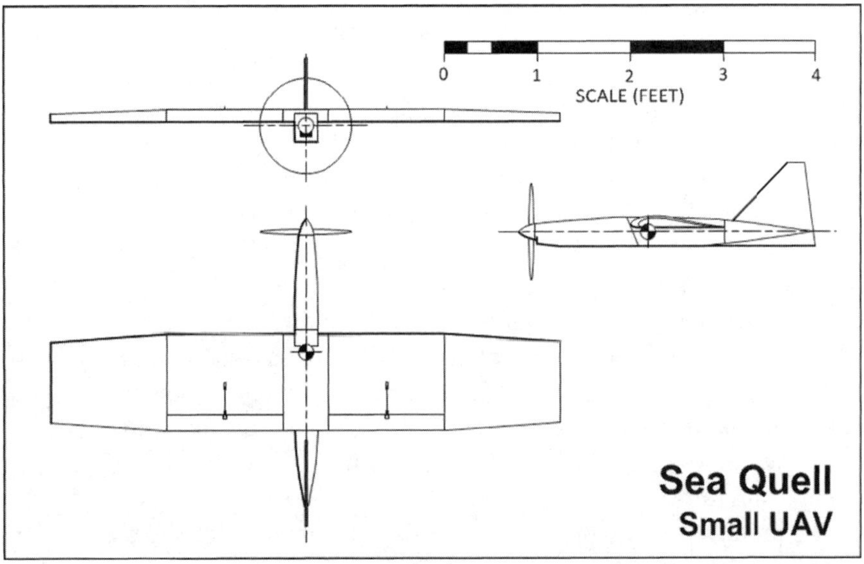

Fig. 5.39 Sea Quell three-view drawing.

Table 5.9 Sea Quell Technical Characteristics

Parameter	Value
Wing span	5.5 ft
Wing area	4.0 ft^2
Length	3.2 ft
Takeoff weight	6.6 lb
Empty weight	3.0 lb
Payload weight	1.5 lb
Battery weight	2.1 lb
Loiter speed	24 KTAS
Dash speed	50 KTAS
Endurance	2.0–2.5 hr
Link range	6 nmi
Airfoils	Liebeck LA 2573A
Launch method	Hand launch
Recovery method	Belly landing
Propulsion type	Battery electric
Power	300 W

SPOTLIGHT, 2005

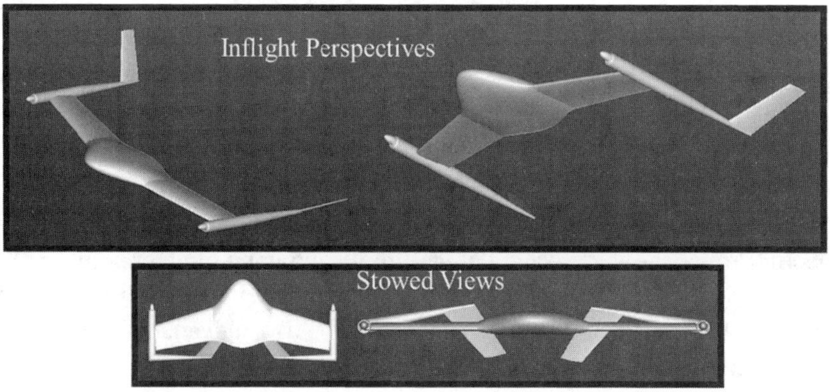

Fig. 5.40 Spotlight unmanned aircraft renderings.

NRL proposed the Spotlight unmanned aircraft for the U.S. Coast Guard. The system was intended to provide capabilities like those of the Dragon Eye, but with longer flight endurance, while still using battery electric propulsion. The system would be operated from a small Coast Guard vessel or a land station. Spotlight would fly out to rapidly interrogate ship targets

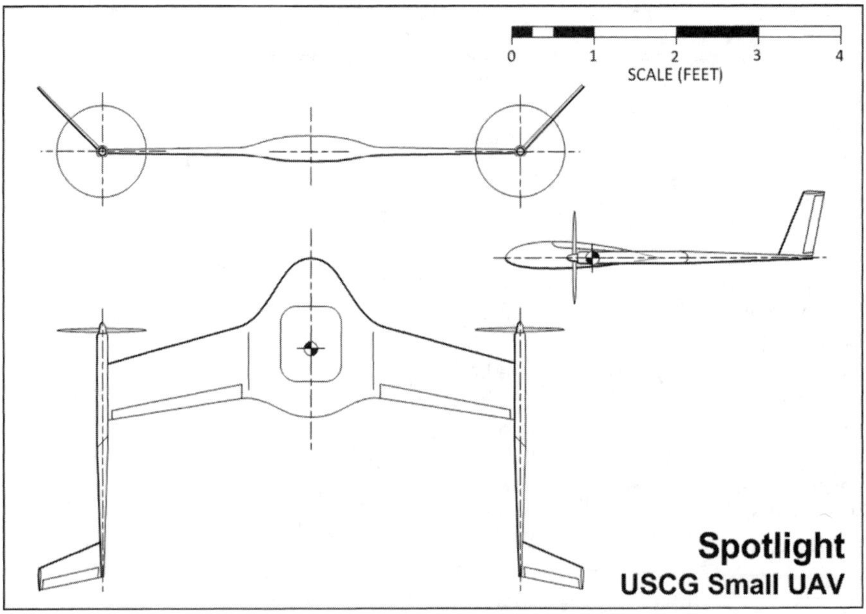

Fig. 5.41 Spotlight three-view drawing.

and return to base. In addition to rapid identification of ship class, the imager would provide real-time video to monitor ship deck activity and to observe the ship's name on its hull.

The Spotlight's displaced tail design was intended to fit within a small container by folding the tail booms on skewed hinges (Fig. 5.40). Unlike Shadow, Extender, and other NRL displaced tail designs, the tails canted outward instead of inward. The tip vortices would have rolled up ahead of the tails thanks to the large tail booms, so flow separation on the tails was not a concern. Propellers were located at the tip of each wing on the booms. Each propeller spun in the opposite direction of the tip vortex to reduce the induced drag. The airframe would have been constructed of advanced composites to save weight.

The Spotlight was designed and proposed, but the Coast Guard opted to buy commercially available systems rather than design a custom system for the mission. No flying aircraft of this interesting configuration were built. Figure 5.41 shows a three-view drawing of this ground-launched aircraft, and Table 5.10 lists its technical characteristics.

Table 5.10 Spotlight Technical Characteristics

Parameter	Value
Wing span	6.7 ft
Wing area	4.2 ft^2
Length	3.9 ft
Takeoff weight	12.0 lb
Payload weight	3.0 lb
Battery weight	5.0 lb
Loiter speed	35 KTAS
Cruise speed	44 KTAS
Endurance	1.0 hr
Link range	8.8 nmi
Max L/D	11 at 44 KTAS
Launch method	Hand launch
Recovery method	Belly landing
Propulsion type	Battery electric
Power	142 W (propulsion) + 50 W (avionics)

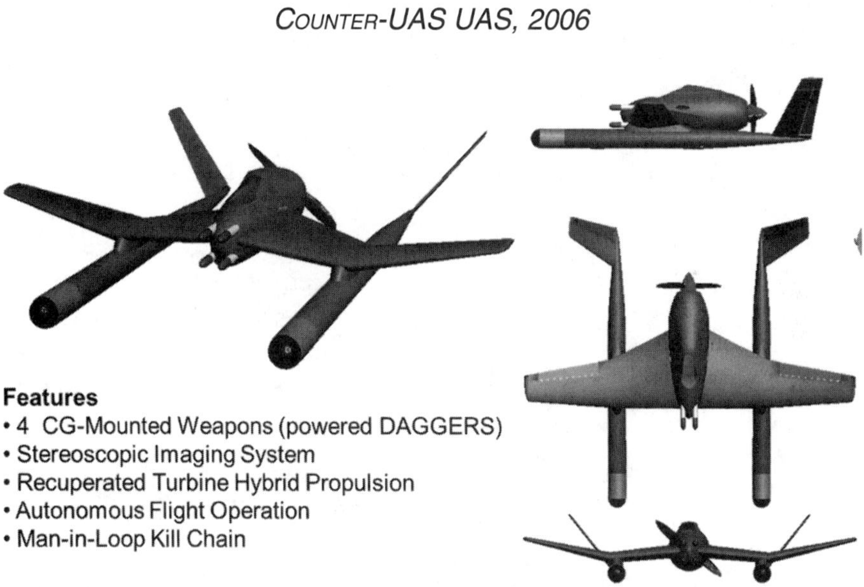

Fig. 5.42 Counter-UAS UAS concept.

The Counter-UAS UAS was a concept design for a small unmanned fighter aircraft intended to shoot down other unmanned aircraft (Fig. 5.42). This was part of a larger study to look at methods of countering unmanned aircraft. Foch believed that the best way to destroy other unmanned aircraft would be with an unmanned interceptor, just like manned fighters are well suited to shooting down other manned aircraft.

To achieve the required flight performance and mission capability, this design had many noteworthy features. Two cameras were mounted on the front of booms, which would provide stereoscopic vision. A central payload pod housed the turbogenerator hybrid electric propulsion system and four small, custom-made, optically guided missiles. The aircraft's CG would remain unchanged as the missiles fired. The configuration had a pusher propeller and a displaced tail layout, with angled tail surfaces at the end of each boom. The aircraft launched and recovered conventionally from an airfield and had retractable landing gear. The flight performance (airspeed and maneuvering) was designed to exceed that of expected target unmanned aircraft.

A stereoscopic vision system was incorporated to aid with target detection and tracking. Moving targets could be detected against the ground background by comparing imagery from the two cameras. The system also provided depth perception. Once the target was detected, the vision-based system would help attain a favorable firing solution. The aircraft would fly a pursuit or a proportional navigation solution. Unmanned aircraft are cooperative targets, because current systems do not have sufficient situational awareness to watch for attacking aircraft and are unable to know when to make effective evasive maneuvers. The Counter-UAS UAS would line up behind the threat aircraft in the six o'clock position, establish a firing solution, and launch its missiles.

NRL designed the initial vehicle concept and established the basic principles, but the design was not taken further. Some believed that the technology was rather far-fetched or even that it was a joke. However, there is no reason the system could not work effectively. Most Uninhabited Combat Aircraft System (UCAS) studies have considered only bomber or attack missions. Others have considered or implemented air-to-air missiles for self-protection. This system concept was the first purpose-designed unmanned fighter for aerial combat with other unmanned aircraft.

SEA WIND, 2008

Fig. 5.43 Sea Wind concept.

NRL performed a study for the Commanding Officer of NAVAIR's Program Executive Office for Unmanned Aircraft and Strike Weapons (PEO [UW]) to investigate the feasibility of a next-generation version of the STUAS. NAVAIR was investigating a new generation of miniaturized payload sensors, which could be combined for a 15-pound mission package. The aircraft flight performance had to be equal to or better than that of STUAS. The payload bays and wings had to be modular to enable mission flexibility. Another requirement was that the system had to use the Insitu SkyHook recovery system that had proven itself on Navy ISR services deployments with the Insitu ScanEagle system. NRL conducted the study for both PEO (UW) and NAVAIR's UAS Program Office (PMA-263) to help plan for future capabilities.

NRL considered multiple configurations in the design study. The displaced tail configuration was selected because it supported the modular design objectives (Fig. 5.43). The tails plugged into the wings rather than the fuselage, so the wing-tail system was independent of the fuselage. With this configuration, a longer-span wing could be used with decreased sweep. The extended-span version would have longer endurance. Similarly, a shorter-span wing with higher wing sweep for maintaining the required longitudinal static stability margin could provide higher dash speed due to the higher wing loading.

The fuselage could be modified as well. The modular fuselage consisted of a nose payload bay, a CG payload bay, a central module above the CG payload bay, and a propulsion module. The fuselage modular arrangement was similar to that of the Insitu Integrator; in fact, the Integrator design of author Gundlach influenced author Foch in his Sea Wind design. Sea Wind's aerodynamically clean camera ball integration was also influenced by the Integrator. The design influence came full circle, since the Integrator design was heavily influenced by Foch's Nexus concept.

The space in the central fuselage could hold fuel or batteries, depending on the propulsion system. The CG payload bay could also be equipped with additional fuel. Alternatively, the entire central fuselage could be modified to hold a pressurized hydrogen tank for fuel cell propulsion, providing 24-hour flight endurance.

The selected airframe geometry allowed the aircraft to fold without removing any components (Fig. 5.44). Electrical and mechanical connectors between airframe components can negatively impact reliability. By having an unfolding aircraft with no component attachments in the field, no connectors are required for routine operations. The ground crew would pull the aircraft out of a container and unfold it prior to flight. The wings were swing wings, with pins in the fuselage. A flexible glove fairing on the upper fuselage locked into place when the wing swung open into the deployed

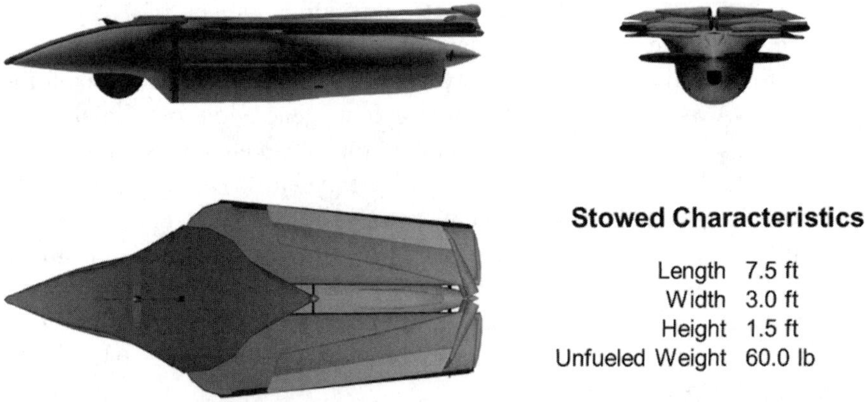

Fig. 5.44 Sea Wind stowed geometry.

position. The outer wings were folded under the inboard panels when stowed. A skewed hinge allowed the wing to rotate through 180 degrees, and then a latch would lock the wing into place. When the aircraft was stowed, each tail rested on top of the inboard wing panel. The tails unfolded using skewed hinges as in the NRL Extender design, and then latched into position.

A novel feature of the Sea Wind was that while it was designed to be manually unfolded on the ground prior to flight, it was also well suited to automated unfolding in flight (Fig. 5.45). This was not a sponsor requirement, but NRL added this feature due to synergies with the self-

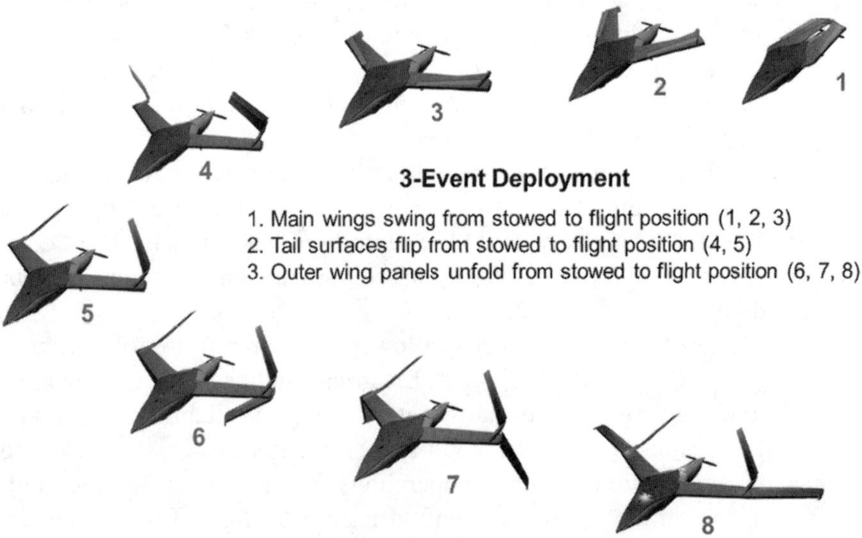

Fig. 5.45 Sea Wind air deployment sequence.

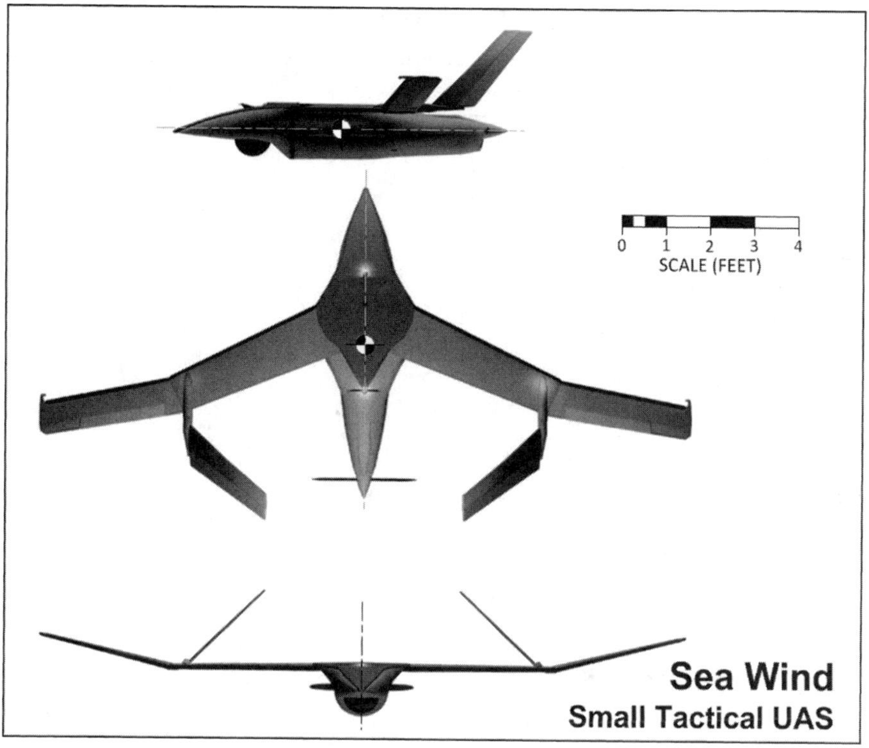

Fig. 5.46 Sea Wind three-view drawing.

erecting arrangement. The configuration had ballistic stability and could be aerodynamically controlled as it unfolded. This would enable the aircraft to be rocket fired or air launched. The inner wing would pivot open using a torsional spring. The outboard panels would aerodynamically deploy using the FLYRT ATD technique. Drag loads would cause the tail to unfold. This airborne deployment approach was very similar to that of Extender. The unfolding was not reversible in flight, so it could only unfold.

Sea Wind was designed with a sophisticated flight control system. The control surfaces included inboard flaps, outboard wing ailerons, and ruddervators. The controls functions could be mixed by the autopilot to optimize control performance by flight phase. Also, the controls could be mixed and scheduled to support in-flight unfolding.

The study was delivered to the sponsor and it was well accepted. No development efforts have yet resulted from this research. The study is being held for future consideration. Figure 5.46 shows a three-view drawing of this ground-launched aircraft, and Table 5.11 lists its technical characteristics.

Table 5.11 Sea Wind Technical Characteristics

Parameter	Value
Wing span	16.7 ft
Wing area	17.6 ft^2
Length	7.5 ft
Takeoff weight	100.0 lb
Empty weight	45.0 lb
Payload weight	15.0 lb
Fuel weight	40.0 lb
Cruise speed	50 KTAS
Dash speed	85 KTAS
Stall speed	35 KTAS
Endurance	32 hr at 50 KTAS with 5% reserves
Launch method	Rail launcher
Recovery method	SkyHook recovery
Propulsion type	Reciprocating heavy fuel engine

TRADE WIND, 2010–2011

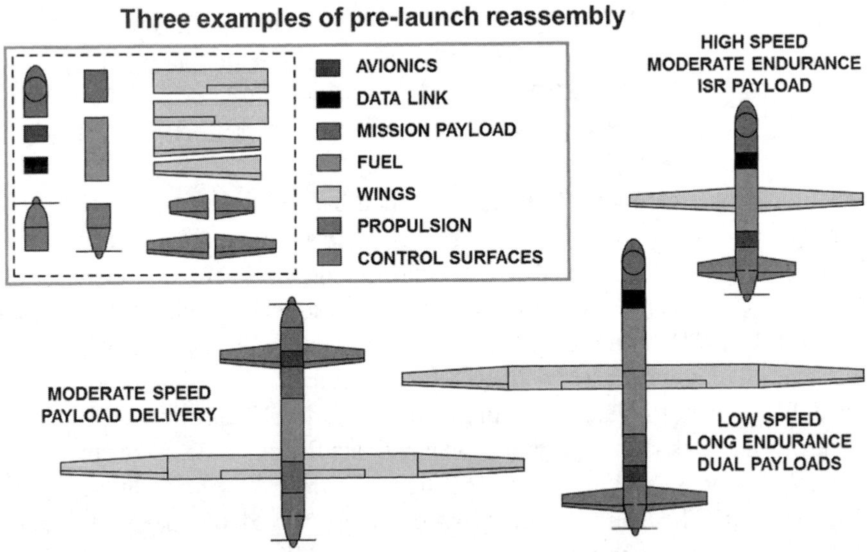

Fig. 5.47 Early reconfigurable modular aircraft concept.

NRL continued to pursue modular unmanned aircraft designs after Nexus. Foch developed a highly modular aircraft concept, called Trade Wind, that could flexibly support a large number of missions. A computer simulation program could determine the module configuration prior to flight to optimize mission effectiveness with minimum system resources. For example, using a robotic system to assemble the aircraft from the modules, a ship could launch a transonic interceptor within 5 minutes of deciding to do so. The program would optimize the vehicle from the available components. The robot would assemble the aircraft and prepare it for launch, and then the aircraft would be launched. In another 5 minutes, the system could launch a long-endurance communications relay aircraft using fuel cells, thus supporting two independent missions using one onboard system that autonomously assembled the needed UASs from a small stockpile of airframe modules.

The system consisted of wing modules, payload modules, fuel or battery modules, avionics modules, and propulsion modules. The computer program selected the components such that the aircraft would be flightworthy and meet the mission requirements. For example, a long-range reconnaissance mission might utilize short-span wings and a turbojet engine. A long-endurance surveillance mission could use a long-span wing module and a reciprocating heavy fuel engine. Both of these examples may use many of the same modules. The modules had carefully controlled geometric, mechanical, electrical, mass properties, fluid, software, and aerodynamic interfaces. The mass properties standards included module CG, total weight, and moments of inertia limitations (Fig. 5.47).

The final Trade Wind design solution used swept tandem wings (Fig. 5.48). Each wing had a dedicated interface module that included a fuselage section. A center module containing the fuel or a payload would be located between each wing module. With the sweep, the CG could always be positioned halfway between the wing roots, regardless of the center module length. In other words, the same wings would work for long or short center modules without requiring ballast to maintain the required CG relative to the neutral point. Weight changes in the center module due to fuel depletion or dispensed payloads would not change the CG. Longer-span wings required less sweep than short wings in order to maintain the desired neutral point location. High-speed vehicles had shorter spans, which also drove higher sweep for balance. The greater sweep also reduced the transonic wave drag, so this relationship was synergistic (Fig. 5.49). Moderate sweep had low Reynolds number profile drag benefits, as was observed in the XFC aerodynamic development.

The front wing had greater incidence than the rear wing, because the front wing required a higher lift coefficient to simultaneously satisfy trim with a positive static margin requirement. The front wing was mounted as a low wing and had approximately 2 degrees of positive incidence. The rear wing module was the same as the front, except it was flipped inverted such that it was a high wing with negative incidence. The common front and rear wings reduced the number of parts that had to be stocked. The electrical interfaces

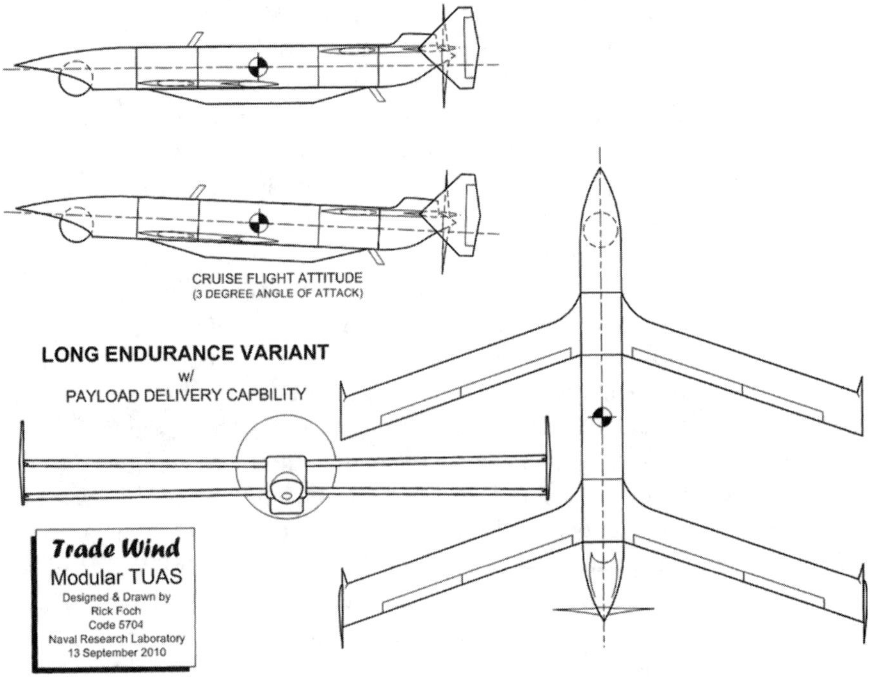

Fig. 5.48 Trade Wind long-endurance configuration.

would be configured to identify when the wings were used as front or rear wings. The system optimization could pick between wing sizes for the front and rear wings as necessary for the mission. The pure tandem wing aircraft would cruise with a 3-degree fuselage angle of attack. The fuselage lift can be beneficial for induced drag, because the vehicle lift distribution with a slightly lifting square fuselage was shown from wind tunnel tests of LODED and LAURA to be close to the elliptical ideal.

Each wing used symmetrical airfoils with full-span control surfaces. These surfaces could be used to adjust the camber on the front and rear wing to minimize drag. The control surfaces all acted as elevons. There were two surfaces per side per wing, and this segmentation helped optimize the camber distribution. The elevons were configured as forward wings or rear wings using information from the mechanical interface orientation to set the configuration. The symmetrical airfoils and tandem wing configuration meant that each wing half was identical. This vehicle had wings that were symmetrical right-left and front-aft. Therefore, the wing halves could be manufactured in large quantities to reduce the unit costs.

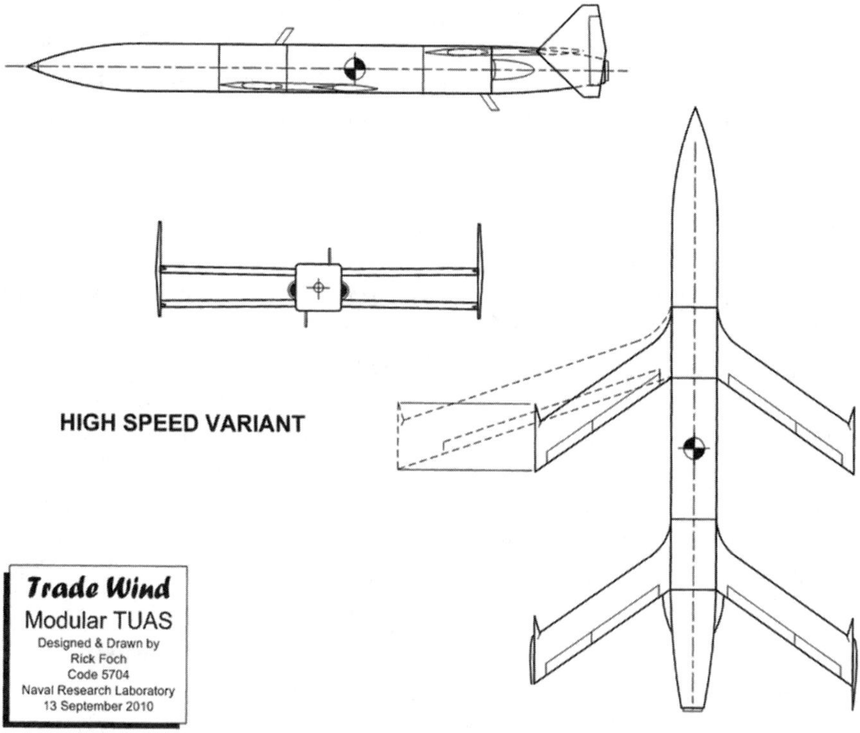

Fig. 5.49 Trade Wind high-speed variant.

Winglet vertical fins were used for directional stability. The tips of each wing had an interface for the vertical surfaces. Various tail sizes were available, based on the vehicle configuration. In general, the winglet vertical stabilizers were applied to the rear wing. However, winglets applied to the forward wing could be used as side force generators for precision flight path tracking.

The wing section interface fuselage modules had a constant rectangular cross section with rounded corners. The center fuselage modules also had a constant cross section. The center modules could house fuel, batteries, or payloads. The nose and propulsion modules that plugged into other modules had variable cross sections to minimize drag. The propulsion modules could be placed at the front or rear of the aircraft, or both. The avionics module was a fuselage slice of constant cross section that could be placed between any two modules.

A number of propulsion system types could be accommodated. The propulsion options included battery electric, fuel cell electric, reciprocating engines, turbojets, and hybrid combinations. Fueled engines would require a fuel tank near the CG, which would be in a center module. Fuel would need to transfer through the rear wing fuselage interface module. The payload could be housed in a nose payload module when a center bay fuel tank was used. Alternatively, a fuel center module and a payload module could be placed between the wings in tandem.

This system would require a unique design approach. A design environment would need to determine the set of components that could most effectively satisfy a large number of missions. The module specifications and interfaces would need to be carefully considered. Once the modules were developed, they could be combined in various ways. The optimization would help define the module requirements and interface specifications.

With this approach, it would be possible for industry to develop noses, wing sets, propulsion modules, and other components to a set of module interface specifications. A single manufacturer would not necessarily develop the entire aircraft. The Navy would generate the specifications for industry, and industry would compete for the module programs. It was hoped that this would greatly decrease development time and cost. In many cases, a new combination of modules or a new module could provide additional mission capabilities without the need for new unmanned aircraft acquisition programs. The Navy would accumulate a tool kit of modules developed under the same specifications. NRL envisioned that such a modular aircraft would not require a full flight qualification for each configuration.

Limited resources were applied to the initial study. While the system-level optimization program was not written, NRL conducted a large number of design iterations. Sample module specifications were written, and modules that could be assembled into viable aircraft were designed. The tandem wing modular design solution found at least one combination of modules that could satisfy each mission.

This robotically assembled modular concept for rapid UAS deployment and its streamlined acquisition process has not yet been fully accepted. However, it was viewed as technically feasible. NRL continues to advocate for reconfigurable unmanned aircraft. Developing and demonstrating the capability would be a welcomed major research program with the potential to establish a new paradigm for future UAS development.

FLYING RADAR REFLECTOR UAS, 2010

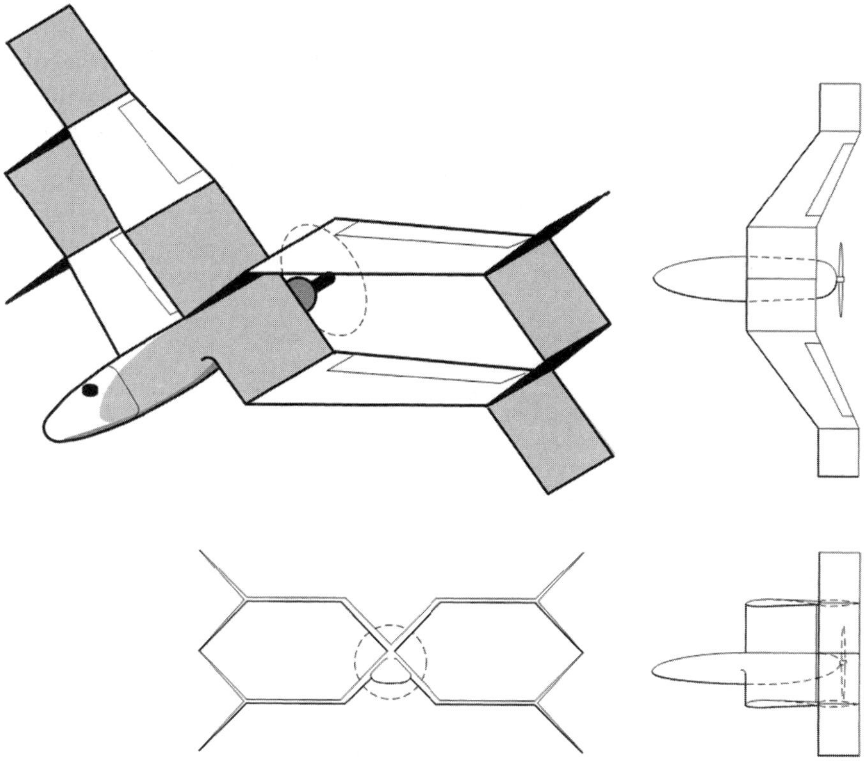

Fig. 5.50 Flying Radar Reflector.

The Flying Radar Reflector aircraft was intended to passively provide an enhanced radar return (Fig. 5.50). This design was the exact opposite of a stealth aircraft, because it was configured to augment its radar cross section rather than to minimize it. Depending on the size chosen for the aircraft and the chosen radar frequencies, the radar return of the aircraft could be similar to that of a small ship. This vehicle would fly at 30 to 40 knots to simulate ships for radars. The aircraft was a tactical-class vehicle that would be affordably expendable. This design was part of a larger effort to explore structural reflectors as passive EW countermeasures.

The wing/body junctures and winglets formed corner reflectors, which directed the radar return back at the emitter over a large range of orientations. A corner reflector is formed when two flat surfaces intersect at a 90-degree angle. This unusual configuration could be described as a joined biplane flying wing. The winglets provided the requisite directional stability. The intersections were arranged such that aerodynamic interference drag was minimized.

Another output of the study was an inflatable structure that formed octahedral intersections. This provided better returns over a broader azimuth angle range. The technology was named the Omni-Directional Inflatable Radar Reflector (ODIRR, pronounced "oh dear"). While not a UAS, and thus outside the scope of this book, ODIRR demonstrator systems are under development at NRL. Based on the current success of the ODIRR project, future UAS designs as follow-ons to the Flying Radar Reflector study could significantly augment their radar signature by installing a pair of ODIRR reflectors within a nonconducting fuselage.

Chapter 6

MICRO AIR VEHICLES

Admiral Owens of the Joint Chiefs of Staff made an appearance on the news program *60 Minutes* one Sunday evening in 1995. He pulled what looked like a toy airplane out of his pocket. It was about 4 inches long, had a canard configuration, and carried a video imager weighing 1 gram. Admiral Owens stated that this was the future of warfare. As Foch watched the show, he remarked to his wife that he would hate to be the person who had to make that-size airplane fly well. Prophetically, Foch received a call from the Chief of Naval Research early Monday morning, tasking him to develop tiny military unmanned aircraft systems (UASs) to carry micro payloads such as that 1-gram video imager, which was made by the Massachusetts Institute of Technology (MIT) Lincoln Laboratory. USAF Colonel Michael Francis of Defense Advanced Research Projects Agency (DARPA) was also tasked with the program.

Foch originally met Francis when the latter was working on his graduate degree at the Air Force Institute of Technology (AFIT) at Wright Patterson Air Force Base, Ohio. Foch visited as a summer researcher in 1974, and the two discussed Francis's research on the low Reynolds number (LRN) aerodynamic coupling of dragonfly wings.

Foch and Francis decided to collaborate on their new tasking. They defined an approach to make for DARPA and NRL programs that were both complementary and supplementary. The two organizations would share their technologies and attempt to minimize redundancy. DARPA was excited by the new miniature imager technology and decided to pursue surveillance missions. DARPA also pursued data link technologies, micro turbine engines, and micro reciprocating engines. NRL was interested in pursuing electronic warfare (EW) applications, vision-based navigation, and electric propulsion. The combined effort, it was hoped, would produce enough output to define the solution space for useful military missions.

Next, the system requirements were defined. The technology challenge had to be "DARPA hard"—meaning revolutionary and "pushing the technology envelope"— rather than an incremental improvement of existing capabilities. On the other hand, the system had to be developed within a reasonable budget and schedule. The pocket-sized aircraft that Admiral Owens described on *60 Minutes* was too small. Foch had experience with free-flight model airplanes, including "peanut scale" models with 12- to 13-inch spans. The largest propeller diameter at small scale was no more than half of the wing span. Foch believed that the smallest prop with reasonable efficiency that could be built out of molded composites would have a 3-inch diameter. Therefore, a reasonable wing span of 6 inches was selected for the DARPA program. The final requirement was for an aircraft that could fit within a 6-inch cube. It was not known if the resulting aircraft would be mission practical, but it was intended to be a technology challenge.

DARPA hosted a workshop to explore the program. Industry, academia, and government labs were challenged with this problem. There was not much industry response, but there was a strong response from academia, including University of Florida, Notre Dame, and MIT. MIT Lincoln Labs presented their 1-gram imager technology. This new class of systems was originally called "Micro Unmanned Aerial Vehicles," but the workshop attendees thought that "micro" and "unmanned" were redundant terms. Therefore, the name Micro Air Vehicle (MAV) was created.

NRL presented a paper on MAV aerodynamics. In higher Reynolds number designs, aspect ratio is generally increased until a structural or packaging limit is reached so that the induced drag can be minimized. Then an airfoil is selected to provide the best performance. If the span is not limited, the chord tends to be small. However, lower-speed aircraft with small chords operate in the low Reynolds number regime. This regime degrades airfoil performance in the form of higher profile drag and reduced maximum lift coefficient. It was found that as the speed was reduced with span constraints, the optimum aspect ratio was reduced. For typical LRN airfoils, the optimum aspect ratio for MAVs was between 1.5 and 3. The better airfoil performance made up for the higher induced drag. MAV airfoils tend to operate at lift coefficients less than 1. The induced drag coefficient is proportional to the square of the lift coefficient. Compared with higher Reynolds number applications, MAV induced drag is less dominant. Some of the other presenters showed much higher aspect ratios, which generated lively debate. Nearly all MAVs developed after this workshop had low aspect ratios based on NRL's empirical equations that were presented to determine optimum MAV aspect ratio.

DARPA proceeded to issue requests for proposals for industry and universities. The areas of interest included vehicle design, liquid-fueled propulsion, payloads, and guidance and navigation. Notable industry-developed systems include the AeroVironment Black Widow and Lockheed Martin MicroStar.

NRL established an internal research program that focused on non-GPS navigation, micro EW payloads, batteries, motors, and LRN vehicle design. NRL developed the airframes and EW payloads internally. The motor and battery development was external via industry contracts.

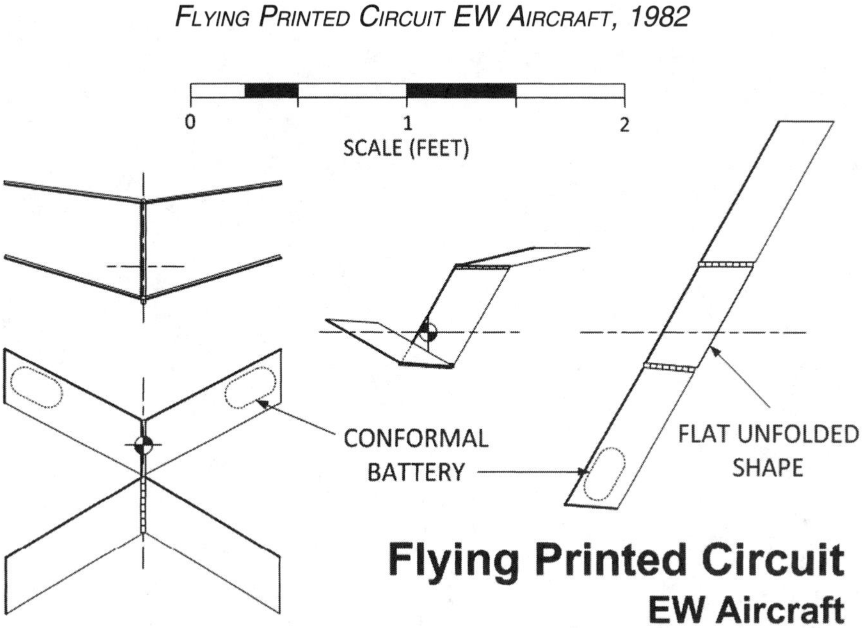

Fig. 6.1 Circuit board aircraft design.

More than one decade prior to the DARPA/Navy MAV programs, NRL developed a design for a very small printed circuit board aircraft for EW (Fig. 6.1). It would serve as a small, inexpensive electronic jammer. The wings and fuselage were constructed of circuit boards. The wings would fold flat in the same plane as the fuselage such that a large number of vehicles could be stacked in a compact volume. Perhaps several tens of thousands could be deployed from a single C-130 sortie. When deployed, the upper and lower wings sprung open with a positive dihedral angle. The front wings were swept forward, and the rear wings were swept aft, producing a tandem wing configuration.

Unpowered models approximately 1 to 1.5 feet long were built and flown from a tower. The vehicles were found to be inherently stable, with good spiral stability. They tended to have a pronounced Dutch roll, which could have been corrected with reduced dihedral. The lift-to-drag ratio (L/D) was approximately 6.

The design was never fully developed at the time. The later CICADA designs reinvestigated the concept and took it further.

MICRO TACTICAL EXPENDABLE (MITE), 1995–2002

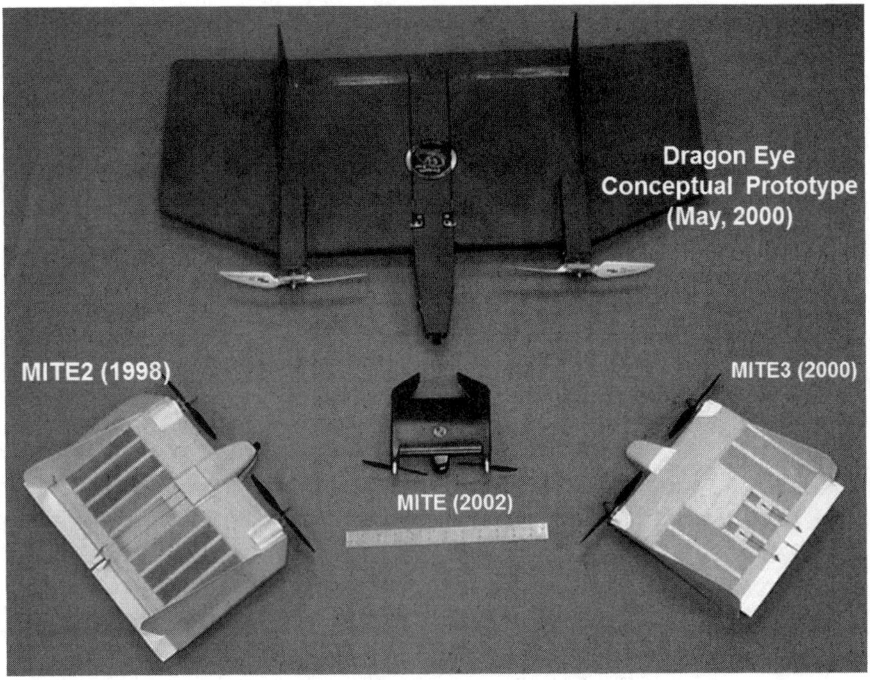

Fig. 6.2 MITE variants and an early Dragon Eye prototype.

Micro Tactical Expendable (MITE) was a family of NRL designs of similar configuration (Fig. 6.2). Foch decided to exploit aerodynamic and propulsion interactions by placing counter-rotation propellers at the wing tips. The propellers spun in the opposite direction of the wing tip vortex, effectively increasing the aspect ratio. This configuration was first developed in the 1940s by Charles Zimmerman, an aeronautical engineer with the National Advisory Committee for Aeronautics (NACA) at Langley Field, Virginia. The blowing over the airframe increases the turbulence and Reynolds number over the wing, both of which improve boundary layer transition characteristics. The MITE configuration was established just prior to the first meetings between NRL and DARPA in 1995 (Fig. 6.3).

NRL engineers realized early on that the 6-inch span constraint was too challenging. NRL was interested in building the smallest unmanned aircraft that could satisfy a useful EW mission rather than developing the

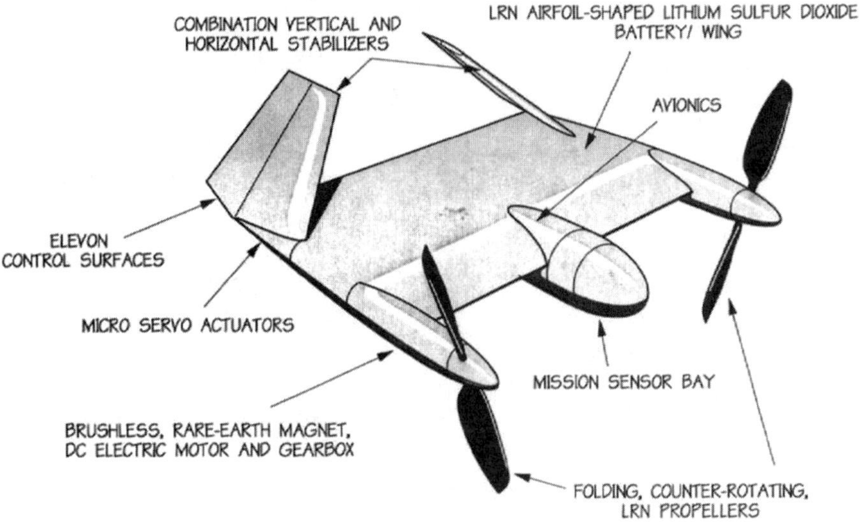

Fig. 6.3 Original MITE concept.

smallest unmanned aircraft possible. The smallest available EW payload was 10 grams, which was an order of magnitude greater than the smallest imagers at the time. The MITE designs were larger airplanes than those developed for DARPA, but they were still small (Fig. 6.4). NRL decided to build the smallest practical aircraft that could be built quickly. The initial MITE had span and length dimensions of 20 inches. As the technology improved, the vehicle shrunk to 10 by 10 inches. NRL never attempted to fully develop a 6-inch-span design (Fig. 6.5).

From the outset, the NRL MITE was considered a swarming system in order to fully realize the tactical advantage. One proposed EW mission was to fly a MITE MAV using optical techniques and to land in the vicinity of a radar.

Fig. 6.4 MITE designs.

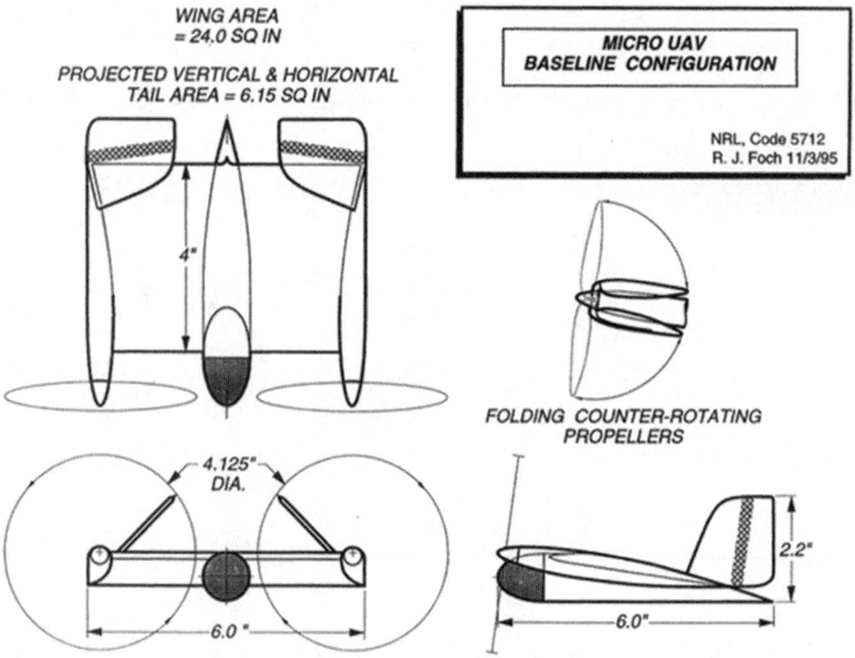

Fig. 6.5 Original 6-inch MITE concept.

These vehicles would then jam the radar with a payload weighing less than 10 grams. The system reliability did not need to be high if a sufficient number of MAVs could land near the radar target. The adversary would need to destroy each vehicle before the radar could be fully operational. Another possible MAV swarm capability was to form a virtual antenna while in flight. MITE vehicles conducted a successful EW mission demonstration in May 2002.

One argument against the effectiveness of a radar-jamming MAV was posed by a visiting U.S. Army colonel. The colonel claimed that as soon as his radar performance was degraded, he would go outside around his radar site and smash the MAV with his boot. During the meeting, Foch had one of the engineers place a MITE behind the Tactical Electronic Warfare (TEW) Division office building. During a break in their meeting, Foch challenged the colonel to find the MITE and smash it! After the colonel spent 1 hour of searching and had walked past the MITE several times, Foch picked up the MITE and handed it to the slightly angered colonel. It was autumn, and the MITE was painted in patches of burnt orange, green, and brown and had been set on top of a pile of fallen leaves.

Multiple MITE vehicles were demonstrated. These used control systems ranging from radio control to advanced miniature autopilots. The latter

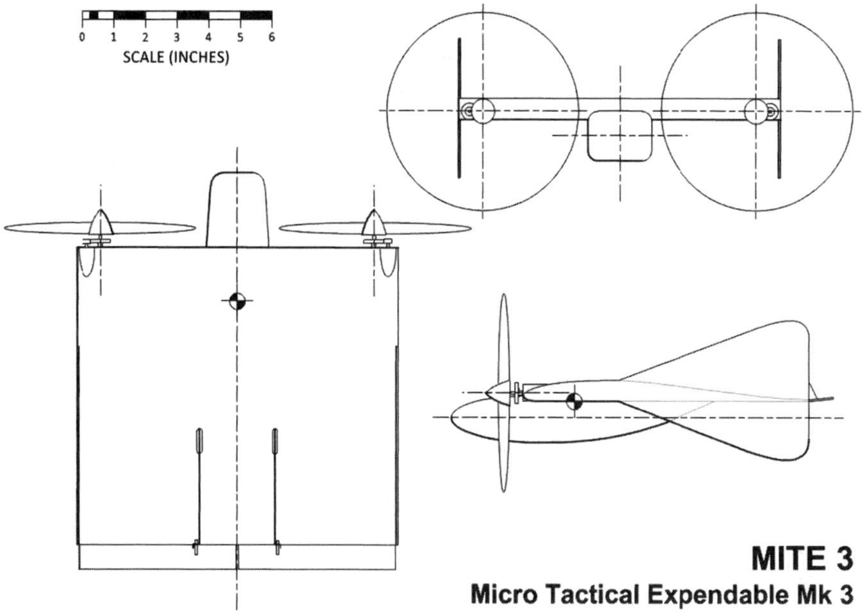

Fig. 6.6 MITE 3 three-view drawing.

employed a single-waypoint GPS navigation system, optical flow terrain avoidance system, heading and altitude hold modes, optical stabilization systems, and inertial stabilization systems. Although none of the MITE family were transitioned to production systems, they were the direct forerunners of the first production military small UAS, the RQ-14A Dragon Eye. Figure 6.6 shows a three-view drawing of MITE 3, and Table 6.1 lists the technical characteristics for MITE 3, MITE 2, and MITE 4.

Table 6.1 MITE Technical Characteristics

Parameter	MITE 3 Value	MITE 2 Value	MITE 4 Value
Wing span	10.0 in.	18.0 in.	24.0 in.
Wing area	97.5 in.^{2}	177 in.^{2}	288 in.^{2}
Length	12.0 in.	12.8 in.	15.0 in.
Takeoff weight	2.5 oz	8.5 oz	10.6 oz
Payload weight	0.5 oz	1.5 oz	3.0 oz
Endurance	10 min	12 min	18 min
Launch method	Hand launch	Hand launch	Hand launch
Recovery method	Belly landing	Belly landing	Belly landing
Propulsion type	Battery electric	Battery electric	Battery electric

MITE: FURTHER READING

Foch, R.J., and Ailinger, K.G., "Research and Development of Micro-Air Vehicles for Electronic Warfare," *1997 NRL Review*, Naval Research Laboratory, Washington, DC, 1997, pp. 86–88.

Kellogg, J., et al., "The NRL Micro Tactical Expendable (MITE) Air Vehicle," *The Aeronautical Journal of the Royal Aeronautical Society*, Vol. 106, No. 1062, Aug. 2002, pp. 431–441.

Kellogg, J., et al., "The NRL MITE Air Vehicle," *Proceedings of the 16th International Conference on Unmanned Air Vehicle Systems*, Bristol, U.K., April 2001.

PINK EYE, 1996–1997

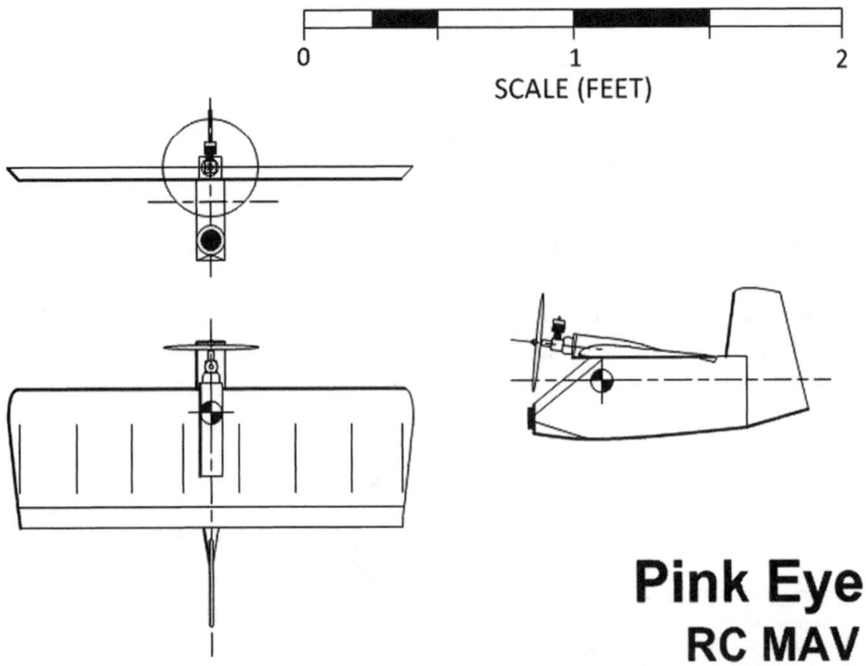

Fig. 6.7 Pink Eye three-view drawing.

The University of Florida hosted an MAV contest for universities. The objective was to image a cloth letter that had been placed in the middle of a field and correctly report it to contest officials. The letter was not directly visible to the pilot. The MAV had to be flown through a camera, and it could be radio controlled (RC). The contest rules were challenging but within the

Table 6.2 Pink Eye Technical Characteristics

Parameter	Value
Wing span	18 in.
Wing area	111 in.^{2}
Length	12.5 in.
Takeoff weight	9.25 oz
Payload weight	1.5 oz
Cruise speed	35 KTAS
Stall speed	17 KTAS
Endurance	4 min
Max L/D	4.5
Airfoils	Custom 9% t/c zero pitching moment
Launch method	Hand launch
Recovery method	Belly landing
Propulsion type	Cox 0.020-cu.-in., 2-stroke engine

state of the art of the time. The contest encouraged students to experiment with MAV technologies.

Government and industry teams could operate MAVs in the contest without directly competing with the university teams. NRL and DARPA volunteered to help with judging and contest support. NRL also developed the Pink Eye vehicle to show students an example MAV that used model aircraft technology.

Pink Eye was developed within the contest rules. It carried a small video imager and was flown via RC. Pink Eye had a plank flying wing configuration with a small fuselage and a single vertical tail. It spanned 18 inches. The engine was a Cox 0.020 single-cylinder reciprocating engine that ran on glow fuel. The airframe was constructed of balsa wood. It was covered with fluorescent pink MonoKote film, inspiring the name "Pink Eye." Figure 6.7 shows a three-view drawing of this MAV, and Table 6.2 lists its technical characteristics.

The NRL team members had positive interactions with the university teams. They practiced with the students and discussed aircraft design. Some of the student projects flew better than Pink Eye, which was encouraging for the future of unmanned flight.

SILICON BUG, 1996

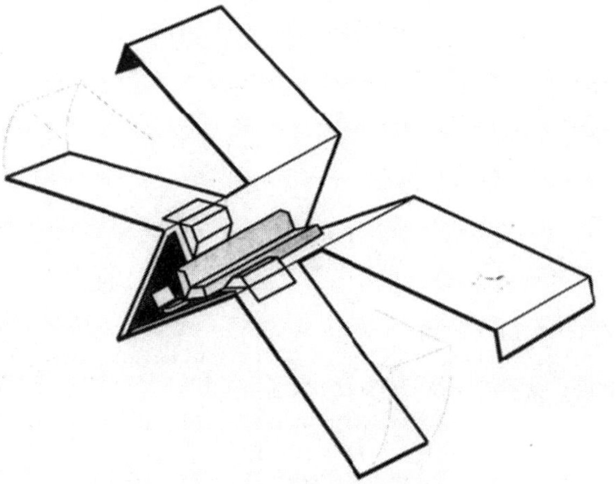

Fig. 6.8 Silicon Bug

The Silicon Bug was a concept for a next-generation MAV that could be mass produced using printed circuit board manufacturing technologies (Fig. 6.8). It would be possible to build millions at very low cost. The wing span could be 1 to 2 inches and the weight would be 50 to 200 milligrams, which was a category now known as Nano Air Vehicles (NAVs). The design enabled 21 vehicles to be constructed from a single 6-inch by 6-inch silicon wafer. The airframe was made of polyamide. The membranes, wiring, battery, payload, and processors would be built directly onto the airframe using integrated circuit manufacturing technology (Fig. 6.9). Thrust was produced by front wing flapping using thin-film piezoelectric bimetallic strip

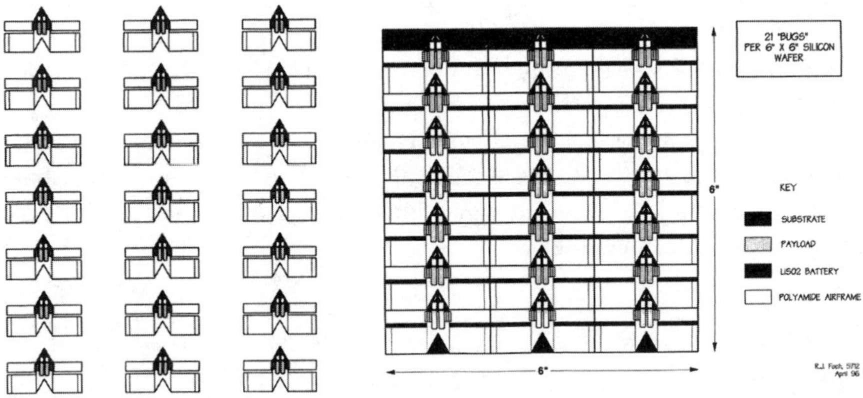

Fig. 6.9 Silicon Bug manufacturing concept.

actuators. Flight speeds would be approximately 2 meters per second. NRL was convinced that such systems could be built, but it was not clear if there were any practical applications. The problem was that the power required to transmit data did not decrease with size, so there was not an effective means of getting the data off the Silicon Bug. One potential mission would be to fly the MAV through a chemical agent plume, return and recover it, and then drop it in solution to analyze the chemicals.

WINGED AUTONOMOUS SILENT PLATFORM (WASP), C. 1997

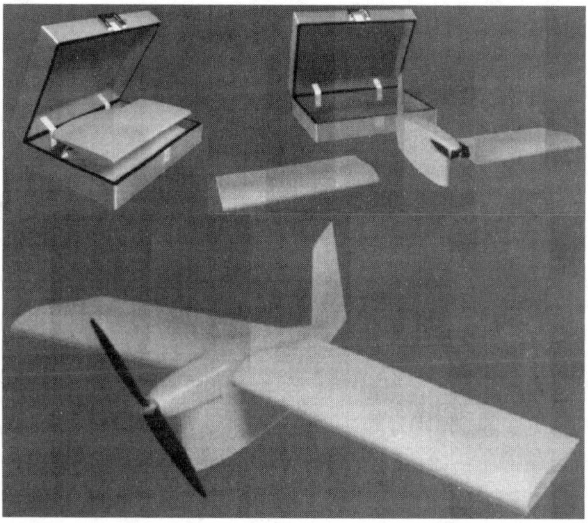

Fig. 6.10 WASP briefcase unmanned aircraft.

The Winged Autonomous Silent Platform (WASP) was a small unmanned aircraft that was designed to fit within a briefcase (Fig. 6.10). The project was a next step from Self-Navigating Drone Expendable/Recoverable (SENDER), which could be carried in a suitcase. The purpose of the briefcase transportation was to enable discrete carriage aboard commercial flights for special operations missions. WASP used electric propulsion to make operations simple. A short flying duration of less than 1 hour was required.

The WASP configuration was similar to that of Pink Eye. It had a plank flying wing with reflexed flying wing airfoils. The wings were slightly swept forward to produce some anhedral (negative dihedral) effect to reduce Dutch roll. The streamlined payload pod protruded vertically below the wing. A motor pod held a single tractor propeller in front of the wing. A single vertical tail extended upward behind the wing. Wing elevons provided pitch and roll control. To store the aircraft within the briefcase, the

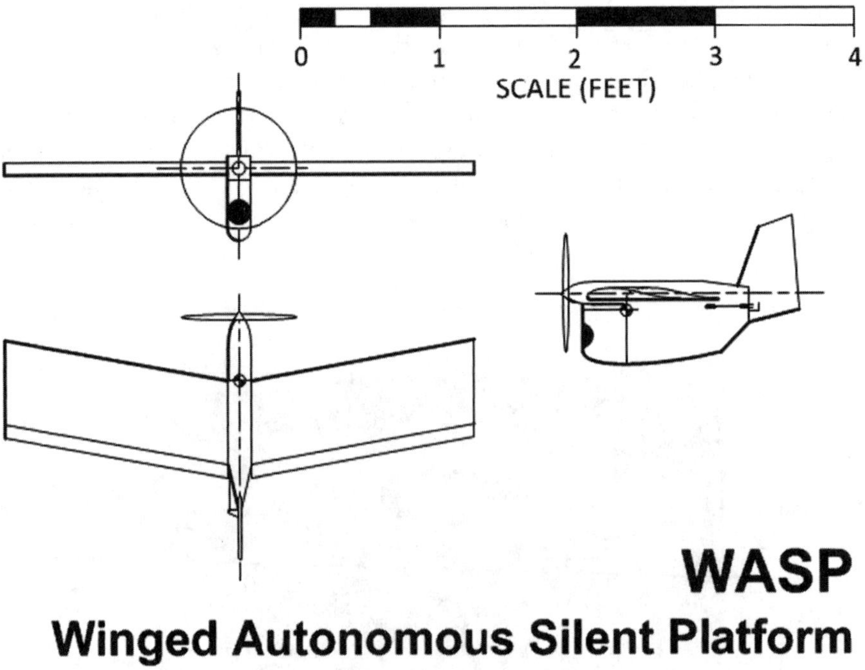

Fig. 6.11 WASP three-view drawing.

wings unplugged from the fuselage, the propeller folded along the fuselage, and the wings and fuselage stacked vertically. Fig. 6.11 shows a three-view drawing of this MAV, and Table 6.3 lists its technical characteristics.

Table 6.3 WASP Technical Characteristics

Parameter	Value
Wing span	3.4 ft
Wing area	2.28 ft^2
Length	1.7 ft
Takeoff weight	2.4 lb
Payload weight	1.0 lb
Dash speed	40 KTAS
Endurance	1 hr
Airfoils	Reflexed flying wing airfoil
Launch method	Hand launch
Recovery method	Belly landing
Propulsion type	Battery electric
Power	180 W

The system did not reach flight status. A full-scale mock-up was built to support a meeting and discussions. Many WASP features were later adopted by the Dragon Eye.

DRAGONFLY, 1997

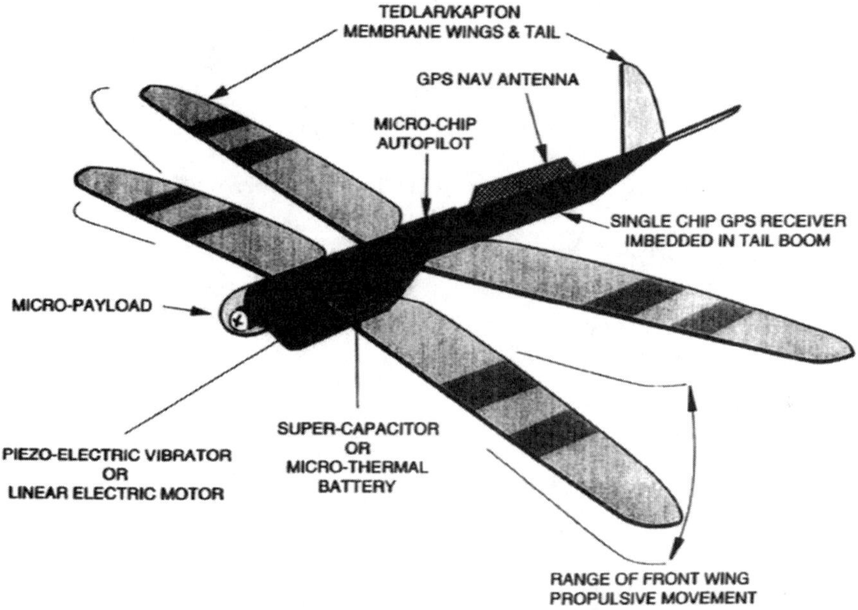

Fig. 6.12 Dragonfly concept.

NRL evaluated a dragonfly design as part of the MAV program. Ornithopters have advantages over propellers at small scales, due to a higher efficiency and reduced acoustic signature. Dragonflies shed a vortex off the front wing, which is then captured by the rear wing to improve efficiency and help maneuvering. The concept Dragonfly MAV would have a span of 3 to 5 inches (Fig. 6.12). This was not taken beyond a conceptual design. Such a system would be easier to build today with nanofabrication technology.

The primary reason that the design was not taken further was that the technology had already been developed by the Central Intelligence Agency (CIA) in the 1970s (now declassified). The life-size Dragonfly "Insectothopter" was built and flown (Fig. 6.13). The mission was to emplace a microphone using laser guidance, though it was never used operationally. This Insectothopter vehicle made its way to Foch's office at the beginning of the MAV program. After seeing this work of art, Foch decided that the concept had already been proven, and he decided to pursue other vehicle designs.

Fig. 6.13 Legacy CIA Insectothopter.

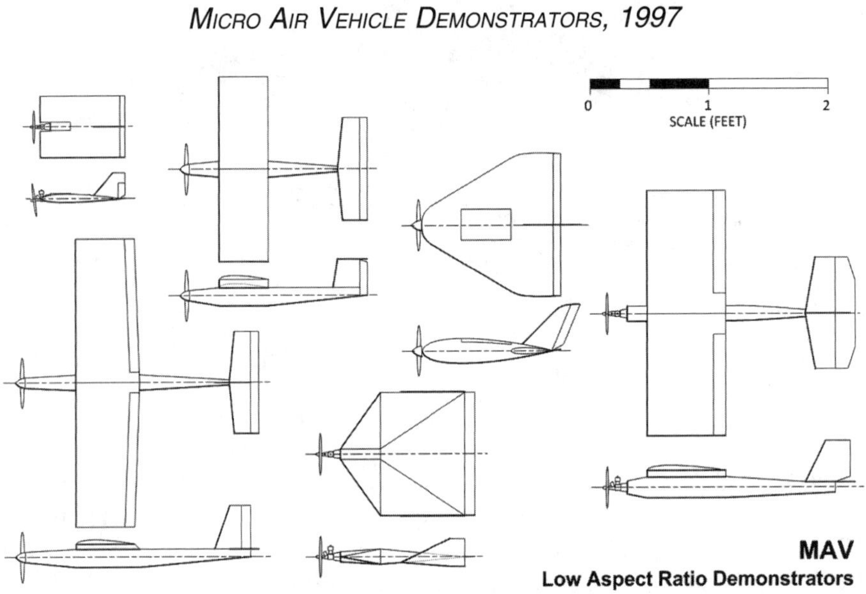

Fig. 6.14 MAV configurations.

In 1997, three VRS summer interns (including coauthor Gundlach) were tasked with building multiple small unmanned aircraft to demonstrate trade-offs in low aspect ratio airframe configurations (Fig. 6.14). The system technologies available at the time were not nearly as mature as those available today. Nickel cadmium and nickel metal hydride batteries were used, rather than the lithium polymer batteries commonly used today. Brushed ferrite motors without gear reduction provided propulsion. The motors were Graupner Speed 200 and Speed 300. Today, miniaturized brushless DC motors or coreless motors would be used. Other designs used

Cox 0.010- and 0.020-cubic-inch, 2-stroke reciprocating engines. The vehicles tested included a deltoid aerobody; three conventional configurations; a faceted design; and a small, low aspect ratio, rectangular-planform vehicle. The vehicles flew successfully, but the heavy propulsion and systems technologies limited the performance.

CUBE OF SKY, 1997

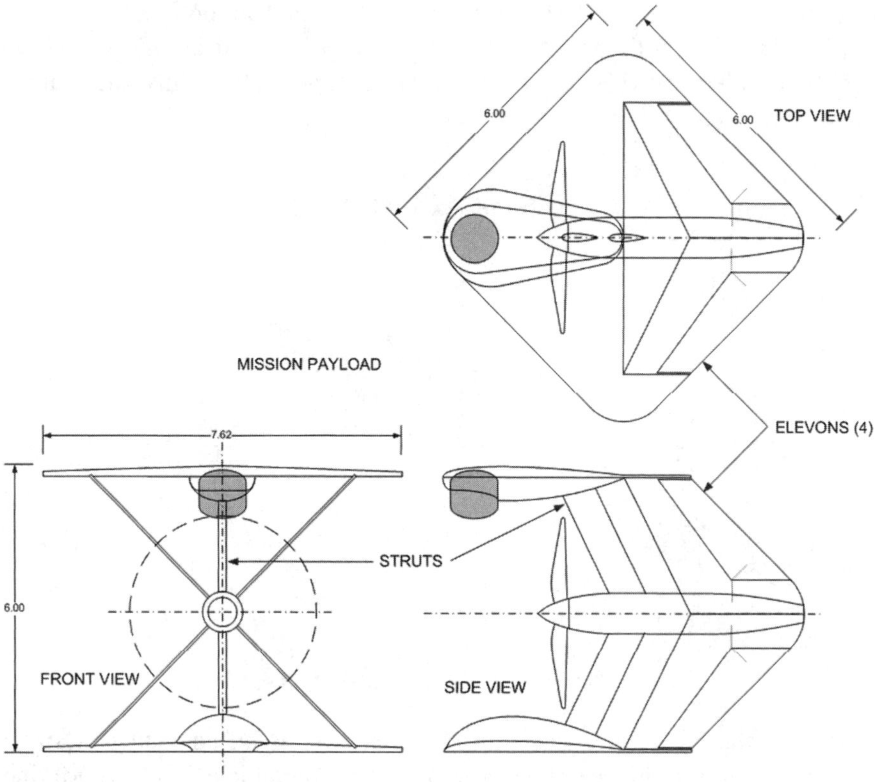

Fig. 6.15 Cube of Sky three-view drawing.

Cube of Sky was a design to maximize MAV performance for a single air vehicle that could be entirely contained within a 6-inch cube, per DARPA requirements (Fig. 6.15). The configuration is challenging to describe within traditional categories. It could be considered a biplane, a joined wing, and a strut-braced wing. The upper and lower wings were moderate aspect ratio delta wings. The mission payload was contained within a bulge below the upper wing. The aft of the upper and lower wings had X-tails that attached to a propulsion module located vertically halfway between the wings. The upper and lower wings also had more substantial vertical pylons that connected the wings with the propulsion module. The central propulsion module had a thrust line that did not generate a pitching moment, since it passed through the vertical center of gravity (CG) and center of drag. The central motor location could also support vertical takeoff and landing with a sufficiently powerful contrarotating propeller. Despite fitting within a 6-inch cube, the wings spanned 7.62 inches. The biplane wings increased the wing area, which reduced the wing loading for longer-endurance flight and greater lifting capacity. Given the low aspect ratio and numerous intersections, the design had a poor L/D. However, most MAVs had poor aerodynamic performance. Free-flight Cube of Sky gliders were built and flown. These flights revealed that this configuration had exceptional stability and control characteristics.

NEXUS SURROGATE MAV, 1999

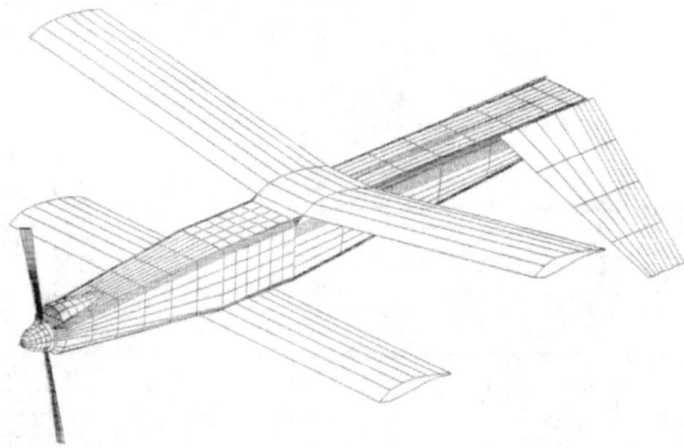

Fig. 6.16 Nexus Surrogate MAV design.

Part of the Nexus design study (Chapter 5) was an air-launched small unmanned aircraft that could be carried by the host vehicle. The U.S. Marine Corps (USMC) later became interested in a variant that could be launched by a grenade launcher. The solution was a simple folding wing design. The wings could be characterized as a highly staggered biplane. Both wings could develop the optimum lift coefficient and be longitudinally trimmed by an inverted V-tail (Fig. 6.16). The vehicle was roughly the size of today's AeroVironment Switchblade. Figure 6.17 shows a three-view drawing of this MAV, and Table 6.4 lists its technical characteristics. This Nexus MAV effort was superseded by the Dragon Eye program.

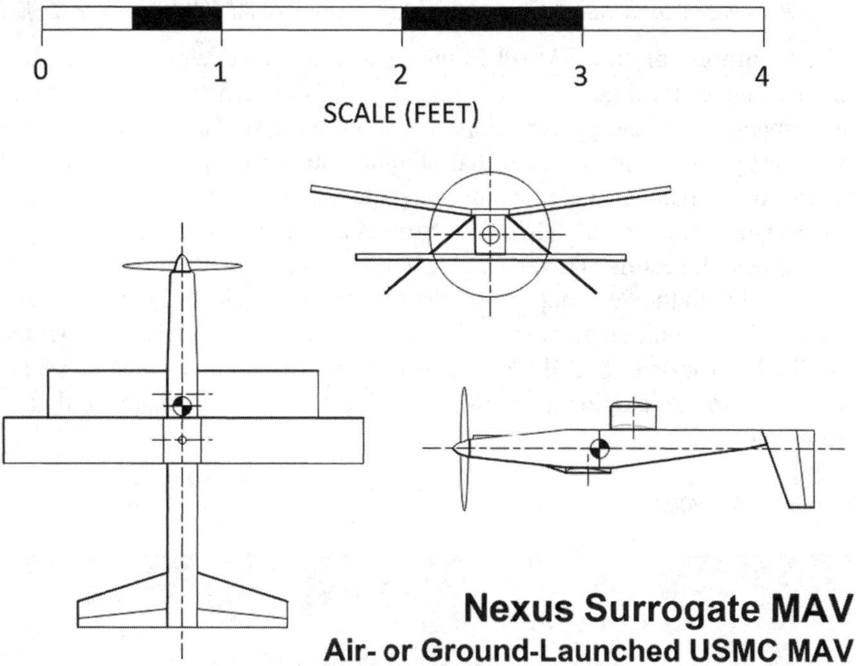

Fig. 6.17 Nexus MAV three-view drawing.

Table 6.4 Nexus Surrogate MAV Technical Characteristics

Parameter	Value
Wing span	2.0 ft
Wing area	0.88 ft^2
Length	2.0 ft
Takeoff weight	1.5 lb
Payload weight	0.3 lb
Battery weight	0.7 lb
Loiter speed	50 KTAS
Dash speed	65 KTAS
Endurance	10 min
Max L/D	6
Airfoils	Wortmann FX 63-137
Launch method	Air drop or grenade catapult
Recovery method	Belly landing
Propulsion type	Battery electric
Power	100 W

SEMI-AUTONOMOUS MOBILE PLATFORM EXPENDABLE (SAMPLE), 2002–2004

The purpose of the SAMPLE program was to develop vehicles that could achieve multiple forms of locomotion. One form would be a high-performance/high energy consumption mode while the second would be a low energy consumption mode that would enable energy replenishment or enhanced survival. A single locomotion system could work in multiple ways. The Samara, which could hop off the ground and transition to forward flight; the Biplane Insectoid Travel Engine (BITE)-Wing, which could fly and crawl; and both the Pectenopter and Delphinopter, which could fly and swim using different, unique methods of locomotion, were all developed under the SAMPLE program. SAMPLE's dual-locomotion premise was also the motivation for developing the Sail-A-Plane demonstrator, designed to fly and sail (Chapter 4).

SAMARA, 2002–2003

Fig. 6.18 Samara.

Samara was a vehicle that could fly horizontally and vertically like a helicopter, fly as a fixed wing aircraft, and perch (Fig. 6.18). The MAV would support operations in a constrained urban environment. It used an innovative stop rotor that could transform between a rotor and a fixed wing. Two single-blade rotors were attached at the root to a coaxial shaft. These contrarotating blades spun in opposite directions for rotary wing flight. The blades could stop in flight to become a conventional wing. The stopped blades formed a right wing and a left wing. By using two single-blade rotors, the stopped rotors were properly oriented as wings such that the leading edges faced forward.

In stop-rotor designs that use two-blade propellers, one of the wings has to operate with airflow reversal over the airfoil. Airfoils do not operate well with flow reversal, due to flow separation along the blunt trailing edge. Active boundary layer control techniques such as suction and blowing are commonly used to reduce drag and improve wing lift. This flow control increases system complexity and depletes energy. The Samara stop-rotor design did not require any active boundary layer control.

The first Samara demonstrator flew under radio control. It weighed 300 grams. It had a conventional tail, with the vertical stabilizer protruding downward. A propeller spun around the tail boom upstream of the tails to provide primary thrust. The blowing from the propeller on the tails provided augmented pitch and coupled yaw-roll control. The downward-oriented vertical tail arrangement provided yaw and roll control in a complementary fashion to turn the aircraft in the same direction.

The Samara 2 design was a tandem wing configuration (Fig. 6.19). The rear wing was a stop rotor, and the forward wing was a tilt wing using two tip propellers. The tip propellers provided thrust for forward flight. In vertical flight, the wing was tilted up, and the propellers provided vertical lift. Another Samara design concept, which included forward flight and hover configurations, was investigated (Fig. 6.20).

Samara's stop-rotor technology was a simple approach for MAVs, but not scalable to tactical and large UASs, which require a much more sophisticated and complex technology; this is discussed in the Stop Rotor vehicle description (Chapter 8).

Fig. 6.19 Samara 2.

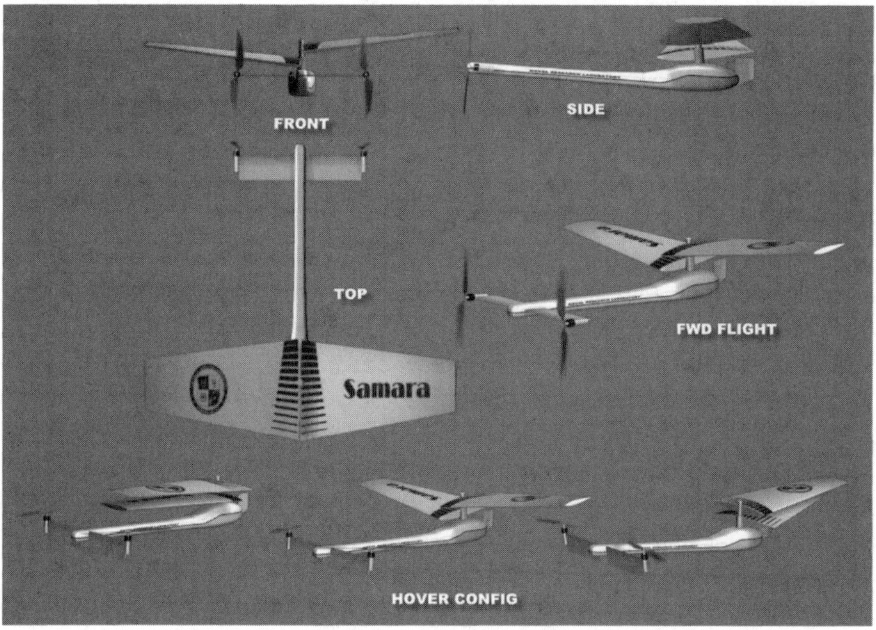

Fig. 6.20 Samara variant.

BIPLANE INSECTOID TRAVEL ENGINE (BITE)-WING, 2003–2004

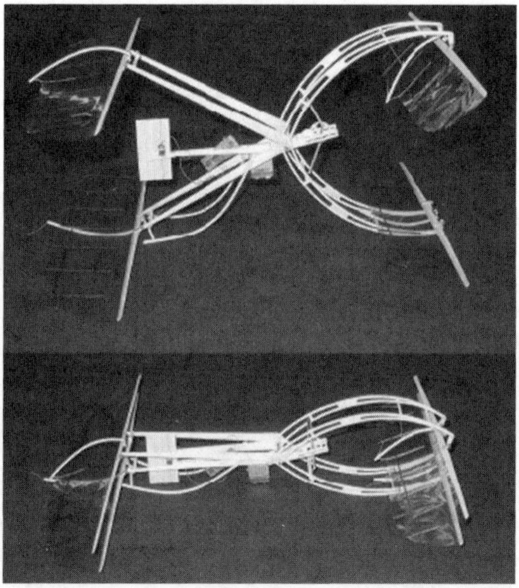

Fig. 6.21 BITE-Wing.

BITE-Wing was developed under the SAMPLE program as a vehicle that could achieve forward flight, hover, crawl on the ground, and hop. BITE-Wing had two pairs of biplane wings, a front pair and a rear pair, arranged in a tandem configuration (Fig. 6.21). Each wing pair had opposite camber. The wings were mounted to beams with a single pin joint in the center, much like a tong. The upper forward wing and lower rear wing were on one arm, and the lower forward wing and upper rear wing were on the other arm. The beams had equal mass and moved in opposition, so the vehicle was dynamically balanced in flight. A motor drove the mechanism and generated the oscillatory motion. Each wing in a wing pair moved toward and away from one another in an oscillatory, or flapping, manner. The forward and rear wing pairs opened and closed at opposite times. The wing pair clapped together when closed and were then separated. This produced a rearward-positive pressure pulse and forward-oriented negative pressure pulse that generated forward thrust. This mode of propulsion was inspired by insect flight (Figs. 6.22 and 6.23). The wings provided both lift and propulsion for flight.

Flapping wing flight has advantages over traditional segregated wing and propulsion flight for low Reynolds number applications. Such vehicles ideally can measure under 4 inches in span or length. The BITE-Wing demonstrators were considerably larger, spanning 15 to 20 inches, driven by available construction techniques for the airframes and mechanisms.

The BITE-Wing demonstrators used wing-flapping techniques similar to those of birds and insects, but the flapping was actuated using mechanical

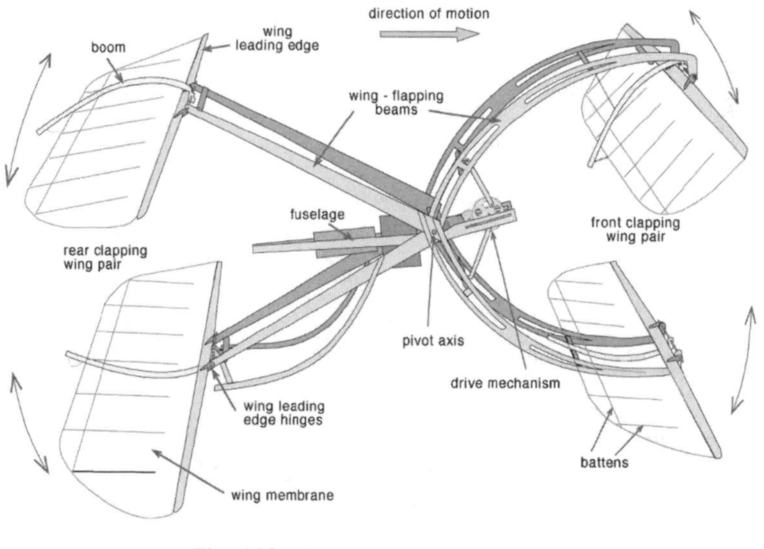

Fig. 6.22 BITE-Wing Mk 1 features.

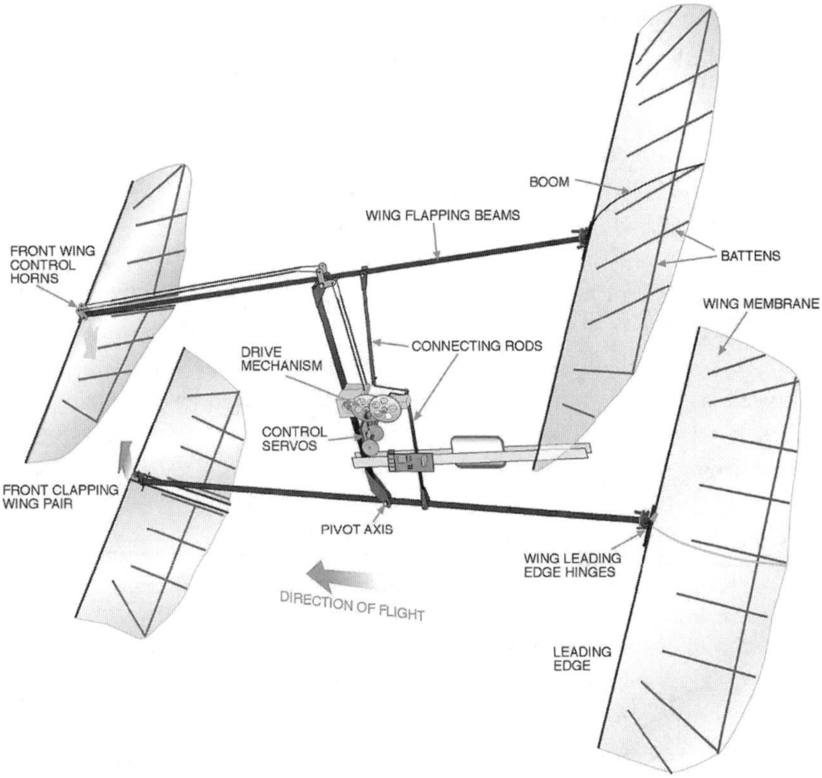

Fig. 6.23 BITE-Wing Mk 3 features.

Fig. 6.24 BITE-Wing moving on the ground.

Fig. 6.25 BITE-Wing Mk 4.

methods such as rotary motion and hinges rather than by mimicking the complex biological structures of bone and muscle.

For ground locomotion, the BITE-Wing vehicle "inchwormed" as the beams oscillated. The wings provided lateral stability to protect against side tip-over. For takeoff, the wing oscillated rapidly to allow the vehicle to hop off the ground. The RC prototype shown in Fig. 6.24 was 38 centimeters long and weighed 19.5 grams. Figure 6.25 shows another version of this vehicle.

Pectenopter, 2004

Fig. 6.26 Pectenopter.

The Pectenopter was a vehicle that could both fly and swim (Fig. 6.26). NRL developed this system in collaboration with Australia's Defence Science and Technology Organisation (DSTO). The forward wing was a fixed lifting surface. The two rear wings oscillated toward and away from one another in unison to provide propulsion, in a manner similar to the BITE-Wing. The Pectenopter successfully demonstrated flying in air and swimming in water. The propulsion system worked the same way in the air and under water, except the oscillation frequency was higher in the air due to the air's lower density.

DELPHINOPTER, 2004

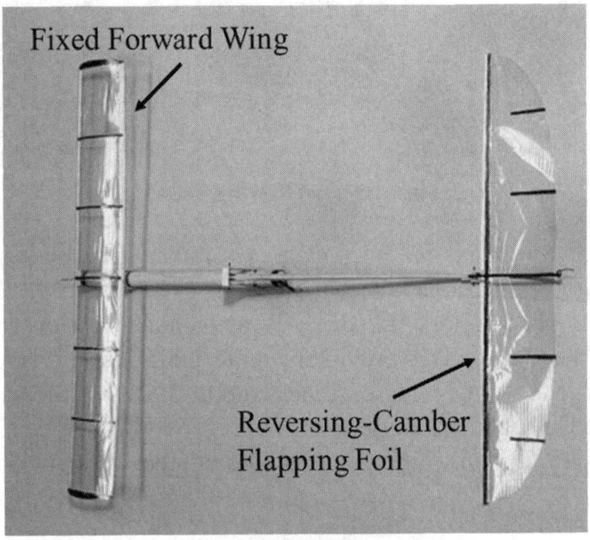

Fig. 6.27 Delphinopter.

Like the Pectenopter, the Delphinopter was a vehicle designed to both fly and swim (Fig. 6.27). The forward wing was a fixed lifting surface. The power source caused the fuselage to bend in the middle, driving the rear wing to oscillate up and down to provide propulsion in the same manner that a dolphin swims. In essence, the Delphinopter swam through the air and under the water in an identical manner, but at a lower oscillation frequency under water. Delphinopter flight testing was successful, but no underwater tests were performed since the main technical objective was to demonstrate swimming propulsion in the air.

SAMPLE: FURTHER READING

Kellogg, J.C., et al., "Design and Development of the Samara Stop-Rotor Hybrid Micro Air Vehicle," *20th International Bristol UAV Systems Conference*, April 2005.

Kellogg, J.C., et al., "Development and Testing of Unconventional Micro Air Vehicle Configurations," AIAA Paper 2003-6656, presented at the *2nd AIAA "Unmanned Unlimited" Systems, Technologies, and Operations Conference*, San Diego, CA, Sept. 15–18, 2003.

Ramamurti, R., Sandberg, W., Vaiana, P., Kellogg, J., and Cylinder, D., "Computational Fluid Dynamics Study of Unconventional Air Vehicle Configurations," presented at the *19th Bristol International Conference on Unmanned Air Vehicle Systems*, Bristol, U.K., March 29–31, 2004.

Close-In Covert Autonomous Disposable Aircraft (CICADA), 2004–2011

Fig. 6.28 CICADA I air-launched swarm concept.

The MITE program showed the potential of MAV systems. Following the MITE program, NRL conducted the CICADA program to explore the potential of these systems for EW and other missions.

A tactical aircraft could rapidly seed an area with CICADA MAVs (Fig. 6.28). Each CICADA would navigate independently using a simple GPS autopilot. The swarm would land in an area and form a dense sensor grid. The vehicles could operate independently or as a network, depending on the mission. The networks could be ad hoc or self-configuring. The sensors could be placed in a preprogrammed geometric pattern. The missions could include forming a communications grid, forming an unattended ground sensor network, or acting as radar jammers. Like the insect namesake, CICADA would be deployed in very large numbers. The host aircraft would not need to fly directly over the target area, thus improving its combat survivability and not alerting the target. This program was likely the first exploitation of the swarm concept.

CICADA I

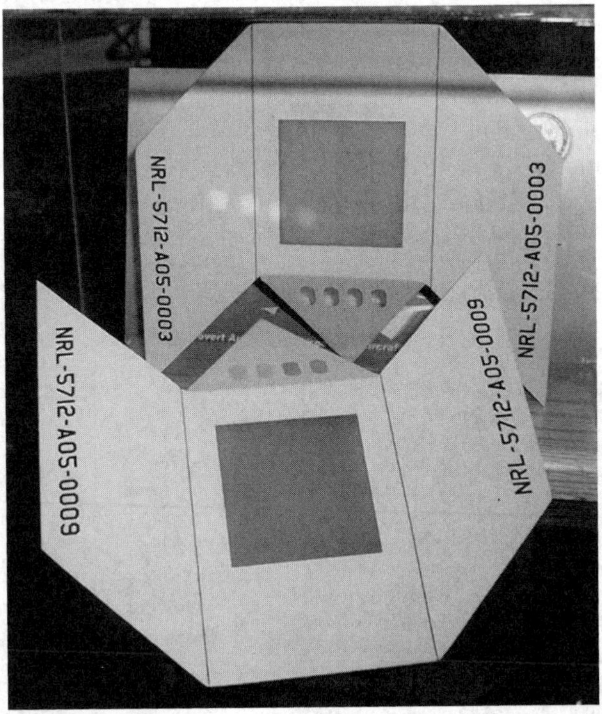

Fig. 6.29 CICADA I vehicle.

In 2004, authors Foch and Gundlach were walking through a parking lot during the 17-year cicada infestation event. The large insects were everywhere, and it was difficult to not step on them. The authors discussed how powerful it would be if these insects had cameras, microphones, and networked communication systems. A few days later, Foch showed Gundlach a flying model of the CICADA I (Fig. 6.29).

Packaging efficiency was important, because CICADA was intended to be deployed as a swarm. A 6-inch wingspan was sufficient for many applications, and if these vehicles were 1 inch tall, then six could fit in a cube with sides measuring 6 inches (Fig. 6.30). Such a design would leave considerable wasted volume within the cube, however, so Foch evaluated concepts that would enable much higher packaging efficiency for these micro gliders.

The CICADA I design had a very high packaging efficiency, such that 18 could fit within the 6-inch cube. This was a factor of 3 improvement over traditional MAV designs. The fuselage had a triangular cross section. The wings folded along the fuselage sides for storage, and sprung open for flight. The wing consisted of a flat plate with a straight inboard section above the

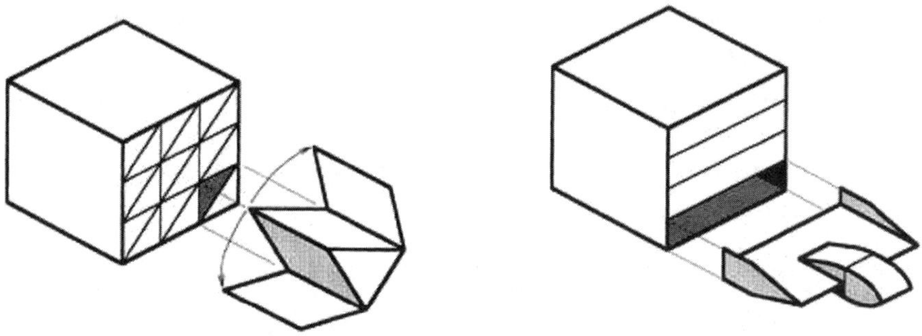

Fig. 6.30 CICADA I (left) and traditional 6-inch MAV (right).

fuselage and swept outboard panels. The front and rear of the fuselage were swept aft such that the vehicles nested longitudinally for storage. The wing outboard panels matched the fuselage side geometry when folded. The triangular fuselage housed the payload.

CICADAs would deploy in the thousands for a single operation, so the systems had to be producible at very low cost. A single Sargent Fletcher pod on a C-130 could disperse more than 9000 CICADAs, for example. Each unit should cost less than $100 to produce. To reduce costs, the wings consisted of a printed circuit board, which was populated with the processors, microelectromechanical system (MEMS) inertial sensors, GPS, battery, payload, and communication system. The fully automated assembly would be performed by robots. Alternatively, the vehicles could be built from a single sheet of metal or paper with simple folds rather than a printed circuit board (Fig. 6.31).

The CICADA I basic flying characteristics were evaluated on small, balsa, free-flight gliders. Initially the fuselage was going to be on top of the wing. However, flight tests revealed that the aircraft consistently rolled inverted. The aircraft trimmed well inverted, so that became the baseline configuration. The gliders could also attain stable flight after tumbling. This behavior greatly simplified the demands placed on the flight control system for air deployment. Also, this self-righting behavior enabled the CIDADAs to recover after a midair collision, so tracking other aircraft in the swarm was not necessary.

The gliders had predictable orientations after landing: either inverted on the flat wing or nose down on the forward face of the fuselage. This enabled antennas and sensors to be placed on the vehicle with a high probability of predictable orientation on the ground. The glide ratio was found to be 3.5:1.

The initial design used flaps on either side of the fuselage for directional control. The vehicle was inherently stable, and it was designed to fly at a

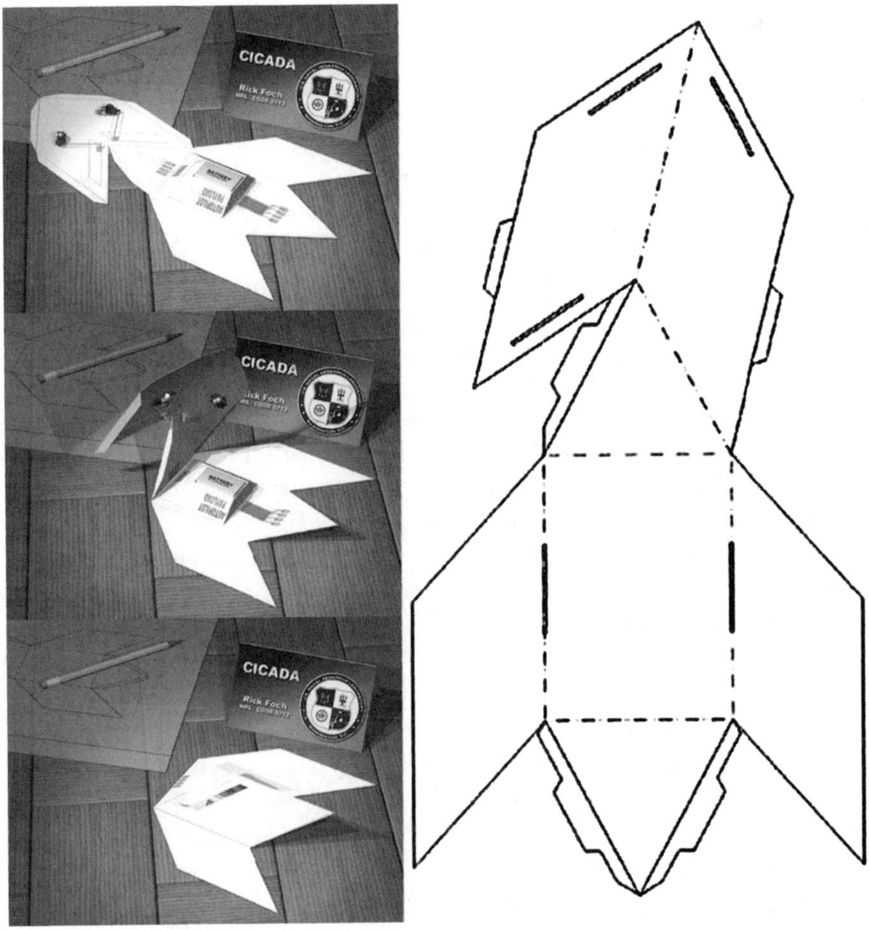

Fig. 6.31 Folded CICADA I arrangement.

single pitch setting. No longitudinal control was necessary. Radio-controlled drop models of 3 and 4 times scale were built to test the control system. The models were equipped with control vanes and split flaps. The flight tests revealed that the CICADA I design was so stable that it was hard to steer. The next approach was to trip the flow over the wings asymmetrically with spoilers to provide roll control. The directional control was vexing, and no suitable solution was found for the CICADA I design. It is possible that a viable directional control solution can be found with further development.

CICADA II

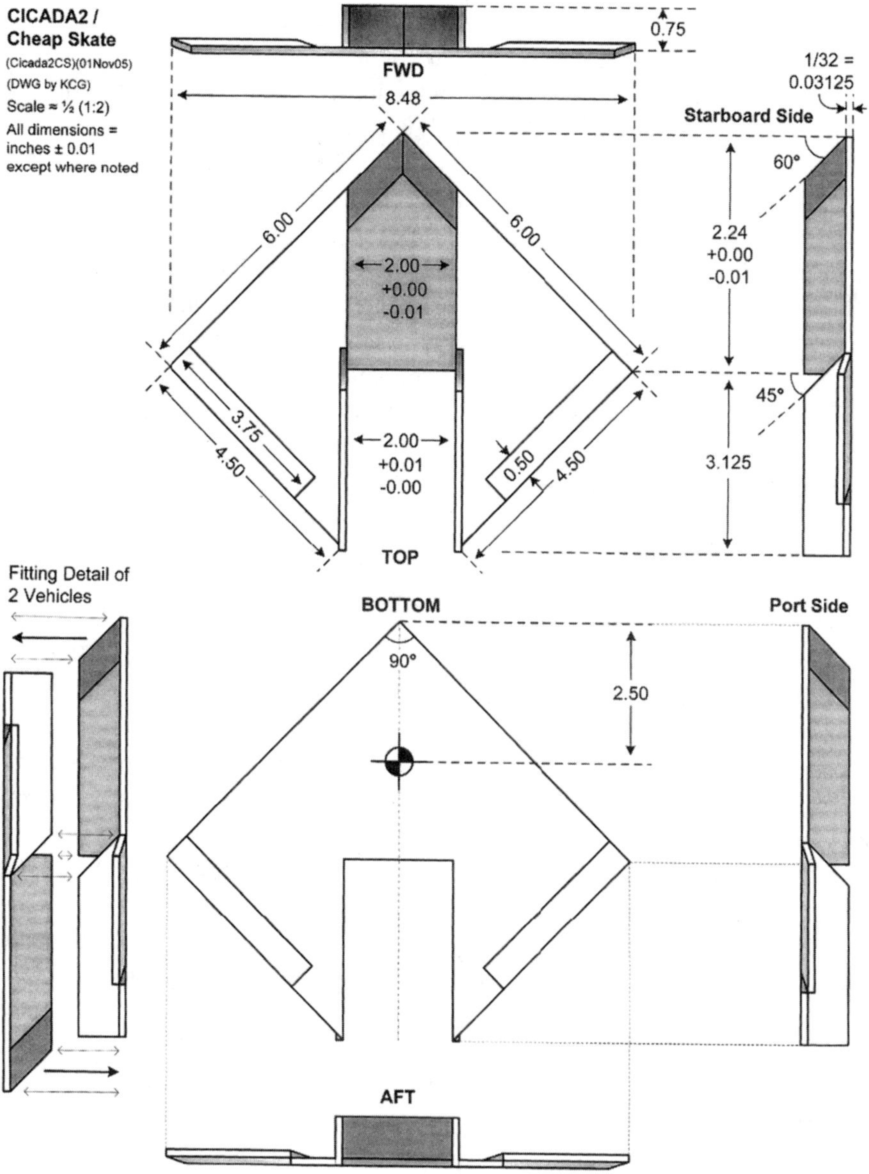

Fig. 6.32 CICADA II.

Armed with lessons learned from CICADA I, designers developed a new configuration. CICADA II was a nonfolding, thin vehicle that resembled a flying plate (Fig. 6.32). The fuselage was thick enough to house the electronics. This new design had control surfaces in the wing but was still constructed of a printed circuit board.

The CICADA II design was progressively tested as free-flight gliders, as RC super-scale drop models, as wind tunnel testing models, and then as autonomous vehicles. The wind tunnel testing was necessary to generate the aerodynamic database for the control law development.

The configuration was designed so that two vehicles nested together back to back and with noses pointing in opposite directions. The planform was a diamond, with 90-degree angles at all corners; this allowed the units to be stacked within a cube. A square notch was cut out of the back of the diamond, which enabled the fuselages to nest together. Vertical plates equal to the fuselage depth extended upward and lengthwise to provide directional stability. An elevon control surface was mounted on each side for roll control.

The CICADA II vehicles were designed to have electrical and data connectivity between vehicles when stored. Each vehicle had a unique IP address. A mission management system in the container would send each vehicle a target coordinate. Prior to launch, the mission planning system would determine if the vehicles had sufficient glide range to reach the objective. Relevant factors included launch altitude, target altitude, flight path, and winds aloft. The mission planner was based on Dragon Eye technology. Once the coordinates were passed to the CICADA swarm, the vehicles would be released. The mission planning capability was not fully implemented for a swarm in this exploratory development program. Vehicle functionality, including self-navigation to a preprogrammed coordinate, was demonstrated with a small number of units, and the swarm behavior was simulated.

CICADA II was considered for a gun-launched application. The round would burst and spew out multiple CICADAs. A key challenge was surviving the 30,000 to 60,000 G shock loads. The servo actuator was considered to be the hardest challenge, since all the avionics were solid-state printed circuit boards. After numerous commercial off-the-shelf servos and more exotic actuator designs were evaluated, a single, low-cost model aircraft micro servo product that survived the shock testing was found. A clipped-wing CICADA II was mounted inside a 155-mm round and fired downrange; the reduced span was necessary to fit within the round without further vehicle development. The vehicle could not fly due to this modification, but it successfully deployed from the round and landed, and the payload and controls functioned appropriately. This may have been the first unmanned aircraft deployed from an artillery round.

CICADA III

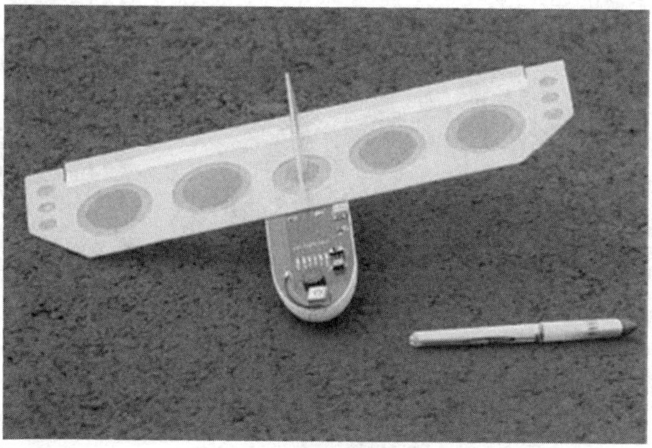

Fig. 6.33 CICADA III.

CICADA III originated from a sponsor request to deploy a small aircraft from high altitude to disperse sensors. This design was larger than previous CICADAs, since there was no size requirement. It did not need to package in large quantities, so the geometry was not constrained. Like its predecessors, CICADA III was constructed of printed circuit boards. The wing configuration was an evolution of the Dragon Eye plank flying wing configuration (Fig. 6.33). Elevons were used for control.

The design had thermal challenges. The electronics and batteries could get too cold after cold soaking at high altitude for an extended time. An insulated fuselage fairing was added, as well as heating for the electronics.

For deployment, two CICADA III vehicles were mounted under the wings of a Tempest unmanned electric sailplane mother ship (made by UASUSA of Boulder, Colorado). Tempest was carried above 50,000 feet by a weather balloon. Tempest released from the balloon, flew for a distance, and then launched the CICADAs (Fig. 6.34). The mother ship had a higher L/D than the CICADAs, so it could penetrate farther before dropping the smaller vehicles. It then served as a communications relay in operations. The CICADAs landed within 15 feet of the programmed recovery point during the demonstrations. Figure 6.35 shows a three-view drawing of the CICADA III, and Table 6.5 lists the technical characteristics for the three CICADAs.

CICADA: FURTHER READING

Anon., "Autonomous Deployment Demonstration Program Completes Flight Testing," *Naval Research Laboratory SPECTRA: The Magazine of the Navy's Corporate Laboratory*, Washington, DC, Spring 2012.

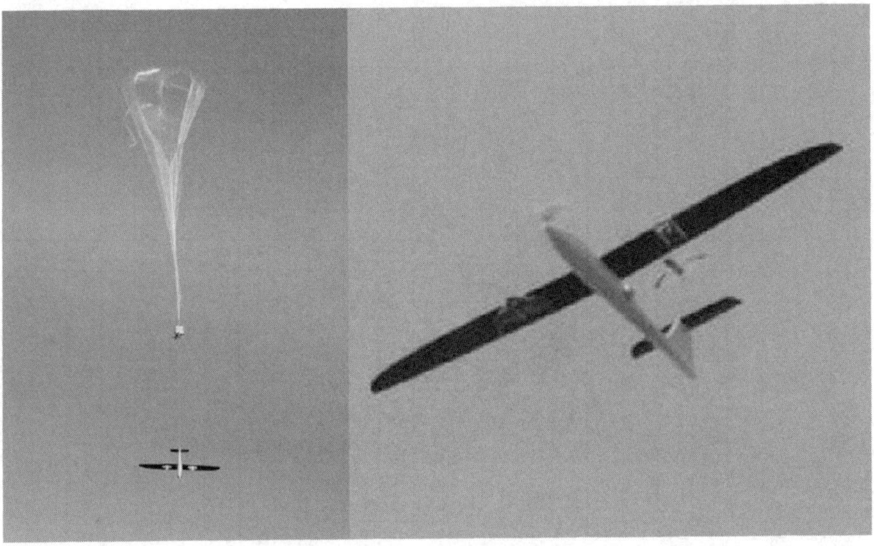

Fig. 6.34 Tempest carried aloft by a weather balloon (left); CICADA III launching from the Tempest mother ship (right).

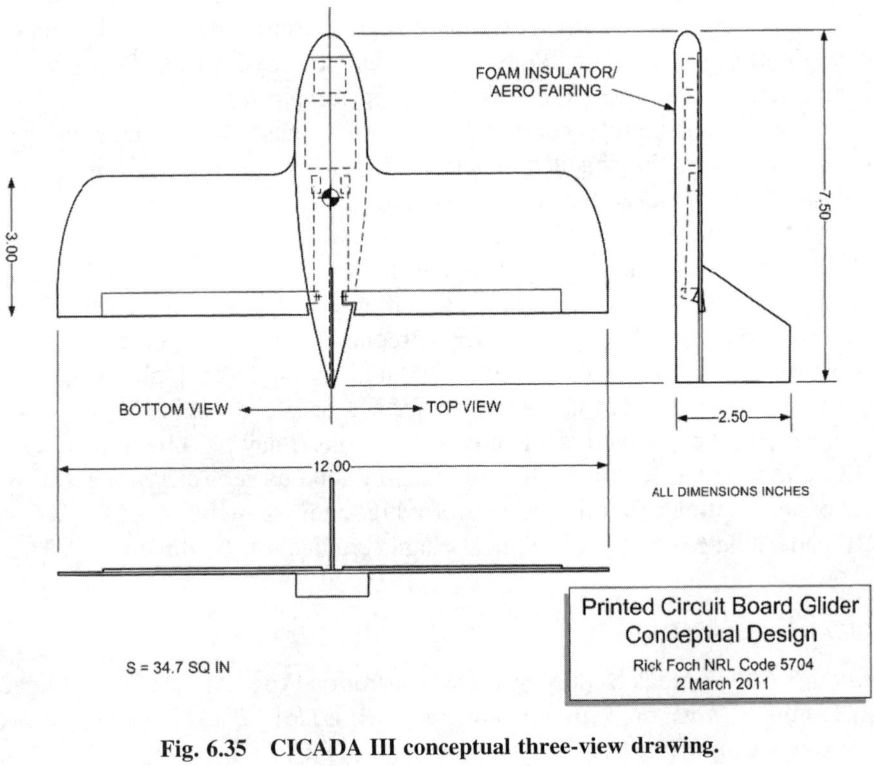

Fig. 6.35 CICADA III conceptual three-view drawing.

Johnson, E.N., Fontaine, S.G., and Kahn, A.D., "Minimum Complexity Uninhabited Air Vehicle Guidance and Flight Control System," *Proceedings of the 20th Digital Avionics Systems Conference*, IEEE, 2001.

Kahn, A.D., and Edwards, D.J., "Design Evolution of a High Packaging Density Micro Air Vehicle for Local-Area Seeding," AIAA Paper 2012-4735, presented at *AIAA Guidance, Navigation, and Control Conference*, Minneapolis, MN, Aug. 13–16, 2012.

Kahn, A.D., and Edwards, D.J., "Navigation, Guidance and Control for the CICADA Expendable Micro Air Vehicle," AIAA Paper 2012-4536, presented at *AIAA Guidance, Navigation, and Control Conference*, Minneapolis, MN, Aug. 13–16, 2012.

Kahn, A.D., and Foch, R.J., "High Packaging Efficiency EAV for Local Area Seeding," *Joint Electronic Warfare Conference*, 2006.

Kahn, A.D., and Kellogg, J.C., "A Low-Cost, Minimum Complexity Altitude/Heading Hold Flight Control System," *21st Digital Avionics Systems Conference*, 2002.

Table 6.5 CICADA Technical Characteristics

Parameter	CICADA I Value	CICADA II Value	CICADA III Value
Wing span	6.8 in.	8.48 in.	12 in.
Wing area	27.2 $in.^2$	29.0 $in.^2$	34.7 $in.^2$
Length	6.0 in.	5.37 in.	7.5 in.
Takeoff weight	4.0–8.0 oz	3.0–4.5 oz	4.0–6.0 oz
Payload weight	0.5–1.5 oz	1.0 oz	1.0–2.0 oz
Max L/D	3.5	3.7	4.5
Airfoils	Flat plate	Flat plate	Flat plate
Launch method	Air launch	Air launch	Air launch
Recovery method	Expendable	Expendable	Expendable
Propulsion type	Glider	Glider	Glider

Chapter 7

Utility Aircraft

These utility aircraft are the least glamorous unmanned aircraft systems (UASs) used by NRL, but they are crucial for rapid technology advancement. They are workhorses that are easy to maintain and operate, and they support the development of key subsystems or new operations that would be too risky to first fly with the prototype high-tech aircraft. Most of NRL's utility aircraft are commercially produced designs that are modified as required. A few have been custom designed and built by NRL when no commercial design was readily available. All are highly cost-effective research tools.

Arsea, 1981–1982

Fig. 7.1 Arsea in flight.

NRL needed a utility aircraft to test avionics and to break in small engines. A plastic Almost Ready to Fly (ARF) kit called a Slo-Comet by Lanier RC was procured and quickly assembled. The Slo-Comet had nice flying characteristics, but its wing proved to be too weak for carrying additional payloads. NRL's VRS decided to design and construct a utility aircraft customized for the envisioned test roles. The result was a quickly built, easily maintained, simple radio-controlled (RC) model that was named Arsea (Fig. 7.1).

"Arsea" is a humorous enunciation of the acronym RC, where the spelling of "C" as "sea" seemed appropriate for a Navy project. The Arsea airframe was constructed of balsa wood and Sitka spruce with MonoKote film model airplane covering. It had a high wing and conventional tail. It featured rudder, elevator, and throttle controls. The wing was given sufficient geometric dihedral to provide strong inherent lateral stability to assist the pilot when flying near the limits of visual range. The dihedral also enabled turns to be easily performed using rudder control, so ailerons were unnecessary. Arsea spanned 6 feet with a 1-foot chord. It used the RF 1165FB airfoil, which has good aerodynamic characteristics in the Arsea's chord Reynolds number range. The landing gear had a tricycle configuration. The tractor propeller was powered by an O.S. 45 FSR 2-cycle engine. The completed aircraft flew as expected and was a reliable test asset.

About halfway through its lifespan, Arsea was modified for radar calibration testing. Two side-looking Lunenburg lens radar reflectors were installed in a fairing beneath the fuselage, with one facing each side, staggered to minimize installed width and depth. Lunenburg lenses are composed of a radome with layers of dielectric foams. These devices focus the radar beam onto a parabolic reflector and have radar cross sections larger than their physical size. The Arsea radar reflectors measured 4 inches in diameter but yielded the Radar Cross Section (RCS) of a 1- or 2-meter-diameter sphere, depending on frequency. The need was to help calibrate test radars at NRL's Chesapeake Bay Detachment (CBD).

CBD radar calibrations had previously been performed by weather balloons equipped with large, lightweight, spun aluminum spheres. The problems with balloons were that their flight path could not be controlled and they had weather launch restrictions. The miniature Lunenburg lens reflectors allowed the mission to be performed by a small unmanned aircraft.

Arsea flew perpendicularly to the radar beams in a racetrack pattern. Since there was a reflector on each side of the aircraft, calibrations could be performed on each leg of the racetrack.

The reflector fairing was a lightweight foam and fiberglass canoe-shaped structure mounted on the lower side of the fuselage, giving the aircraft an ungainly appearance. The landing gear had to be extended to allow ground clearance. The landing gear configuration was changed from a nose wheel to a tail wheel to allow operation from unprepared grass fields near the radar site. The additional weight and drag of these modifications made the aircraft a marginal performer, even though the engine was upgraded from the O.S. 45 FSR to an Enya 90 4-cycle. To remedy this, the wing span was increased to 8 feet using plug-in wing tip panels. This decreased the wing loading,

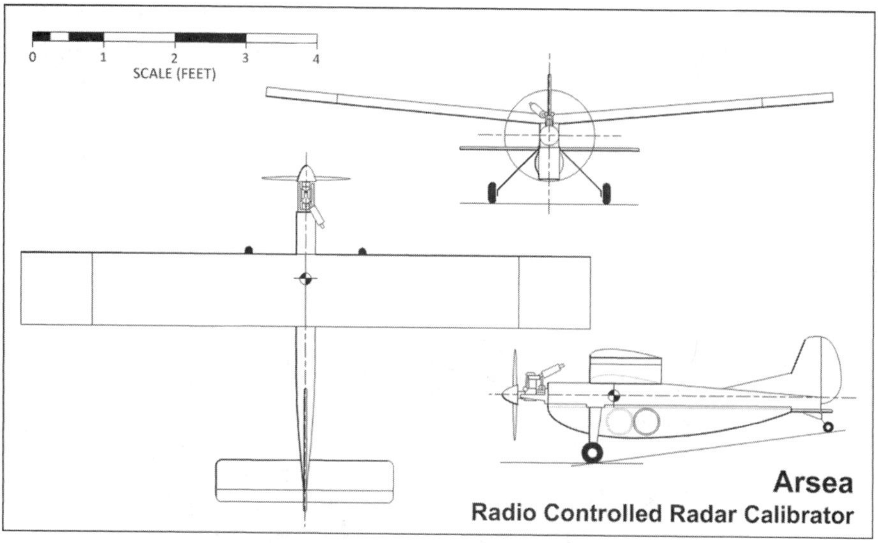

Fig. 7.2 Arsea three-view drawing.

reduced the span loading, and improved the lift-to-drag ratio (L/D). The vertical fin area was also enlarged to compensate for the longer wingspan and increased fuselage side area.

Radar calibration flight operations were made difficult by the airfield dependency. The aircraft was flown by RC, so the pilot needed to maintain visual contact. However, the airfield was located some distance from the radar range. The pilot had to be driven between the sites while maintaining control and keeping visual contact with the flying aircraft.

Table 7.1 Arsea Technical Characteristics

Parameter	Value
Wing span	8.0 ft
Wing area	8.0 ft^2
Length	4.7 ft
Takeoff weight	17.3 lb
Empty weight	9.3 lb
Cruise speed	40 KTAS
Endurance	1.0 hr
Launch method	Conventional
Recovery method	Conventional
Propulsion type	Reciprocating engine O.S. 45 FSR 2-cycle and Enya 90 4-cycle
Power	0.9 hp and 1.3 hp

The Arsea periodically performed radar calibrations, but it had a smaller radar signature than desired. As an upgraded radar calibration capability, the Arsea was replaced by the DARC, a larger and more capable system, discussed next. The Arsea pilots all have fond memories of flying this dependable aircraft. Figure 7.2 shows a three-view drawing of this utility aircraft, and Table 7.1 lists its technical characteristics.

DRONE AIRCRAFT RADIO CONTROLLED (DARC), 1982–1984

Fig. 7.3 Final DARC configuration.

Drone Aircraft Radio Controlled (DARC) was a dedicated radar calibration unmanned aircraft designed specifically for the mission (Fig. 7.3). It was a follow-on project to Arsea. DARC carried a larger Lunenburg lens payload measuring 10 inches in diameter, which generated a 10-square-meter radar cross section to simulate tactical aircraft. This radar reflector payload weighed 12 pounds. Unlike the Arsea, DARC carried a single reflector in the nose rather than one on each side. DARC flew parallel to the beam rather than perpendicular to it. This flight profile allowed the radar to calibrate to different ranges. The operational limitation of this payload arrangement was that calibration could occur only when the aircraft flew toward the radar. The wings spanned 10 feet, so this aircraft was substantially larger than its predecessor.

The fuselage was tubular from the wing forward, and the payload radome formed a hemisphere at the nose (Fig. 7.4). The fuselage aft of the wing tapered back to the tail. The engine and pusher propeller were located at the back of the fuselage. The stabilizers had a T-tail configuration. The vertical tail was swept back such that the horizontal stabilizer rested over the

Fig. 7.4 Original DARC configuration.

propeller. The configuration closely resembled the Prescott Pusher general aviation design.

The fuselage had a fiberglass skin and aircraft plywood bulkheads. The fiberglass was wrapped around a 0.5-inch-thick foam shell that was hot-wired from a larger block. The wings had foam cores with 1/64-inch bonded plywood skins. The tail booms were made of the type of poplar plywood used for hollow core interior house doors. The tails were foam core with balsa covering. The aircraft was finished with polyurethane enamel paint.

There were few suitable reciprocating engines for the program. The first power plant selection was the O.S. 61 2-stroke model aircraft engine, which is in the 2-horsepower class. It used high-nitro glow fuel, and it was fed by an electric fuel pump to get maximum performance. The aircraft experienced continuous engine problems during taxi tests. The problems were related to fuel oscillations in the long fuel lines between the central fuel tank and the pusher engine, which were partially caused by the fuel pump. The aircraft was also underpowered.

A larger engine was required for adequate performance. The final engine choice was the O.S. 1.08, which produced about 3 horsepower. However, this engine was significantly heavier than the O.S. 61. The additional weight at the rear of the fuselage created an unacceptable aft center of gravity (CG) shift. The location of the payload in the nose could not be altered, and the most forward ballast location did not have a large moment arm from the desired CG location. The short nose arm would have required so much ballast that the aircraft would no longer be able to fly.

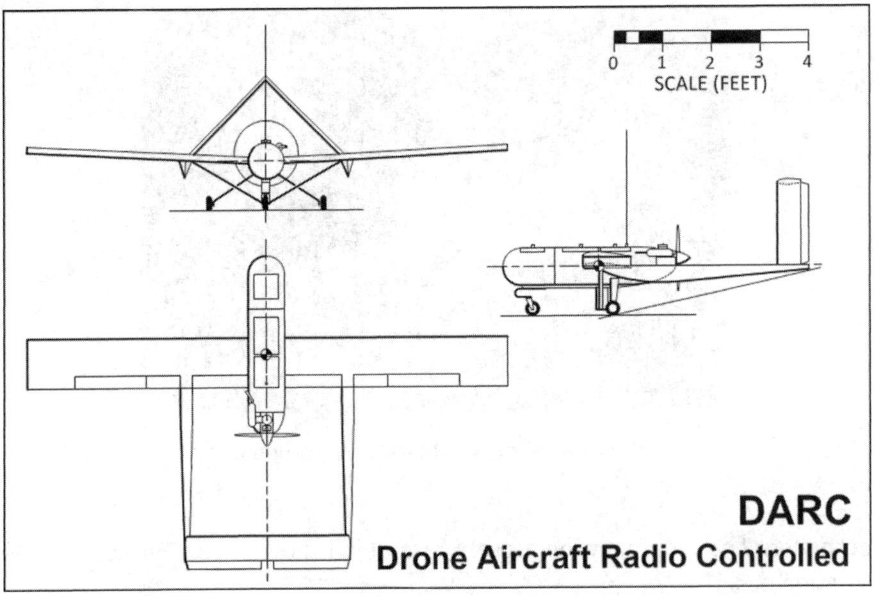

Fig. 7.5 DARC three-view drawing.

Drastic measures were in order to solve the aft CG problem. The fuselage had a bulkhead just aft of the wing trailing edge. The fuselage was sawed off just aft of the bulkhead, and the bulkhead then served as a firewall for the engine. This location had a much shorter moment arm from the CG to the engine, which solved the balance problem. In addition, the larger engine located closer to the fuel tank did not require a fuel pump. Two booms extended aft from the wings. The tail surfaces connected to the booms in an inverted V-tail configuration, with 45-degree anhedral angles. Thus modified, DARC had a configuration similar to today's AAI Shadow 200.

The twin-boom variant flew successfully. The final aircraft weighed 40 to 45 pounds, so it had a higher wing loading than similarly sized model aircraft. Like Arsea, this aircraft was flown under standard radio control. A simple Kraft FM radio system was used.

DARC never flew any radar calibration missions. The radars were deactivated for maintenance for several years. When they were reactivated, alternative calibration techniques employing ships and aircraft were available. The DARC sponsor also retired during this time, so the program no longer had a champion. Figure 7.5 shows a three-view drawing of this utility aircraft, and Table 7.2 lists its technical characteristics.

Table 7.2 DARC Technical Characteristics

Parameter	Value
Wing span	10 ft
Wing area	10 ft^2
Length	6.3 ft
Takeoff weight	33 lb
Payload weight	12 lb
Minimum speed	28 KTAS
Maximum speed	50 KTAS
Launch method	Conventional
Recovery method	Conventional
Propulsion type	O.S. 1.08-cu.-in., 2-cycle
Power	3 hp

Baseline Aircraft Radio Controlled (BARC), 1985–1990

Fig. 7.6 BARC.

The purpose of the Baseline Aircraft Radio Controlled (BARC) utility aircraft was to serve as a propulsion test bed for the LAURA program (Chapter 4). The BARC airframe was a modified, commercially available, giant-size RC aircraft (Fig. 7.6). The aircraft was purchased Almost Ready to Cover (ARC), which was a novelty in the RC model airplane hobby business at the time. Today, numerous model aircraft of all sizes are sold as ARF, Ready To Fly (RTF), and various levels in between, including ARC. The commercially available kit was named "Big Dog," but after extensive modifications, NRL decided that BARC was a more appropriate name.

The aircraft had a low wing and a conventional tail. The wings had a nearly flat bottom semisymmetrical airfoil and were made of foam cores with balsa covering. The kit had a canopy on top, but NRL removed this and added a large fairing to house larger fuel tanks and instrumentation. The aircraft had a tail wheel–type landing gear arrangement.

The airframe was modified to carry a larger engine. The nose length was reduced to maintain the required CG position with the increased engine weight. The wing center section structure was strengthened. The fuselage was covered with fiberglass cloth. The tails were given aluminum struts because the original tail structure would not have been able to handle the increased loads of the now much heavier and faster aircraft. To avoid flutter, all control surfaces were statically balanced. The landing gear and its mounting locations were strengthened as well.

Two candidate LAURA engines were initially considered for flight experimentation. The first was World Engines' Tartan Twin 2-cycle, but this engine was in short supply and the distributor could not provide a guaranteed delivery schedule. The next engine was the O.S. FT-300 twin-cylinder 4-stroke, which had lower fuel consumption but was slightly heavier. The O.S. engine was selected to be the first engine flown on the BARC.

The aircraft had a relatively high wing loading (by model airplane standards) of about 35 to 40 ounces per square foot. The airframe was originally designed for an engine with one third of the power. BARC had flying characteristics much like a giant-scale RC warbird such as a P-51 Mustang. This was due to the combination of high wing loading, high power-to-weight ratio, and tail wheel landing gear arrangement. The takeoff and landing speeds were fairly high. Plenty of right rudder deflection was required during takeoff to counter the slipstream effects. The aircraft had an abrupt stall with a strong left-wing drop that occurred with both power-on and power-off stalls, so care was necessary to avoid stalling during the landing approach. The NRL test pilots were experienced RC pilots and had little difficulty flying the BARC with its exciting model fighter-like handling characteristics.

The first flight was made on October 1985. It was equipped with a telemetry system and sensors that measured airspeed, fuel flow rate, and engine rpm. The test program characterized the O.S. FT-300 twin-cylinder 4-cycle engine and led to its adoption on the LAURA program. There were several dozen flights made by multiple pilots.

The aircraft continued to fly after the initial LAURA propulsion test flights. The Saito FA-300T twin-cylinder engine was considered as a backup LAURA power plant, so one was installed in the BARC. BARC continued to provide utility flights for several more years with the Saito engine before being retired. BARC was an inexpensive research aircraft that expeditiously utilized a modified model aircraft. Figure 7.7 shows a three-view drawing of this utility aircraft, and Table 7.3 lists its technical characteristics.

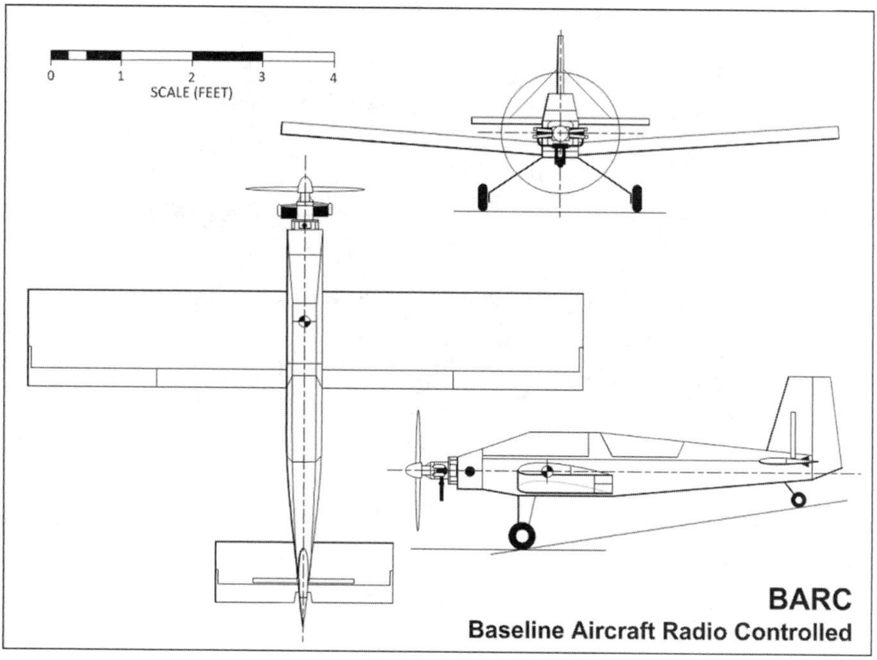

Fig. 7.7 BARC three-view drawing.

Table 7.3 BARC Technical Characteristics

Parameter	Value
Wing span	7.8 ft
Wing area	11.7 ft^2
Length	6.2 ft
Takeoff weight	29.5 lb
Empty weight	26.8 lb
Cruise speed	65 KTAS
Endurance	0.5 hr
Launch method	Conventional
Recovery method	Conventional
Propulsion type	Reciprocating engine O.S. FT-300 and Saito FA-300T
Power	4.0 hp

NDM I, II, AND III, 1991–PRESENT

Fig. 7.8 NDM I with FLYRT ATD subscale model.

The NDM aircraft were utility aircraft intended for testing subsystems and air-launching research aircraft. The name NDM was inspired by the popular sitcom "Dinosaurs," in which a notable line from one of the characters was "Not Da Momma." The humorous name was carried forward for three separate utility aircraft, each able to carry aircraft on its back or below its belly.

The NDM I was built in 1991 (Fig. 7.8). It was designed to be a stable aircraft with a 10-pound test payload capacity. A key feature of this vehicle was docile handling characteristics to minimize RC pilot workload. It had a three-piece wing with generous dihedral on the outboard panels. The wing had no control surfaces. The wing was mounted as a low wing arrangement. The conventional tail had a rudder and an elevator. A fixed tricycle landing gear was used. The design was spirally stable and used yaw control for turning. The large horizontal tail provided a large static margin and a large permissible CG range.

NDM I could launch drop models (Figs. 7.9 and 7.10). A glider release mechanism was mounted to the top of the fuselage. This device was mounted

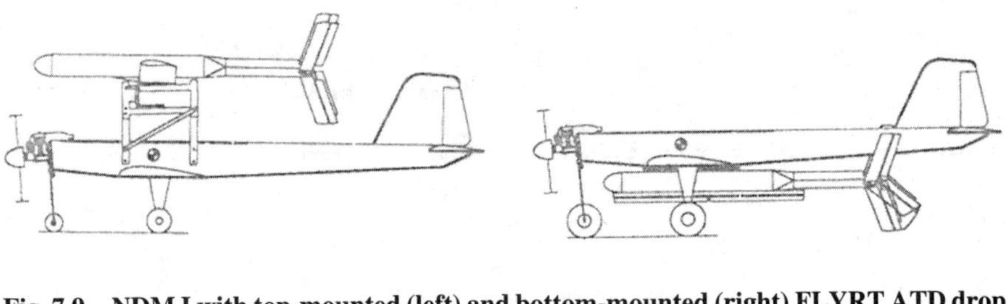

Fig. 7.9 NDM I with top-mounted (left) and bottom-mounted (right) FLYRT ATD drop models.

Fig. 7.10 NDM I air launch of FLYRT ATD models.

above the vertical tail height to ensure adequate separation from the drop model. Drop models could also be released from the bottom of the fuselage below the wing. A large landing gear height provided the necessary ground clearance for the model. The glider release mechanisms were actuated by a servo. To release a top-mounted drop model, the NDM I would enter a steep dive. It would achieve less than zero-*g* while still accelerating. The drop model would release and pull up. NDM I and the drop model then turned in opposite directions to minimize the probability of a mid-air collision.

In addition to drop model launch support, NDM I was also used for autopilot testing. A box with horizontal stabilizers and a downward vertical stabilizer was fitted to the lower fuselage aft halfway between the wing and tail. The horizontal and vertical surfaces could rotate, and these are known as "full flying stabilizers." The main tail's rudder and elevator were controlled via radio control to enable a safe landing if the autopilot malfunctioned. The additional surfaces were controlled by the Light Weight Autopilot prototype units. This autopilot was later added to the SENDER and other systems.

NDM I crashed on a windy day due to structural problems. The spar joiner was made of brass tubing, which was considered sufficient for the intended flight operations. However, a large gust load while in a turn caused the wing panel joiner to fail and led to its destruction. Figure 7.11 shows a three-view drawing of NDM I.

NDM II (Fig. 7.12) replaced NDM I. The fuselage had geometric characteristics similar to those of FLYRT ATD, with a rectangular forward fuselage that transitioned to a diamond rear cross section. The fuselage

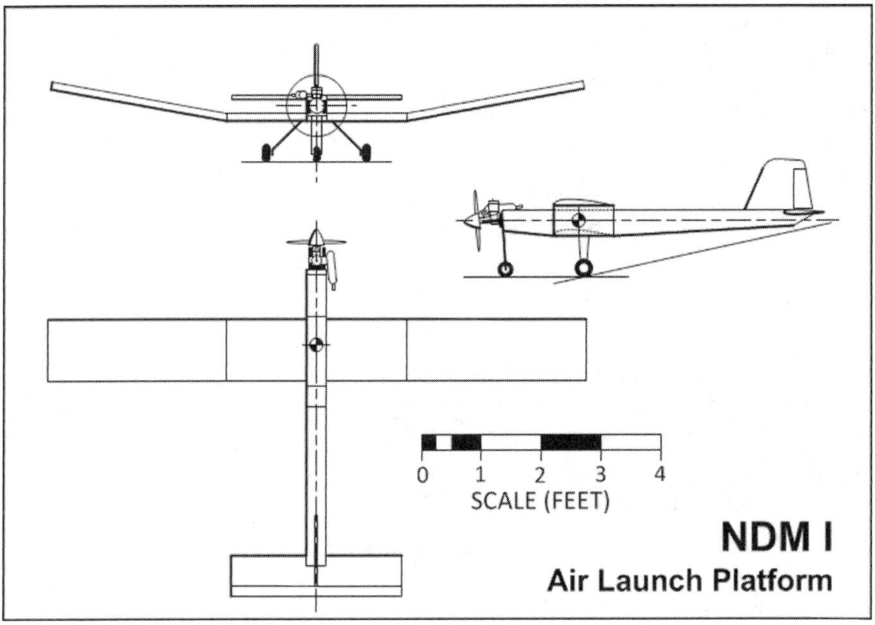

Fig. 7.11 NDM I three-view drawing.

was first used for the Ramwing demonstrator. After the Ramwing testing, the fuselage and tails were salvaged, and conventional wings were added. The NDM II had tapered foam core wings with fiberglass skins and featured ailerons and plain flaps. The conventional tail had rudder and elevator controls. A fixed tricycle landing gear with large ground clearance was used. After many successful tests, NDM II was destroyed during the first launch of the CG Shift Bird demonstrator, which struck and sheared off NDM II's vertical tail, causing it to crash. Figure 7.13 shows a three-view drawing of NDM II.

Fig. 7.12 NDM II.

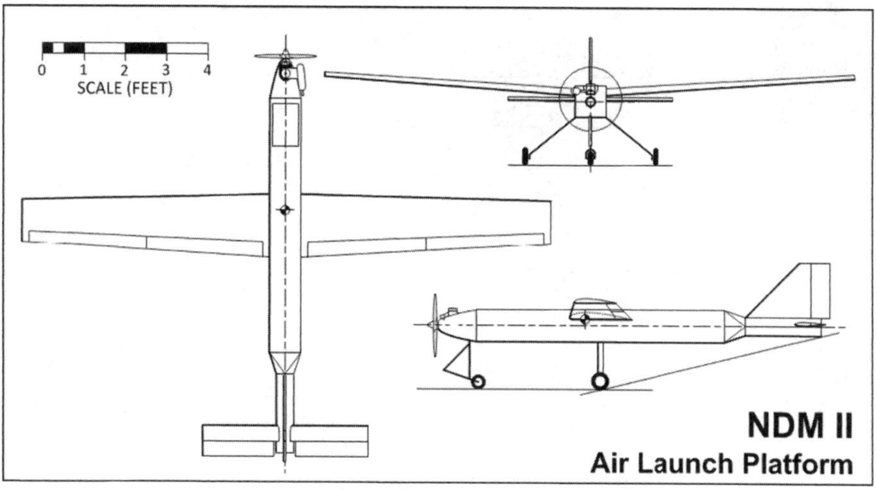

Fig. 7.13 NDM II three-view drawing.

The NDM III was built in 1994. It was designed with an H-tail arrangement (Fig. 7.14), so that the loss of a single vertical tail would not result in the loss of the aircraft. In one of its later drop tests, the drop model knocked off one of the vertical tails and the NDM III safely landed. The H-tail arrangement allowed the release mechanism to be placed close to the top of the fuselage without concern over vertical tail strike. Like its predecessors, NDM III had a fixed tricycle landing gear with large ground clearance. NDM III had rudders, elevator, ailerons, flaps, throttle, and wheel brake controls. The wings and tails were constructed of foam core with fiberglass skins. The fuselage used balsa strips for corners with thin Rohacell foam sheeting panels that were covered with a fiberglass skin. Figure 7.15 shows a three-view drawing of NDM III, and Table 7.4 lists the technical characteristics for the NDM series.

Fig. 7.14 NDM III with CG Shift Bird.

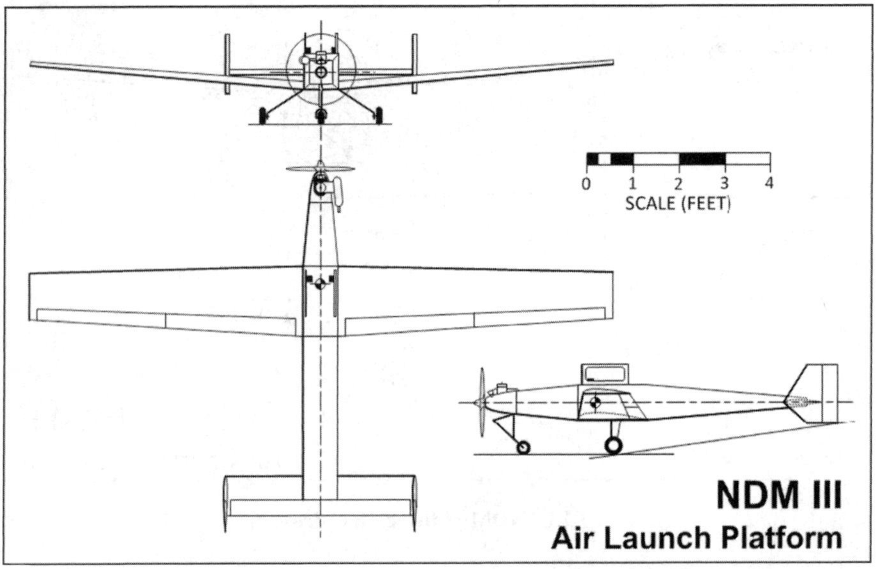

Fig. 7.15 NDM III three-view drawing.

Table 7.4 NDM Series Technical Characteristics

Parameter	NDM I Value	NDM II Value	NDM III Value
Wing span	9.0 ft	12.75 ft	12.75 ft
Wing area	9.0 ft^2	16.1 ft^2	16.1 ft^2
Length	6.0 ft	9.75 ft	7.9 ft
Takeoff weight	16.0 lb	36.0 lb	34.0 lb
Empty weight	8.0 lb	24.0 lb	22.0 lb
Payload weight	7.0 lb	10.0 lb	10.0 lb
Cruise speed	35 KTAS	40 KTAS	45 KTAS
Endurance	0.5 hr	0.5 hr	0.5 hr
Airfoils	Wing RF 1165FB Flat plate (tail)	Wing RF 1165FB Flat plate (tail)	Wing RF 1165FB Flat plate (tail)
Launch method	Conventional	Conventional	Conventional
Recovery method	Conventional	Conventional	Conventional
Propulsion type	O.S. 61 FSR 0.61 cu. in.	O.S. BGX 2.1 cu. in.	O.S. BGX 2.1 cu. in.
Power	1.6 hp	3.8 hp	3.8 hp

NDM: Further Reading

Bovais, C.S., Haupt, M.E., and Toot, P.L., "Flight Testing with Sub-Scale Remotely Piloted Drop Models," AIAA Paper 92-4079-CP, 1992.

Test Radar Calibration Experiment (TRACE), 1993

Fig. 7.16 Renderings of TRACE.

Leading up to the FLYRT ATD flight test program, there was concern that the airframe could not be tracked by radar due to its small physical size should its electronic warfare (EW) payload fail to operate. The flight test plan would not be approved until this issue was resolved. The risk to the schedule and the additional program cost could have put the FLYRT ATD program in jeopardy. TRACE was an extremely rapid risk reduction effort to test the radar calibration of the test range (Fig. 7.16). It also proved that a FLYRT-size aircraft could be monitored with remote, long-range video cameras.

Three NRL researchers built three TRACE aircraft in a total elapsed time of 3 hours, so the entire design and fabrication portion of this effort was a mere 9 man-hours! It was designed and built concurrently. TRACE was a very simple monoplane of conventional configuration with a single fuselage, V-tail, and three-piece polyhedral wing. No landing gear was fitted, and it was reminiscent of a powered version of the Second CG Shift design. It used an O.S. 25 FSR, 0.25-cubic-inch, 2-stroke model aircraft engine in a tractor configuration. It was constructed of hot-wire-cut foam core wings, sheet foam fuselage, sheet balsa wood tail surfaces, and birch plywood hardwood parts at critical high-stress areas. No drawings or plans were generated,

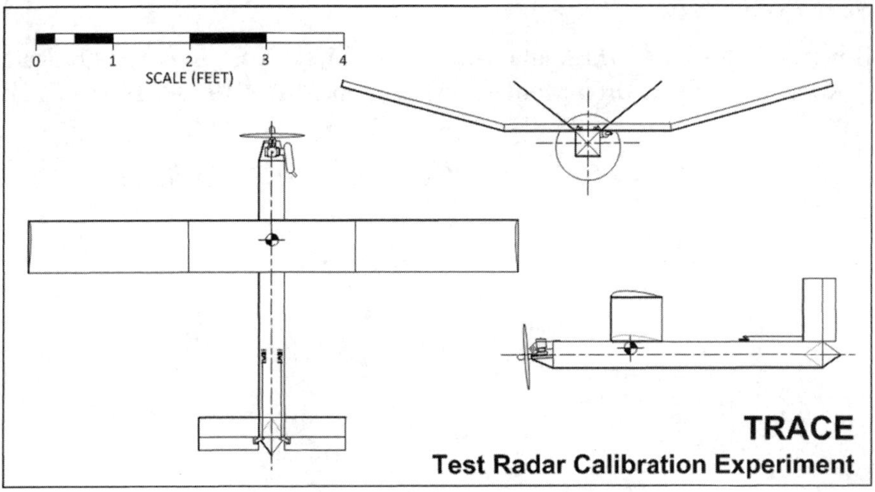

Fig. 7.17 TRACE three-view drawing.

only the most rudimentary sketches and two airfoil cutting templates. Figures 7.16 and 7.17, generated for this book, exceed the amount of documentation that was developed for the project! The team built a small octahedral set of corner reflectors out of a foam sheet and conducting tape attached to the back of the fuselage to create a FLYRT-like radar signature (the FLYRT ATD had undergone extensive RCS measurements in an anechoic chamber, and its radar characteristics were well known). The airframes were spray painted with one quick coat of Rustoleum high-visibility orange paint.

The team test-flew the TRACEs at Blossom Point field the same day they were constructed. The next day, one was flown from a small boat in the Chesapeake Bay. It proved to be challenging to safely start a 2-stroke engine and then hand-launch the vehicle within the confines of the small open boat. The aircraft landed in the water after the flight, as designed, since these were one-time-use systems. The fuselage had a torsional aeroelastic issue that was resolved by the field-expedient upgrade of diagonally wrapped fiberglass strapping tape. The aircraft had benign handling qualities and was fun to fly.

TRACE proved that FLYRT ATD could be tracked by the test radars. This entire effort consumed a total of less than 40 man-hours over a two-day period, but it saved the program an estimated $250,000 and a significant amount of time. The rapid development and testing of TRACE proved the benefits of having a well-stocked prototyping facility and a team of dedicated engineers who were also skilled model airplane enthusiasts. Unfortunately, no photos were ever taken of this unique project. Figure 7.17

Table 7.5 TRACE Technical Characteristics

Parameter	Value
Wing span	6.3 ft
Wing area	4.2 ft^2
Length	4.1 ft
Takeoff weight	3.9 lb
Empty weight	3.5 lb
Cruise speed	25 KTAS
Endurance	0.4 hr
Launch method	Hand toss
Recovery method	Water ditching
Propulsion type	Reciprocating engine
	O.S. 25 FSR, 0.25-cu.-in., 2-cycle
Power	0.7 hp

shows a three-view drawing of this utility aircraft, and Table 7.5 lists its technical characteristics.

Telemaster, c. 1993–Present

Fig. 7.18 Twelve-foot Telemaster.

NRL has utilized the Telemaster model aircraft series for low-cost flight research and payload experimentation over several years (Fig. 7.18). Kits are sold by Hobby Lobby RC Inc. Telemasters are docile, high wing trainers that place low demands on RC pilots. The aircraft are constructed of balsa wood, spruce, and aircraft plywood. The airframes are covered with MonoKote. NRL either builds the kits in-house using the aerospace technician staff or outsources the work.

The early, 8-foot-wingspan Senior Telemaster was developed in Germany as an aid in stringing telephone lines across wide valleys, leading to the "tele" in the name. This activity predates any NRL involvement with the aircraft. In this operation, the aircraft carried a lightweight cable and laid it across a valley. The heavier telephone cable was then affixed to the end of the cable and hoisted across.

NRL uses various scales of the Telemaster, including the 8-foot Senior Telemaster and 12-foot variants. The 12-foot Telemaster has a cavernous payload bay and is more like a traditional unmanned aircraft than a model aircraft. NRL widened the standard fuselage from 6 inches to 12 inches to better facilitate payload integration. All of the Telemasters have lightweight airframes for their sizes, which is partially due to the built-up wood and covering construction that does not suffer from minimum gage skin thickness as composite aircraft do. All variants are easy to fly, which reduces pilot workload for tests. NRL has operated four Senior Telemasters and six 12-foot Telemasters. Two more 12-foot Telemasters are under construction at the time of this writing.

NIGHT FLIGHTS

NRL Senior Telemasters were employed in a series of night flight experiments in support of an airborne biological-agent-detection project. The final demonstration flights of the detection system were to be performed at night at the Army's Dugway Proving Ground. The demonstration aircraft were not fully autonomous at that point, so they needed to be controlled manually through radio control. However, external RC night-flying techniques for unmanned aircraft were not well established. NRL wished to pursue the opportunity so committed to solving the problem. NRL developed the necessary night-flying techniques at an Edgewood Arsenal airfield at Aberdeen Proving Ground.

The first night-flying tests used one of Foch's small RC aircraft with lights taped to each wing tip and the top of the vertical tail. This proved viable, so the night-flying experiments then shifted to a Senior Telemaster.

The test flights had a number of provisions to aid the pilot. The pilot used a pair of Gen III Night Vision Goggles (NVGs). The aircraft had "grain of wheat" bulbs on the wingtips, which were covered by seven layers of white vinyl tape so they were barely visible to the unaided eye and not excessively bright for the NVGs. Finally, the edges of the runway were illuminated with Cyalume light sticks spaced every 8 to 10 feet. The NVGs had a very narrow field of view, which created great challenges for situational awareness, and the pilot had no depth perception. Clearly, new flying techniques had to be developed.

Landing was a particular challenge. Due to the NVG's narrow field of view, the pilot could not detect the end of the illuminated runway unless the aircraft flew over it at low altitude and close to the horizon. Foch developed a landing technique called the "button hook." The aircraft crossed the runway at an angle of 60 to 90 degrees relative to the runway length direction. This gave the pilot a position reference relative to the end of the runway. The aircraft passed and then conducted a 180-degree turn, arcing away from the runway in the turn. Then the aircraft executed a 90-degree turn toward the runway for final approach. The maneuver shape resembled a question mark.

Dozens of missions were flown at Dugway Proving Ground. The 6-, 8-, and 12-foot Telemaster variants were all flown. The 6-foot Telemaster was a personal model loaned by one of the NRL technicians, and it flew the first missions from unprepared surfaces and a one-lane road. The 8-foot Senior Telemaster was flown with a larger payload, also from a paved road. Later, an unmanned aircraft airfield named Ligler Field was built by the U.S. Army Corps of Engineers. The airfield was approximately 1000 feet long by 100 feet wide. The 12-foot Telemaster operated from Ligler Field. The airfield was named after NRL's Dr. Fran Ligler, who developed biological-agent-detection payloads and participated in the tests.

A humorous episode came out of the test program. The tests took place at about 5000 feet above sea level in the high desert, and the nights were very cold. One night the flight test crew decided to rest by lying down on the warm road in the hours between test events. They were awoken by an Army vehicle racing down the road. An Army Colonel jumped out of the vehicle in visible distress. The facility had been remotely monitoring the crew, and they grew concerned by what appeared to be casualties strewn across the road. The fear was that the NRL test crew was inadvertently exposed to harmful substances and was sick or dying!

Prior to the program, Foch had read that another government agency was flying RC aircraft at night. This helped build his confidence in the feasibility of the endeavor. After the project was well under way, he contacted that organization and was told that night flying had been attempted by them but that it had proved unsuccessful. At the conclusion of the program, that agency asked NRL to train them in night-flying techniques. NRL was successful because they did not know it could not be done, and the Telemasters are very forgiving aircraft to fly. Had NRL learned of the earlier failures, the night-flying project might never have started. Table 7.6 lists the technical characteristics of this utility aircraft.

Table 7.6 Telemaster Technical Characteristics

Parameter	Senior Telemaster Value	12-Foot Telemaster Value
Wing span	8.0 ft	12.0 ft
Wing area	9.3 ft^2	21.1 ft^2
Length	5.3 ft	7.5 ft
Takeoff weight	20.0 lb	55.0 lb
Empty weight	8.5 lb	38.0 lb
Payload weight	10.0 lb	14.0 lb
Cruise speed	40 KTAS	40 KTAS
Endurance	60 min	60 min
Launch method	Conventional	Conventional
Recovery method	Conventional	Conventional
Propulsion type	O.S. 61 FSR 0.61-cu.-in., 2-cycle	Xenoah G62 3.8-cu.-in., 2-cycle
Power	1.6 hp	5.5 hp

DAKOTA, 1995–1998

Fig. 7.19 Second Dakota aircraft.

The Dakota aircraft served as large, affordable, payload testbed vehicles (Fig. 7.19). Two aircraft were built by Daedalus Research Inc. (DRI) based on an NRL conceptual design and requirements. This was the largest fixed-wing UAS that NRL had used up to that time.

The original platform was called "Truck," but the name was changed to "Dakota" because radio communications on military test ranges were confusing when an unmanned aircraft was referred to as a truck. The original name Truck was intended to convey that this aircraft was a payload-hauling truck of an aircraft. The name Dakota was an alternative along the same theme, as the name is also shared by a popular pickup truck model.

The low-cost payload demonstrator aircraft was designed by DRI. Foch generated the initial vehicle sizing and specifications. Preliminary design included the major vehicle dimensions such as the wing span and aspect ratio and the tail geometry. The aircraft was intended to have handling qualities similar to a general aviation aircraft. Ned Allen of DRI conducted the detail design. NRL employees assisted with the developmental flight test program at DRI's facilities in Logan, Utah, at Webster Outlying Field in Maryland, and at a privately owned airfield in the Washington, DC, area.

The basic aircraft was originally designed to carry a 20-pound payload and stay aloft for up to 3 hours on 12 pounds of fuel. The original empty weight was 71 pounds and it had a design gross weight of 103 pounds. However, a 50-pound payload and other avionics were required for a chemical agent detection mission, and the aircraft was ultimately flown at a gross weight of 155 pounds. The large structural margins enabled this overload case without excessive risk of structural failure. This aircraft had an unusually high payload mass fraction.

The composite airframe was simple, lightweight, and inexpensive. The fuselage was constructed of flat panels made of carbon fiber cloth and a Nomex (by DuPont) core sandwich. The fuselage cross section was rectangular, which was ideal for payload integration. The first aircraft had sharp corners where the panels met. The second aircraft had rounded corners made from molded carbon and foam, with a constant radius for the entire fuselage. The fuselage composite panels could be removed and replaced when damaged or for modification. The wings and tails were foam core covered with fiberglass and carbon fiber. The wings had a carbon fiber spar to take the bending loads.

Beyond the carrying capacity, Dakota had generous payload accommodation. Up to 400 watts of payload power was available from an engine-driven generator. The payload bay measured 17 inches by 11.5 inches by 12 inches. This volume was located under the wings near the CG, so light or heavy payloads could be installed without adding ballast for balance. The payload mechanical interface was the same as a 19-inch industrial computer rack. A hole could be cut in the bottom of the fuselage for payload apertures.

NRL flew Dakota as an RC aircraft. The aircraft had ailerons, flaps, a rudder, and elevator control surfaces. The rudder was aerodynamically and statically unbalanced and used a servo tab to reduce the actuator loads. All other control surfaces had two servos for reliability. Servos directly controlled the surfaces without any pulleys or cables. Dakota had a fixed tricycle landing gear with pneumatic brakes and a steerable nose gear.

Dakota flew very well, as expected. The generous tail volumes gave the aircraft gentle handling characteristics. The platform had similar stability derivatives and coefficients as a general aviation aircraft. The RF 1165FB airfoil and constant chord wing provided gentle stalling characteristics. This enabled the pilot to focus on the mission rather than on flying the aircraft. The

Fig. 7.20 The NRL team posing in protective gear with Dakota at Dugway Proving Ground.

light wing loading and powerful engine allowed takeoff from a stop to clearing a 50-foot height in only 250 feet when not overloaded. The landing from a 50-foot height to a stop took only 770 feet.

The first Dakota was fitted with 50 pounds of payload including a chemical agent detector, a biological agent sampler, and meteorological sensors. The aircraft was to be flown at Dugway Proving Ground, which is at an elevation of over 4000 feet above sea level (Fig. 7.20). The additional weight plus higher altitude would have resulted in a poor climb rate. Suitable larger engines were not available at the time. NRL added wing span extensions at the wing tips to reduce the span loading. The combination of higher weight and increased span reduced the structural margins, but the margins were still acceptable. The aircraft successfully conducted night operations. Only one incident occurred: when the pilot missed the one-lane road and set the Dakota down into a ditch parallel to the road. Fortunately, the only damage was a broken propeller.

The second airframe was fitted with two floats. The balsa wood and fiberglass floats were made by Stream Inc. and later slightly modified by NRL. The aircraft made two flights from the water. The first flight was fully successful. On the second flight as a floatplane, Dakota experienced a rudder failure with a hard-over deflection. The rudder servo may have been damaged by water during the first flight. The aircraft had a strong yaw-roll coupling (dihedral effect), so large aileron deflections were necessary to maintain straight flight. However, this resulted in a large sideslip angle and a bank angle (much like crossing the controls to slip a general aviation aircraft). The

Fig. 7.21 V-tail modification of Dakota.

pilot explored his options until the fuel was nearly exhausted. He decided to level the wings and alight with a crab angle. The floats sheared off the airframe due to excessive side loads, and the aircraft was destroyed. The other option would have been to alight without crabbing but with a turn rate, but it is not clear if this would have produced a better outcome.

Dakota served as a launch aircraft for SENDER before the NDM III was ready, and during one incident, a release mechanism failure caused the SENDER to fly back into the horizontal tail, resulting in a Dakota crash, which was repairable. Further details are in the description of SENDER (Chapter 5).

The first Dakota was modified with a V-tail (Fig. 7.21). This aircraft served as an autopilot test bed for the NRL Ghost project. It was flying in this configuration when it clipped a guy wire, which broke off a wing panel. The resulting crash destroyed this well-flown prototype.

DRI was purchased by Geneva Aerospace, which in turn was purchased by L-3. Dakota continued to be developed after the NRL involvement ceased. Later versions had a conventional tail rather than the T-tail of the NRL variants. The new wing was made of aluminum. Another group at NRL acquired two of these aircraft for research. Years later, these assets were sent to the VRS and are still occasionally performing tests in support of mission payload developments. Figure 7.22 shows a three-view drawing of this utility aircraft, and Table 7.7 lists its technical characteristics.

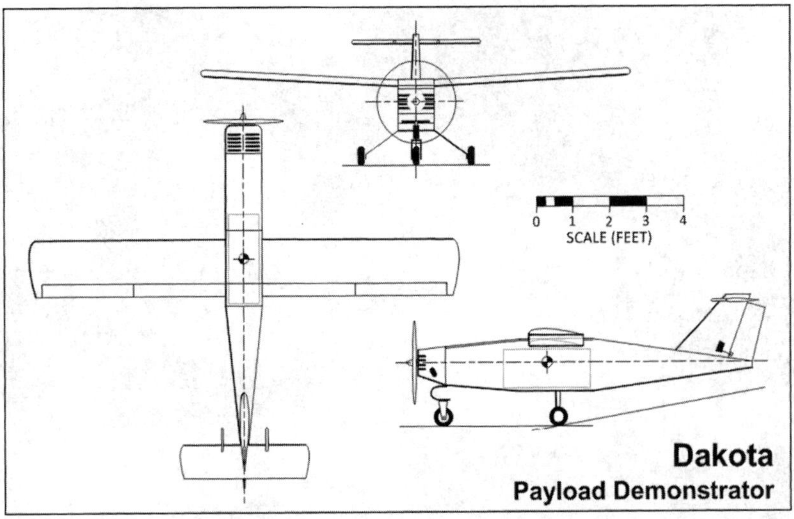

Fig. 7.22 Dakota three-view drawing.

Table 7.7 Dakota Technical Characteristics

Parameter	Value
Wing span	12.7 ft (original)
Wing area	18.8 ft^2
Length	9.7 ft
Takeoff weight	103 lb (1st airframe)
	133 lb (2nd airframe)
	155 lb (1st airframe w/extended wing)
Empty weight	71 lb
Payload weight	20 lb (design threshold)
	65 lb (max flown)
Fuel weight	12.0 lb
Loiter speed	73 KTAS
Dash speed	105 KTAS
Stall speed	36 KTAS (without flaps)
	28 KTAS (with flaps)
Endurance	3.3 hr
Airfoils	RF 1165FB (wing)
	Eppler E520 (vertical tail)
	NACA 0009 (horizontal tail)
Launch method	Conventional
Recovery method	Conventional
Propulsion type	Reciprocating gas engine
	Aerrow A200
Power	17 hp

Dakota 2, 1998–2002

Fig. 7.23 Dakota 2 prototype.

The Dakota 2 (Fig. 7.23) was designed as a payload test bed to replace the original DRI Dakota. This version was larger yet engineered to be easier to transport. Theiss Aviation was selected to build the aircraft because of their proprietary, novel, low-cost manufacturing technique. Foch provided a sketch of the airframe and some requirements, and Theiss Aviation conducted the detail design and construction. Dakota 2 had to be transportable in a small vehicle and carry more than 100 pounds of payload. The manufacturer's internal name for the aircraft was Tarzan TD-1c.

Dakota 2 had multiple hinges to enable it to be stowed in a shape that could be rolled into and out of a panel van without disassembly. The 18-foot-span aircraft folded into a volume measuring only 6 feet long by 6.5 feet wide by 4.5 feet high. The fuselage had a hinge on top just behind the wing, so the aft fuselage folded over the forward half. The tails were fixed to the aft fuselage. The wings had a middle panel that folded up and an outer panel that folded inward such that the wing formed a box when viewed from the front. The width of the box was approximately the same as the width between the V-tail tips. The aircraft rested on the landing gear when folded so that it could be rolled without ground handling equipment (Fig. 7.24). Latches secured the structure for flight. The aircraft could be configured for flight from a folded state in approximately 15 minutes.

Dakota 2 featured an innovative structure. The construction closely resembled a classic wood and fabric structure. However, uprights and ribs were replaced with sheet foam. Spruce spars and longerons were used. The airframe was covered with Dacron polyester fabric. Latex paint was applied in place of nitrate dope to protect the foam. A later version had plywood skins. The structure was much faster to build than traditional wood or composite airframes. One objective of the program was to evaluate the practicality of this novel Theiss construction method. The aircraft could carry 105 pounds of payload with an empty weight of only 150 pounds, thanks to the lightweight structure. The cavernous fuselage had a payload

Fig. 7.24 Dakota 2 in the folded configuration.

volume of 9 cubic feet (Fig. 7.25). The airframe was very inexpensive due to the low-cost materials and low labor hours needed for construction. Another advantage of the foam structure was that the payload was well isolated from shock and vibration.

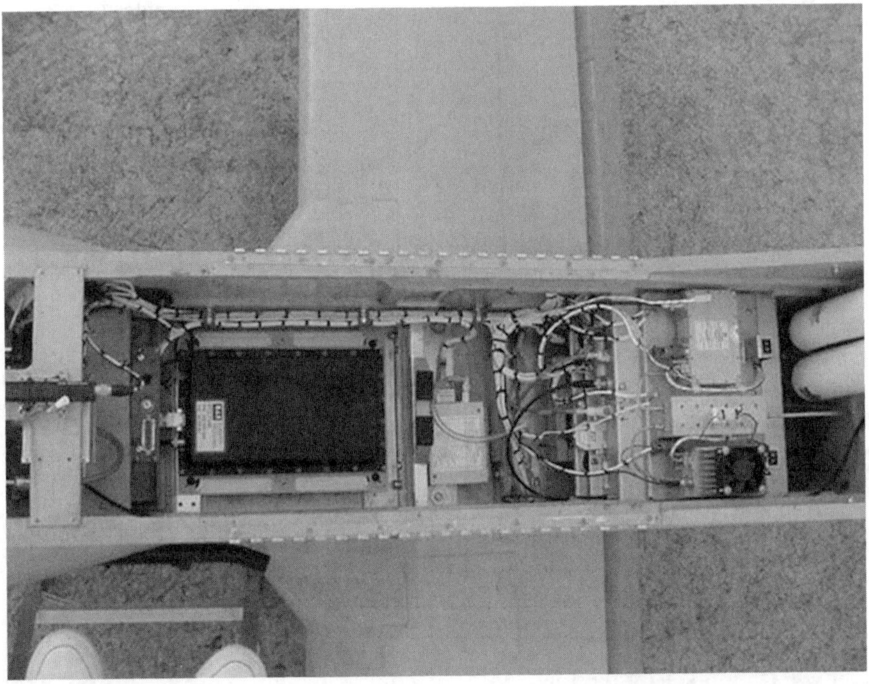

Fig. 7.25 Dakota 2 internal fuselage bay with a payload installed.

Fig. 7.26 Modified version of Dakota 2 with side folding fuselage hinges for hangar storage.

Several aircraft were built. The Dakota 2 aircraft flew at Camp Roberts in California, performed experiments with the Naval Postgraduate School, and supported numerous NRL experiments (Fig. 7.26). The airframes gave good service, but they eventually wore out through frequent use. The airframes had a limited lifespan in order to leverage this low-cost construction method. Dakota 2 had gentle handling qualities to reduce pilot work load. Figure 7.27

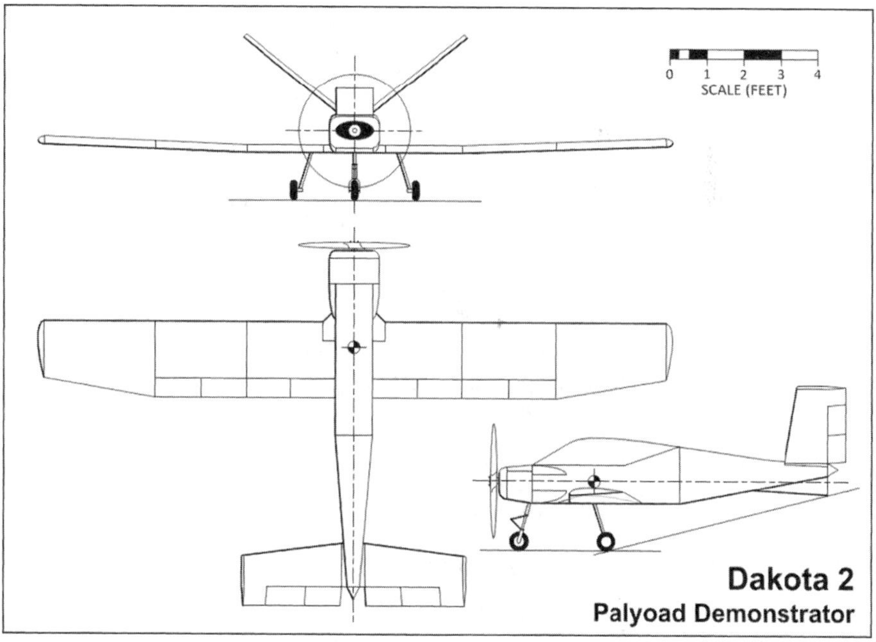

Fig. 7.27 Dakota 2 three-view drawing.

Table 7.8 Dakota 2 Technical Characteristics

Parameter	Value
Wing span	17.2 ft
Wing area	34.0 ft^2
Length	10.75 ft
Takeoff weight	255.0 lb
Empty weight	150.0 lb
Payload weight	105.0 lb (useful load)
Fuel capacity	5.4 gal
Loiter speed	43 KTAS
Stall speed	31 KTAS
Endurance	>4 hr
Airfoils	Wing: RF 1165FB Tails: Flat plate w/tapered trailing edge
Launch method	Conventional
Recovery method	Conventional
Propulsion type	Gas 2-stroke reciprocating engine Lightning Aircraft Engines 302cc
Power	28 hp

shows a three-view drawing of this utility aircraft, and Table 7.8 lists its technical characteristics.

UNMANNED TITAN TORNADO, 1995–1996

Fig. 7.28 NRL's Titan Tornado.

The Unmanned Titan Tornado program (Fig. 7.28) was the result of a white paper that Foch wrote to the Secretary of the Navy, the Honorable Mr. Richard Danzig. The Secretary of the Navy was troubled by the expense of the TRW Hunter tactical UAS program. The Hunter development and acquisition costs were higher than would be expected for a similar-sized general aviation aircraft. Rather than using general aviation hardware, many Hunter components were custom built for the application. For example, two heavily modified Moto Guzzi motorcycle engines were used rather than aviation engines. These adapted engines had development problems and were expensive.

Foch suggested that general aviation components and airframes could be applied to tactical UASs for lower-cost and faster development. The widespread Department of Defense (DoD) perspective was that general aviation aircraft were not sufficiently robust for unmanned aircraft missions and field conditions. Foch pointed out that the Cessna Bird Dog and O-2 aircraft served over several decades as forward air controller aircraft under harsh wartime conditions. These aircraft were also very affordable.

The white paper described the concept of modifying existing general aviation aircraft to become unmanned aircraft. While aviation maintenance training would be required for operations, the systems would be reliable and affordable. It was suggested that a small manned aircraft could fulfill the Hunter mission.

Rather than considering high-end general aviation aircraft, NRL evaluated two-place light aircraft. NRL surveyed the field for the most airworthy and aerodynamically efficient aircraft. The Rotax 503-powered Titan Tornado was the best candidate. This aviation engine was an improvement upon motorcycle engines. The pusher engine configuration would enable unobstructed payload integration in the front of the aircraft.

The NRL team visited Titan Aircraft to evaluate the aircraft. The Titan Tornado was sold as a homebuilt kit. The FAA requirement for this class of aircraft was that the builder had to perform 51% of the airframe construction. NRL labor costs were higher than those of Titan Aircraft, so Titan Aircraft was awarded a contract to assemble 100% of the aircraft. It was decided that since it would fly as an unmanned aircraft, the homebuilt rules were not applicable. NRL purchased a single assembled Titan Tornado aircraft.

The aircraft was transported to NRL facilities in a customized Wells cargo trailer that was 24 feet long. The single-piece wing, which was held on by only four bolts, and the horizontal tail were taken off the fuselage for transport. The aircraft was reassembled at the lab workshop for display. Numerous visitors came to see it. Some visitors supported the concept while others found it controversial. Most came to the conclusion that the aircraft was adequate for the mission.

NRL requested additional funds to convert the manned aircraft to an unmanned aircraft. The seats would be left in the aircraft to allow a pilot to

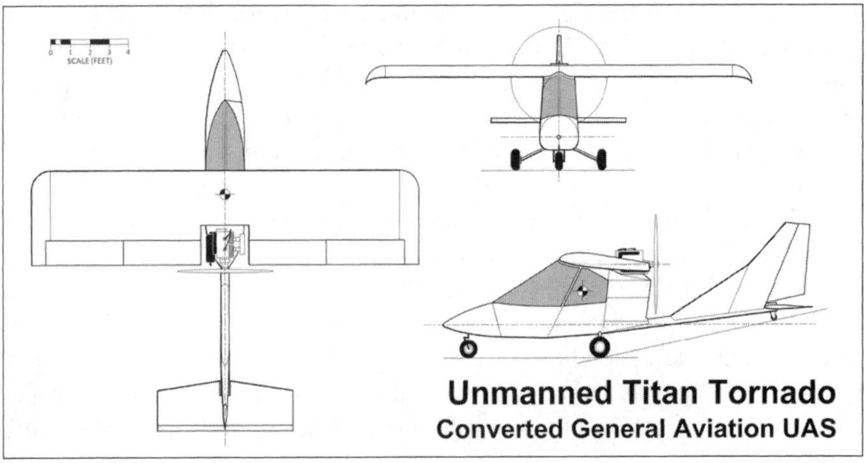

Fig. 7.29 Titan Tornado three-view drawing.

ferry the aircraft. A chassis was bolted to the steel structure to house the actuators and other systems. The servo actuators would move the control stick, throttle column, rudder pedals, and toe brakes. Today such as system is called an Optionally Piloted Aircraft (OPA).

The field of OPAs is currently quite active. Aurora Flight Sciences produces the Centaur OPA, which is a converted Diamond DA42 twin-engine general aviation aircraft. The Gulfstream G550 jet was proposed as a contender for the Navy Broad Area Maritime Surveillance program. Other manufacturers have developed OPAs as well. The rationales for such systems were affordability, supportability, and the ability to fly in a manned mode for transit.

The Hunter program was cancelled at a time that coincided with NRL starting the development of the utility UAS version of the Titan Tornado. The DoD then reconsidered UAS requirements and decided that there was no immediate need for a Hunter-like unmanned aircraft capable of carrying 200 pounds of payload with general aviation performance. Consequently, the unmanned Titan Tornado was placed into storage at NRL and was never flown. Figure 7.29 shows a three-view drawing of this utility aircraft, and Table 7.9 lists its technical characteristics.

Table 7.9 Titan Tornado Technical Characteristics

Parameter	Value
Wing span	20.0 ft
Wing area	98.0 ft^2
Length	19.0 ft
Takeoff weight	1000 lb
Empty weight	500 lb
Payload weight	500 lb (maximum)
Fuel weight	60 lb
Loiter speed	72 KTAS
Cruise speed	87 KTAS
Dash speed	105 KTAS
Stall speed	35 KTAS (with flaps)
Endurance	2.5 hr
Launch method	Conventional
Recovery method	Conventional
Propulsion type	Rotax 503 reciprocating engine
Power	50 hp

Science Instrumentation Environmental Remote Research Aircraft (SIERRA), 2002–Present

Fig. 7.30 SIERRA.

NRL developed the Science Instrumentation Environmental Remote Research Aircraft (SIERRA) unmanned aircraft (Fig. 7.30) for NASA Ames Research Center to provide a platform for scientific missions. The 20-foot-span utility aircraft can carry up to 100 pounds of payload. SIERRA is on the NASA approved platforms list.

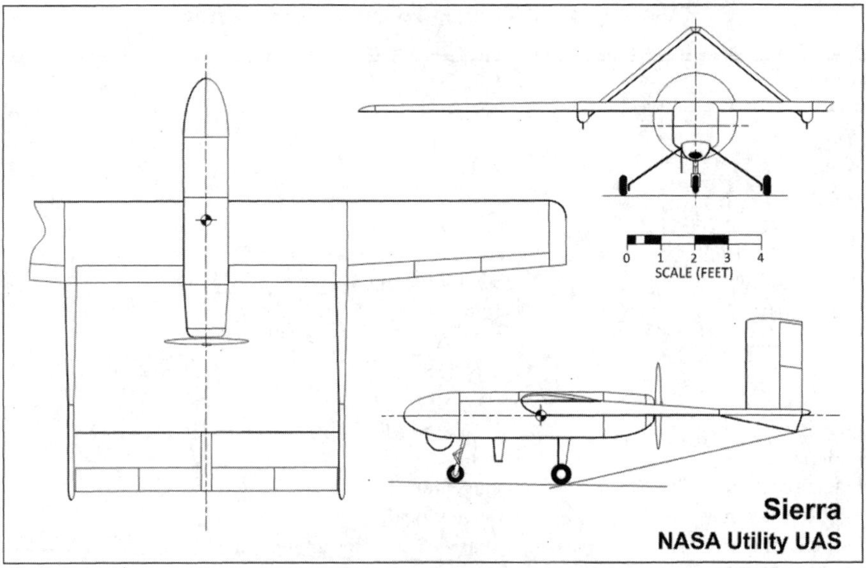

Fig. 7.31 SIERRA three-view drawing.

The aircraft is a conventional twin-boom pusher configuration with an inverted V-tail. It has a fixed tricycle landing gear for runway operations. Payloads can be mounted in the nose or in under-wing pods. The modular design allows the propulsion system to be mounted in the front or rear of the fuselage. The airframe was constructed from composite materials. The design leveraged experience from the Dakota, Dakota 2, and NDM vehicles. Figure 7.31 shows a three-view drawing of SIERRA, and Table 7.10 lists its technical characteristics.

SIERRA has performed multiple NASA science missions. A tropospheric atmospheric chemistry mission carried a Picarro gas analyzer to measure carbon dioxide, methane, and water vapor in real time, and a wind probe to measure vertical fluxes. NASA conducted a joint operation with the Japan Aerospace Exploration Agency (JAXA) to calibrate and validate the GOSAT satellite greenhouse gas remote sensing capability. SIERRA's arctic ice reconnaissance mission, conducted out of Svalbard, Norway, used a payload suite consisting of laser altimeters, synthetic aperture radar, microspectrometers, still and video cameras, and pyrometers.

Table 7.10 SIERRA Technical Characteristics

Parameter	Value
Wing span	20.0 ft
Wing area	42.4 ft^2
Length	11.8 ft
Takeoff weight	310.0 lb
Empty weight	180.0 lb
Payload weight	100.0 lb maximum
Cruise speed	55 KTAS
Dash speed	79 KTAS
Stall speed	42 KTAS (clean)
Rate of climb	545 ft/min
Endurance	8–10 hr
	2.5 hr with 100-lb payload
Range	500–600 mi
Max L/D	11.5
Maximum altitude	12,000 ft
Launch method	Conventional
Recovery method	Conventional
Propulsion type	Gas reciprocating engine
Power	25 hp

ARCTURUS T-15, T-16, AND T-20, C. 2009–PRESENT

Fig. 7.32 Arcturus T-16 in flight.

Fig. 7.33 Arcturus T-20 with under-wing pods.

The wing of the NRL Swallow was built by Mark Allen, who later became an employee of Arcturus-UAV, LLC. He designed the T-15 unmanned aircraft, which used the outer 5 feet of the Swallow wings built from the Swallow molds. NRL became interested in the T-15's capability and eventually bought larger T-16 and T-20 follow-on versions.

NRL used the Arcturus T-16 for an autonomous payload directed flight program (Fig. 7.32). The aircraft was a good design for a utility vehicle, so NRL purchased the vehicle for use as a demonstrator. The T-16 can be equipped with a retractable full-motion video gimbal.

NRL used the larger Arcturus T-20 for continued flight research (Fig. 7.33). This unmanned aircraft had a larger span than the T-16. NRL designed and built wing payload pods that can also be used on many different aircraft for flexible integration. NRL integrated into one of these pods a payload that is now being integrated onto a Shadow 200 UAS (AAI Corp.). The T-20 research included payload directed flight, altitude hold at 100 feet above ground level, and auto-land. Table 7.11 lists the technical characteristics.

Table 7.11 Arcturus T-16 and T-20 Technical Characteristics

Parameter	T-16 Value	T-20 Value
Wing span	13.0 ft	17.0 ft
Takeoff gross weight	85.0 lb	150.0 lb
Payload weight	30.0 lb	>30.0 lb
Cruise speed	50 KTAS	60 KTAS
Endurance	16 hr	>16 hr
Propulsion	Honda 60cc 4-stroke reciprocating engine	10 hp 4-stroke reciprocating engine using AVGAS

BAI TERN, C. 2001

Fig. 7.34 BAI Tern during flight test operations.

NRL used the BAI Tern for payload test operations (Fig. 7.34). (BAI Aerosystems Inc. was subsequently acquired by L-3 Communications.) The Tern was an outgrowth of a Maynard Hill design. It could be fitted with a payload canoe beneath the fuselage for simplified payload integration. NRL flew chemical/biological detection payloads and five other payloads and also used it for checkouts of the FINDER autopilot system. The aircraft had a narrow speed range but good handling qualities. Table 7.12 lists its technical characteristics.

Table 7.12 BAI Tern Technical Characteristics

Parameter	Value
Wing span	11.3 ft
Fuselage length	8.8 ft
Fuselage width	10.0 in.
Takeoff gross weight	110.0 lb
Payload weight	18.0 lb (useful load)
Propulsion	Reciprocating engine

THORPE P10B, C. 2005

Fig. 7.35 Thorpe P10B engine start.

NRL used the Thorpe P10B as a payload utility test aircraft (Fig. 7.35). This simple, large aircraft had similar characteristics to an ultralight manned aircraft. Unlike many NRL unmanned aircraft, the P10B had a tail dragger landing gear. The P10B was constructed of inexpensive composite materials. This vehicle enabled more affordable payload testing operations than alternative operational unmanned aircraft such as the Shadow or Pioneer. It was purchased for the purpose of air-launching the 60-pound FINDER for flight testing. The contemporary NDM III was too small for this role. Table 7.13 lists its technical characteristics.

Table 7.13 Thorpe P10B Technical Characteristics

Parameter	Value
Wing span	21.0 ft
Length	12.5 ft
Payload volume	4.0 cu. ft
Empty weight	150–160 lb
Payload weight	50 lb
Airspeed	30–100 mph
Propulsion	Reciprocating engine

MAGPIE, C. 2008

Fig. 7.36 Magpie.

The Magpie AP (autopilot) airframe was a simple electric model aircraft (Fig. 7.36) manufactured by Mountain Models (Appleton, WI). The commercial platforms were designed to carry cameras for aerial photography. In 2008, NRL integrated autopilots into these models to create a low-cost, autonomous unmanned aircraft. This is a good example of leveraging off-the-shelf model aircraft for unmanned aircraft applications to save cost and to accelerate program schedules. Modern small autopilots can be rapidly integrated. The Magpies served as multipurpose test platforms. NRL flew multiple autonomous Magpies in formation. This platform demonstrated extended non-GPS navigation with autonomous dead reckoning and auto-land capabilities. After numerous modifications, the Magpies were used as aerial targets during exercises. NRL frequently used the unmanned Magpies as test beds on new flight test ranges, where they conducted Radio Frequency (RF) communications link testing and helped establish procedures with the facilities. Table 7.14 lists its technical characteristics.

Table 7.14 Magpie Technical Characteristics

Parameter	Value
Wing span	4.0 ft
Takeoff gross weight	3.0 lb
Payload weight	1.5 lb
Cruise speed	28 KTAS
Endurance	30 min with 5-min reserve
Propulsion	Battery electric
Power	200 W

UASUSA TEMPEST, 2011

Fig. 7.37 Tempest unmanned aircraft (left) and Tempest carried aloft by balloon (right).

NRL used the UASUSA Tempest as a research aircraft and launch platform for CICADA III MAVs. The aircraft is based on high-performance, electric-powered model sailplane technology. NRL needed a robust airframe for high-altitude flights. The Tempest was used by University of Colorado at Boulder for tornado research, so the airframe was clearly robust in high winds. Furthermore, it was very inexpensive for the capability. NRL launched the Tempest from an Aerostar high-altitude helium weather balloon as part of the CICADA III program (Fig. 7.37). These tests were performed at Yuma Proving Ground, Arizona, and drops were successfully made from altitudes as high as 57,000 feet. Three Tempest systems were purchased for the CICADA III program.

Chapter 8

VERTICAL TAKEOFF AND LANDING (VTOL) AND ROTORCRAFT

Rotary wing unmanned vehicles are highly attractive for many electronic warfare (EW) and surveillance missions. By virtue of being able to perform vertical takeoffs and landings, they are runway independent, which enables operation from ships and small ground vehicles. In addition, they store in compact packages compared with fixed wing aircraft and have the ability to operate in poor weather conditions. These attributes come at a price. They are not aerodynamically efficient, generally carry a smaller payload mass fraction, and are more difficult to fly autonomously by virtue of being inherently unstable. Nevertheless, NRL rose to the challenge and developed several rotary wing small unmanned aircraft systems (UASs) and small tactical UASs. Early projects were especially challenging and not fully successful. The perseverance of the vehicle research engineers eventually paid off with several highly successful programs that developed the requisite technologies and set the standards for autonomous VTOL UASs.

SHIP'S TETHERED AUTOGYRO PLATFORM (STAPL), 1975–1980

Fig. 8.1 STAPL prior to flight.

The Ship's Tethered Autogyro Platform (STAPL) was a tethered unmanned aircraft based on a highly modified Benson B8M Gyrocopter rotorcraft (Fig. 8.1). The purpose of the platform was to place an electronic payload above a ship to extend the Radio Frequency (RF) horizon. NRL oversaw the contract with Kaman Aircraft in Winsor Locks, Connecticut. The program began in 1975 and ended in 1980. The first flight was made in 1977. This was the first program at NRL to which Foch was assigned. As a newly graduated engineer, he joined the effort in 1979, when it was well under way.

The gyroplane flew like a tethered kite when there was relative wind. However, ship operations could not guarantee suitable wind, so the STAPL had a propulsion system (visible in Fig. 8.2). The autogyro would fly figure eight patterns when the wind was low to stay aloft.

The autopilot, battery, and actuators were placed ahead of the rotor mast in the former cockpit area. The combined unmanned provisions weighed approximately 150 pounds. The actuators directly interfaced to the rotor head and rudder. Without the original control stick or rudder pedals, STAPL was not an early example of an optionally piloted aircraft. Because it started as a manned aircraft, it could have been optionally piloted had there been any interest in dual operations. The autopilot used analog electronics, and its operation was sensitive to temperature changes and other factors. The engine remained the standard Benson B8M power plant, a 72-horsepower 2-cycle McCulloch 4318 drone engine. This power plant drove a small-diameter, wide-blade "club" prop and had a high vibration level compared to standard 4-cycle aviation engines. It was also quite loud.

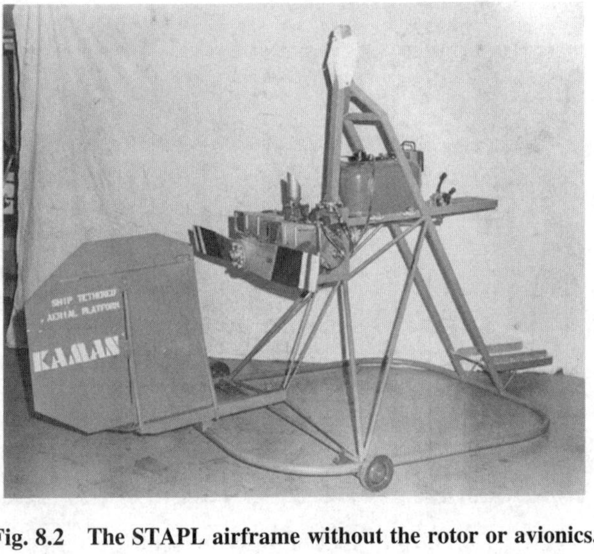

Fig. 8.2 The STAPL airframe without the rotor or avionics.

Fig. 8.3 STAPL on the trailer in the test configuration.

Initially the aircraft used a fixed-pitch wooden prop for forward thrust. However, a more aggressive thrust control response was desired, which could be met by a controllable pitch propeller. Kaman Aircraft was a rotorcraft company, so they adapted a tail rotor rather than using a more traditional aircraft propeller solution. The 2-cycle drone engine had very high vibration, but no dynamic analysis was conducted prior to testing. The aircraft experienced significant damage when both propeller blades shed during a test. Fortunately, no personnel were injured. After that, only the original fixed-pitch wooden propeller was used on STAPL.

To represent the wind over the deck of a moving ship, the STAPL aircraft was launched from a flatbed trailer towed behind a truck (Fig. 8.3). The truck drove along a runway and could stop at the end of the runway when STAPL was aloft in powered flight. STAPL flight tests were conducted at Barnes Municipal Airport in Westfield, Massachusetts, a short distance from the Kaman plant. This airfield was not overly busy and had a 10,000-foot runway aligned with the predominant winds (north/south). During these pioneering years of UAS research and development, it was quite easy to obtain permission from both the FAA and the military to test unmanned vehicles at active airfields. Obtaining airspace is much more challenging today. The NRL/Kaman test team simply worked with the control tower and safely conducted tests between general aviation and Massachusetts Air National Guard air traffic. There were no interruptions in normal airport operations.

The trailer contained a winch and a control hut (Fig. 8.4). An operator and test engineer stood behind a steel mesh barrier in the hut to control and monitor the aircraft. The truck would drive at approximately 50 mph, which was a chilly experience for the engineers during the winter months. Prior to moving, the drone engine was hand started and set to idle rpm. The truck then

Fig. 8.4 STAPL winch system (left) and operator control station (right).

moved forward down the long military runway to generate sufficient wind over the flight deck to spin up the rotor. When sufficient rotor speed was attained, the aircraft would lift off. The winch let out line and controlled the tension, much like lining out a kite. The engine rpm was adjusted as required. The winch would reel in the aircraft for landing on the trailer platform.

Initially the system experienced significant instability problems. There were short period pendulum problems at short line lengths. Letting the cable out more rapidly solved that problem. The more significant problems happened at greater line length. In one flight, the aircraft began to yaw to the right. Rather than correcting to the left, the aircraft turned sharply to the right and made a hard-over turn into the ground. Dr. Klemm reviewed the control law block diagrams at the Kaman plant after the incident. The diagram showed feedback loops and sensors. He discovered that the directional rate feedback loop should have had a positive feedback rather than a positive feedback. This problem was corrected prior to the next flight.

The most spectacular crash was caused by a pitch problem. The winch control system experienced a problem when an oscillation occurred. The autopilot system had pitch gain schedules based on line tension and the amount of line paid out. The aircraft pitched forward and crashed.

Little documentation exists for the STAPL program. The interim reports contain limited information. The final report would have provided hundreds of pages of details. Unfortunately, the Kaman lead engineer kept much of the development history in his head and, tragically, passed away from a heart attack before completing the report. A key lesson is that it is best to document key program information as it happens. Figure 8.5 shows side and top views of this rotorcraft, and Table 8.1 lists its technical characteristics.

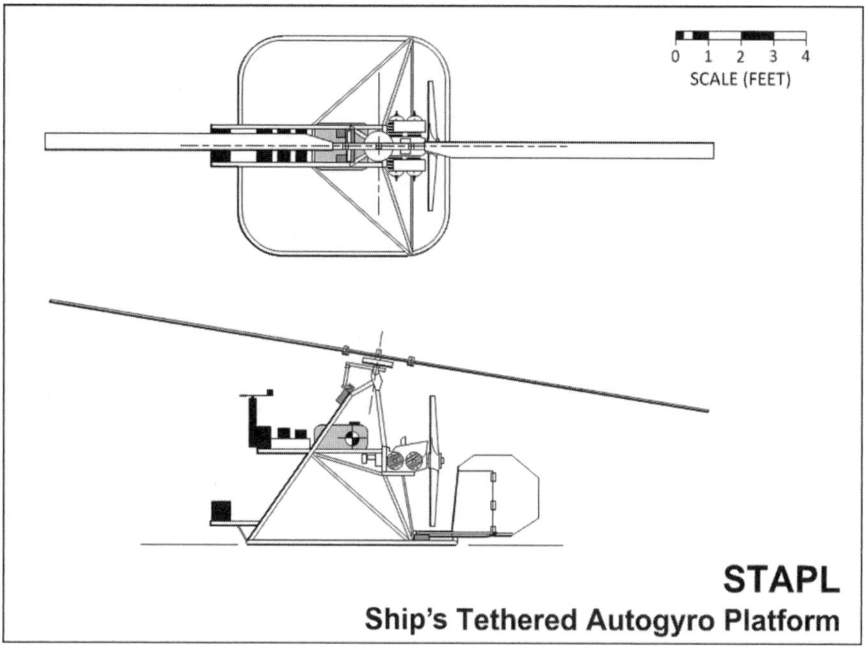

Fig. 8.5 STAPL side and top views.

Table 8.1 STAPL Technical Characteristics

Parameter	Value
Rotor diameter	20.0 ft
Length	10.1 ft
Rotor disk area	314 ft^2
Takeoff weight	550 lb
Empty weight	247 lb
Flight speed	45 to 50 KTAS
Endurance	2 hr
Launch method	Tethered launch with winch
Recovery method	Tethered recovery with winch
Propulsion type	McCulloch 4318, 2-stroke reciprocating engine
Power	72 hp

LIBELLE, 1986

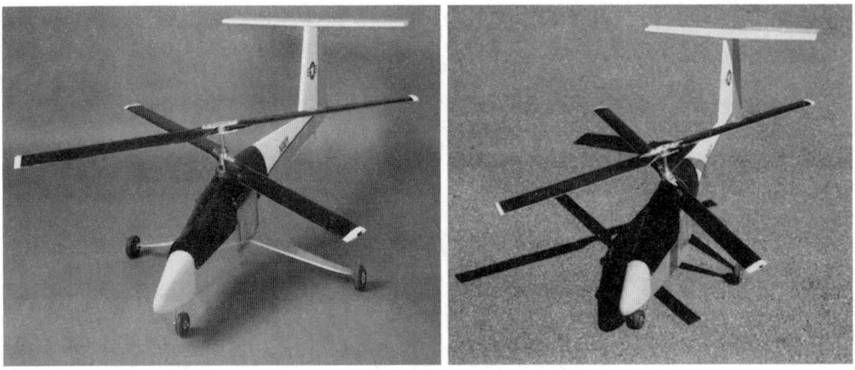

Fig. 8.6 Libelle demonstrator.

The purpose of the Libelle program was to evaluate a simple rotary wing EW decoy for ship protection missions (Fig. 8.6). An effective decoy must fly at ship-like speeds independent of wind direction, threat direction, or ship direction. Fixed wing aircraft can have difficulty achieving low enough airspeeds, so a rotorcraft with vertical lift has distinct advantages.

Unmanned helicopter technology was immature when the program began in 1986. NRL had never developed a helicopter autopilot, and it was difficult enough to make a stable, fixed wing aircraft autonomous at the time. The Libelle design concept was an attempt to leverage the advantages of rotorcraft but reduce the dynamics complexity to allow simple control laws. Cyclic and collective control of unmanned and model rotorcraft is routine today but was not widely available in the 1980s.

NRL already had experience with the STAPL tethered autogyro, but there were disadvantages to that approach. The rotor had to be prespun prior to flight. The rotor required forward airspeed for sustained rotor operation, which restricted its flight envelope. STAPL needed a low disk loading, which drove a large rotor that was difficult to stow and had a large deck footprint.

Libelle had a coaxial powered rotor. The opposite spin direction of the rotors was intended to trim out asymmetric lift of the advancing and retreating blades. Also, the coaxial arrangement canceled the torque of the rotors, so no tail rotor was needed. To simplify the design, Libelle had rigid rotor heads with no teeter or flap. There were no rotor controls other than power. The rotors were driven by a single electric motor that drove the rotors in opposite directions via a series of spur and bevel gears. The performance of the intended payload was not sensitive to vibration.

The Libelle demonstrator was intended to prove out the rotor design and control arrangement for an operational system. The demonstrator had a

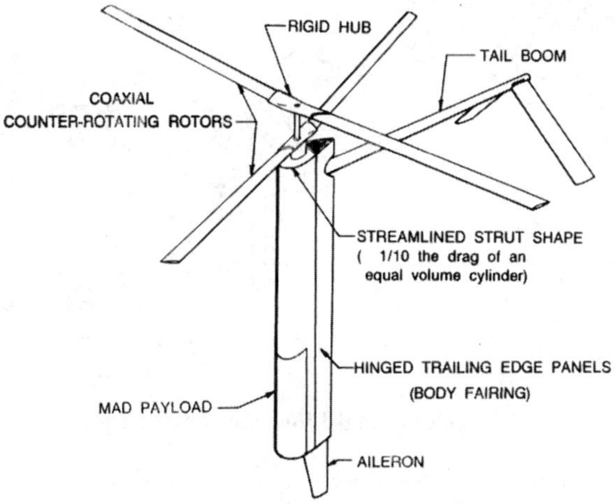

Fig. 8.7 Operational Libelle concept.

fuselage configured much like an airplane. However, the final operational design would fit within an MK 36 DLS and have a vertical cylindrical fuselage (Fig. 8.7). The Libelle would be fired bottom first with a rocket with the rotors folded along the fuselage sides (Fig. 8.8). The acceleration and drag would open the two sets of rotors. The rotors would begin to rotate as they deployed due to the blade pitch. The rotors would be allowed to over-deploy to reduce stresses during deceleration to 50 knots. The rotors stabilized the fuselage during deployment such that the fuselage was aligned

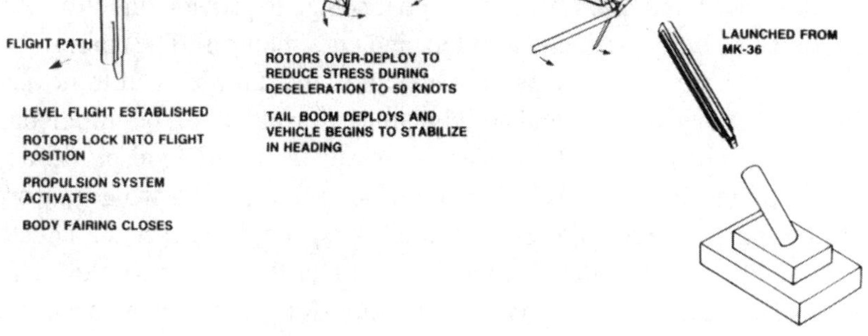

Fig. 8.8 Operational Libelle ship deployment concept.

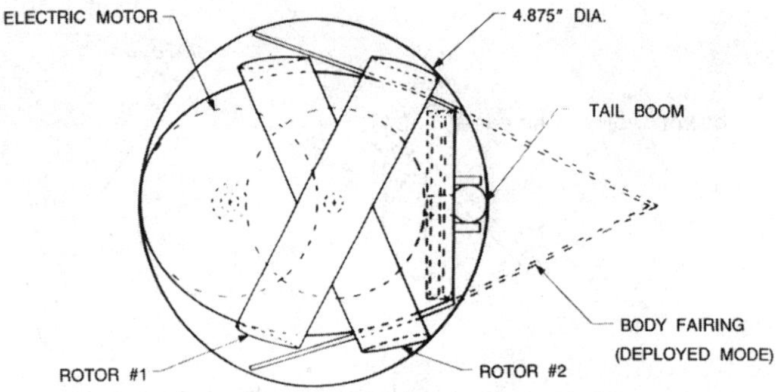

Fig. 8.9 Operational Libelle packaging concept.

with the airflow. Then a tail boom with an inverted V-tail deployed to provide directional and longitudinal stability. After the rotors and tail boom deployed, an aft fuselage fairing with a vertical aileron deployed by sliding along the fuselage (Fig. 8.9). The vertical fuselage acted as a pendulum for lateral stability. Once in level flight, the rotors would lock into the flight position, and power would be applied to the coaxial rotors.

The operational system could also be deployed from a manned aircraft such as a P-3. Rather than using a rocket launch, the Libelle would fall vertically with the rotors stowed. It would fall with the bottom of the fuselage pointed downwards. The 150- to 250-knot descent would force the rotors to deploy and start to spin. Other aspects of the deployment were similar to those of the ship-based rocket launch.

The demonstrator airframe was built of balsa and aircraft plywood to minimize the weight. The balsa wood was wet-formed to achieve the compound curves. Balsa planking provided rigidity. The structure was covered in MonoKote film model airplane covering. The hatch was made from thin sheet aluminum. The landing gear was made of aluminum and had commercial model aircraft wheels.

Libelle had a T-tail at the aft end of the fuselage to provide directional and longitudinal stability as well as pitch and yaw control. The aircraft was intended to have positive airspeed at all times and did not need to hover. It was hoped that the conventional aerodynamic controls would be sufficient for control during all phases of flight. The minimum designed flight speed was 10 to 15 knots, which compares favorably to alternative fixed wing platforms that could only achieve 25-knots minimum airspeed. Libelle had a negative $C_{l,\beta}$ (dihedral effect), so the designers hoped that the rudder would provide sufficient lateral-directional control. The elevator would pitch the aircraft forward such that the rotors would provide a thrust component. This is similar to a fixed wing aircraft on which the elevator controls airspeed.

The rotor had a high disk loading to support compact storage. The rotors were driven by a single cobalt brushed motor that was connected via a series of gears. Nickel cadmium batteries were intended to provide 3 minutes of endurance, though this was never realized due to crashes at takeoff.

Libelle was designed to have a rolling takeoff and landing. It had a tricycle landing gear configured such that the rotor had a forward tilted orientation while on the ground. The rotor would generate a thrust component to build airspeed. After sufficient airspeed was generated to provide rudder and elevator control authority, the aircraft would lift off.

Before starting the radio controlled (RC) prototypes, NRL built a rubber-band-powered balsa wood version. It had a coaxial rotor that was powered by a single rubber band with a torque reaction tube separating the rotors. The model helped establish the appropriate center of gravity (CG) position and trim settings. The model flew well enough to encourage the program to proceed. The rubber model's balsa wood rotors were flexible compared to the larger prototype's rigid rotors, which masked some of the relevant dynamics at this early stage.

Two prototypes were built for flight testing, but only one was fully completed. On the first flight, the prototype lifted off before reaching the minimum controllable airspeed. The aircraft crashed but was repairable. The control surface size was increased to reduce the required airspeed for control. Also, the nose gear length was decreased to give the prototype a more nose-down orientation, which gave Libelle a greater forward thrust component for the ground roll.

Rotor dynamics problems arose during later flights. After takeoff, the pilot pitched Libelle nose down with elevator command to gain more airspeed. The rotors acted as a gyroscope, which generated moments 90 degrees out of phase with the orientation change. The nose-down pitch generated a rolling moment that could not be overcome with the rudder. In other takeoff attempts, the aircraft rolled and pitched forward on the tricycle landing gear. This problem was aggravated by the nose-down landing gear stance.

The flight test program ended after six flights. The problems were judged to be insurmountable. The rotor dynamics were far more complex than anticipated. NRL engineers gained important rotorcraft experience from the program that helped with future rotorcraft programs such as Eager, Dragon Warrior, and SPIDER. Lessons learned include

- Rotorcraft should be controlled with rotor head controls. The elevator and rudder control system did not work at low speeds. Perhaps having the control surfaces operate in the higher dynamic pressure of the rotor wake would have improved controllability.

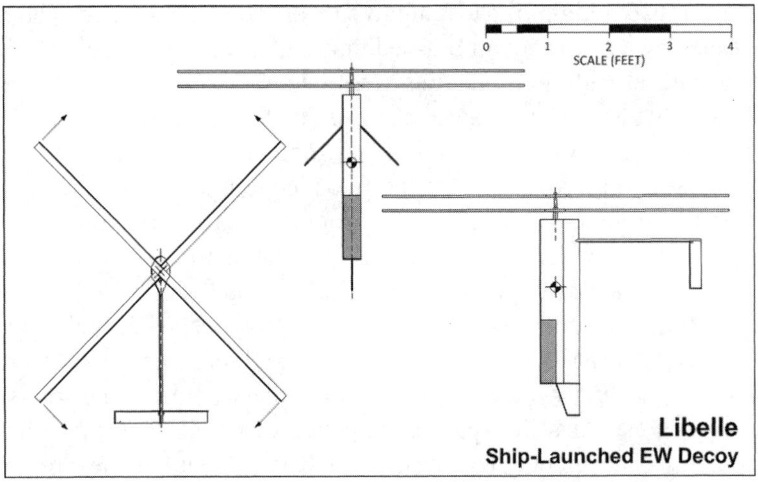

Fig. 8.10 Libelle three-view drawing.

- Elevator pitch commands resulted in a rolling moment due to gyroscopic precession. A modern control system could leverage this coupling phenomenon by commanding roll with the elevator.
- Coaxial rotors are subject to vibrations that must be balanced. The rigid rotor system did not appropriately manage the imbalances. Even slight inertia imbalances or flow differences created significant vibrations. A teetering system or delta hinge would have greatly improved the vibration problems.

Figure 8.10 shows a three-view drawing, and Table 8.2 lists Libelle's technical characteristics.

Table 8.2 Libelle Technical Characteristics

Parameter	Value
Rotor diameter	5.67 ft
Length	2.6 ft
Height	3.6 ft
Empty weight	10 lb
Payload weight	2 lb
Battery weight	3 lb
Loiter speed	25 KTAS
Dash speed	50 KTAS
Endurance	5 min
Launch method	Ballistic rocket launch from the MK 36 DLS
Recovery method	Expendable
Propulsion type	Battery electric coaxial rotor

COMMANDER'S OVERHEAD VEHICLE FOR ELEVATED RECONNAISSANCE (COVER), 1986–1989

Fig. 8.11 COVER vehicle.

The purpose of the Defense Advanced Research Projects Agency (DARPA) Commander's Overhead Vehicle for Elevated Reconnaissance (COVER) program was to develop a tethered hovering platform to support tank operations (Figs. 8.11 and 8.12). The elevated platform was intended to carry cameras and synthetic aperture radar (SAR) payloads for protection and targeting. DARPA ran contracts with Tracor and NRL; Moller was a subcontractor to TRACOR.

DARPA wanted to leverage an existing Moller four-duct electric vehicle design. This Paul Moller design built upon his flying car research. The vehicle had four ducted fans housed within a single body. The power was generated on the ground and was brought to the vehicle via a high-voltage cable. The 10.5-inch-diameter fans had fixed stators, moving control vanes, and two blades. The fan blades were 12-inch model aircraft propellers with tips clipped to the reduced diameter. NRL designed a custom fan with molded composite blades. Testing proved that this custom fan was significantly more efficient than the model airplane propellers, but they proved to be extremely brittle and vulnerable to damage. The payload was stored in the center of the vehicle. Alternative payload locations were located around the body perimeter.

The electric propulsion system was distributed between the air and ground elements. The system nominally operated at around 150 volts DC. The high voltage yielded low resistive losses and allowed a small cable diameter for the tether. The onboard electric motors were rare earth samarium cobalt brushed motors that were approximately 60% efficient. The low efficiency created heating problems due to losses, which decreased the peak motor power. Later

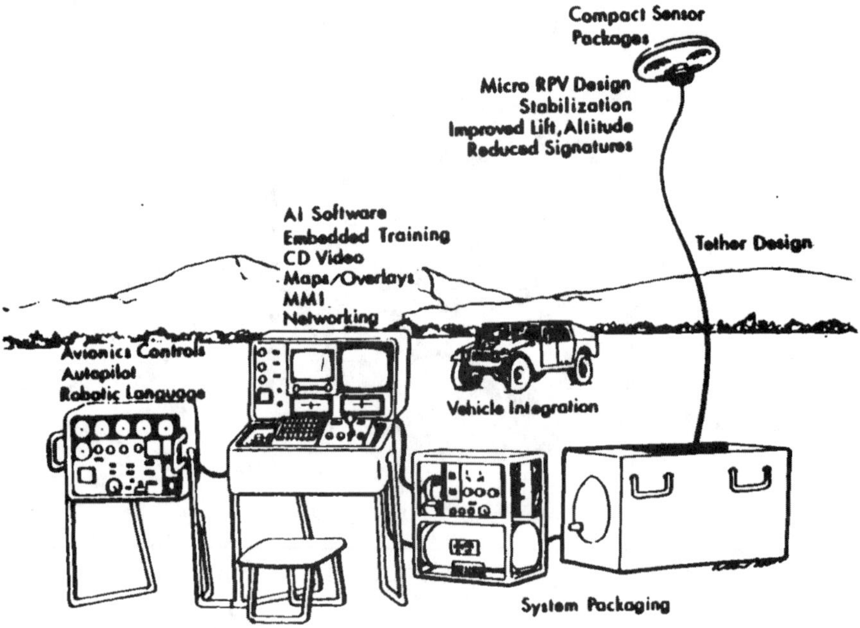

Fig. 8.12 COVER system.

in the program, the voltage was increased to 400 volts DC to reduce the tether weight and to further reduce resistive losses. The new tether had electrical cables of 1/8-inch diameter. The higher voltage hit a motor voltage limit, where the brushes continuously arced, giving the appearance of blue fire.

NRL's roles in the program were to make the COVER platform autonomous and to support systems design. The platform was radio controlled at the beginning of the program. With its high disk loading, neutral stability or instability, lack of damping, and fast control responses, it was challenging for the pilots, but satisfying when mastered. The end objectives were to make the platform deployable with a keyboard command, to automatically station-keep with the tank, and to be oriented based on the desired sensor field of view.

NRL developed the flight control system. First, NRL pilots and engineers learned to manually fly the vehicle via radio control at Moller's facilities. NRL then wrote the control laws and evaluated them in the six-degree-of-freedom (6-DOF) simulator. Once NRL refined the control laws, the autopilot was integrated into the COVER vehicle. Control gains were adjusted between flights, and many of the flights consisted of short hops.

Camera pointing proved to be challenging. Initially the camera was fixed to the platform with no ability to gimbal. Camera pointing was accomplished by adjusting the vehicle heading. To move the aircraft forward, the vehicle had to pitch nose down to generate a forward thrust vector component. The ducts could not be tilted independently of the body. To help remedy this problem, NRL added a pan-tilt camera gimbal to permit camera pointing independent of the vehicle orientation. NRL did not fully close the camera pointing control loops by the end of the program.

DARPA also wanted COVER to support a synthetic aperture radar payload that was conformal to the body. The system had to estimate the vehicle position within 1 centimeter relative to the base with a tethered distance of 300 feet. The system utilized an RF kinematic relative navigation system (pre-GPS) to provide precise estimations of the vehicle position relative to the base. Tight attitude estimation information was also required. This vehicle state information was needed to enable SAR processing for an accurate image. Ultimately the system was not able to station-keep adequately or provide sufficiently precise state information for SAR imagery.

The tether had some effects on the flight dynamics. It had weight and drag impacts on the vehicle. As the vehicle moved forward, the tether formed a catenary angle due to drag and inertia. The angle of the tether away from vertical generated a retarding force that required additional nose-down pitch to overcome. The test program did not use a winch with a tether tension control system. The tether passively spooled out and then loosely rested on the ground. Ground crew manually wound in the tether after the flight. Full system development would require a more sophisticated controlled winch system with a level winding reel that would provide figure-eight spooling to prevent wire twisting.

The flying duct system had large pitching moments due to horizontal flow over the ducts. The vehicle was found to be unstable when the airspeed was approximately one-third of the duct velocity. This problem was never fully solved. In hindsight, a tail would have made the platform much more controllable. The neutral point in forward flight was close to the quarter chord, which was ahead of the center of the vehicle. The CG was at the center of the circular planform, so the vehicle had a negative static margin in forward flight. This resulted in unstable pitch stability and a strong pitch-up tendency.

NRL conducted numerous flight tests. The control system worked reasonably well when hovering over a fixed base with a static air vehicle position. The system had more difficulty with a moving base. This was the first time that NRL made a neutrally stable or unstable platform autonomous. All previous NRL autonomous platforms were inherently stable.

The flight test program experienced a notable pilot-induced oscillation (PIO) event. The test took place within a warehouse. Pilots were controlling the vehicle via radio control, and they were becoming familiar with the flight behavior. One pilot flew COVER too high, and it hit the ceiling. The vehicle then fell rapidly to the ground as the pilot attempted to apply full power. It impacted the ground, bounced, and flew back up to impact the ceiling once more under full power. After the second ceiling impact, COVER fell back to the ground. Despite the trauma, the rugged COVER survived the incident.

COVER had a distinctive acoustic signature. The system had four ducts with multiblade rotors, stators, and control vanes. The fans had a high disk loading, which produced significant noise. The motor brushes tended to arc due to the high voltage. Also, the duct acoustics were highly dependent upon vehicle orientation relative to the observer. Some characterized it as sounding like a nest of angry hornets.

The program ended without fully accomplishing the objective of providing a practical payload platform for tank operations. It turned out to be a considerably more difficult problem than anticipated by the sponsor, and the program did not have sufficient resources to address all the challenges. The mission remains valid today, and a modern version of COVER would benefit from the availability of high-efficiency brushless motors, GPS navigation, and many other newer UAS technologies. Several companies are today pursuing similar tethered platforms, which generally employ a quad rotor configuration. Figure 8.13 shows a three-view drawing of COVER, and Table 8.3 lists its technical characteristics.

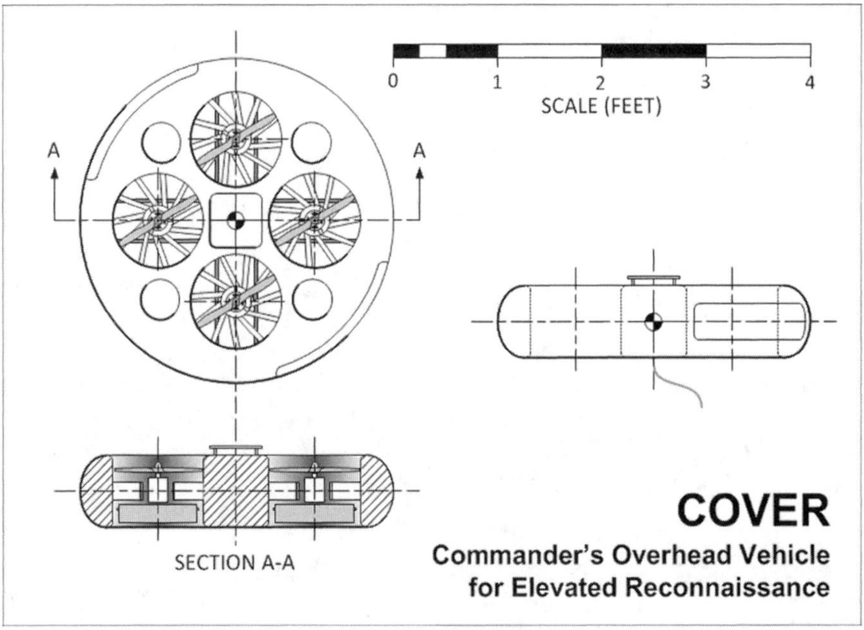

Fig. 8.13 COVER three-view drawing.

Table 8.3 COVER Technical Characteristics

Parameter	Value
Diameter	2.8 ft
Takeoff weight	10.0 lb
Payload weight	3.0 lb
Altitude	300 ft (limited by tether length)
Endurance	Vehicle lifespan (100 hr approx.)
Launch method	Vertical launch
Recovery method	Vertical landing
Propulsion type	Electric motors and ducted fans Ground power source transmitted via tether

SMALL SHIPS COMPATIBLE DECOY (SSCD) AUTOGYRO, 1992–1993

Fig. 8.14 M-wing SSCD Autogyro drop model.

The purpose of the Small Ships Compatible Decoy (SSCD) Autogyro was to defend small ships against anti-ship missiles with an EW payload (Fig. 8.14). The 10-pound payload was a radar repeater that generated a false target, which leveraged the AAED payload technology (Chapter 3). Since the system defended small ships such as destroyers, the payload antenna could be relatively compact, and so the vehicle could be small, as well. The program was an exploratory development program to advance the vehicle design.

SSCD was a powered vehicle capable of flight for 5 to 10 minutes. It would be launched from a ship and fly a nonballistic flight. Initially, rigid wing unfolding aircraft were considered. However, rotary wings could fold up more compactly, so the rotary wing approach was selected.

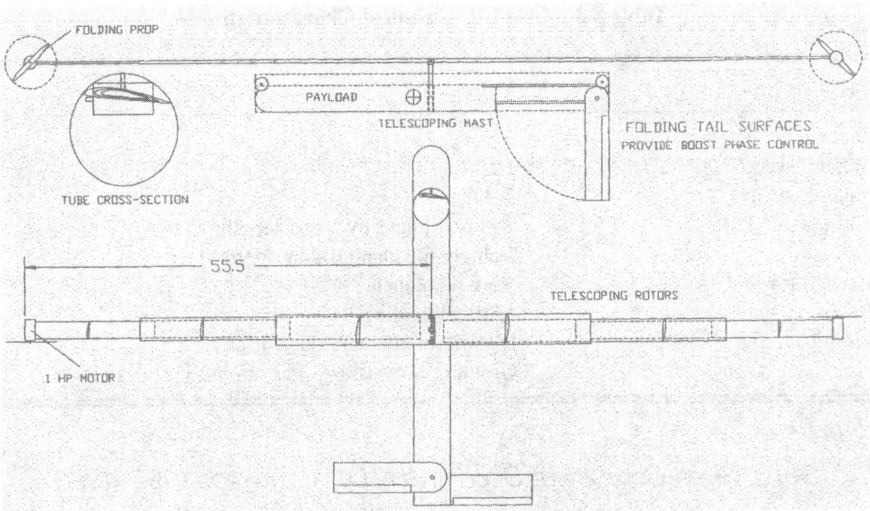

Fig. 8.15 Tip propeller SSCD concept.

Several rotary wing configurations were considered at the conceptual stage. One design was a dual-rotor system with electric tip propellers (Fig. 8.15). The rotors telescoped out to provide a low disk loading. Another design was a telescoping monocopter with an electrically powered propeller on the counterbalance side (Fig. 8.16). Centripetal acceleration would cause the rotors to extend, though telescoping rotors could produce large vibration if the deployment was uneven. Finally, an autogyro with an electric propeller for vehicle thrust was considered (Fig. 8.17). All vehicles had pivoting tail surfaces.

The operational vehicle concept would use a solid propellant rocket motor for a ballistic launch from an MK 36 DLS (Fig. 8.18). The system would fit within a 60-inch-long, 5-inch-diameter canister. The booster included fins to provide stability during the boost phase. After booster burnout and separation, the rotors pivoted back from the stowed position along the fuselage. The rotors began to spin up as the vehicle approached apogee. The rotors also provided weather vane stability during spin-up and helped orient the vehicle, which eliminated the need for a stabilizing parachute. The vehicle then had an autorotative descent with the nose pointed along the flight path, and the body swung about the rotor mast. The rotor head was wide enough that the blades did not interfere with the body as they spun up. The tails deployed and the propeller unfolded after the rotor mast was elevated. Once the rotors generated lift and the propulsion system was operational, the vehicle pulled out of the dive into level powered flight.

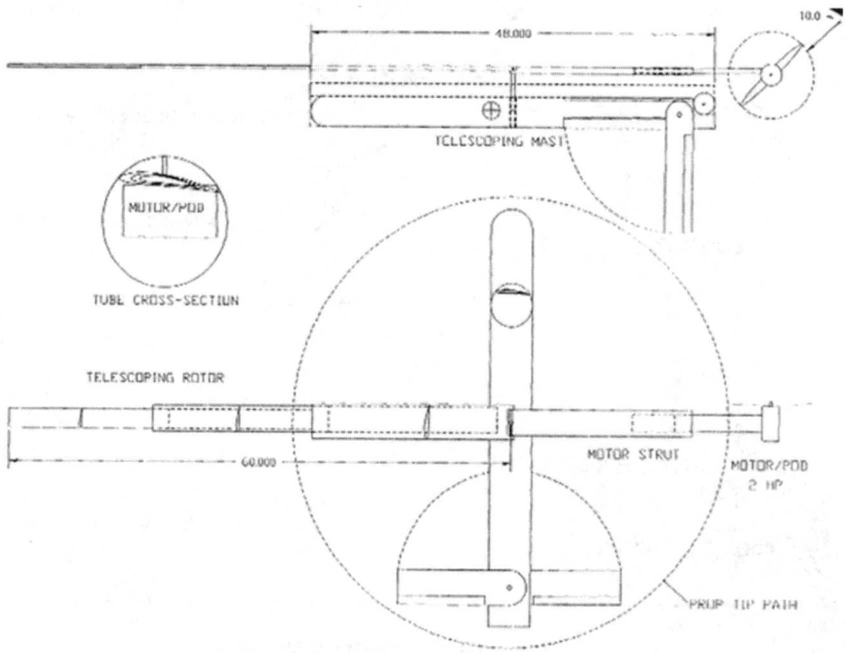

Fig. 8.16 Single-blade rotorcraft SSCD concept (Samara concept).

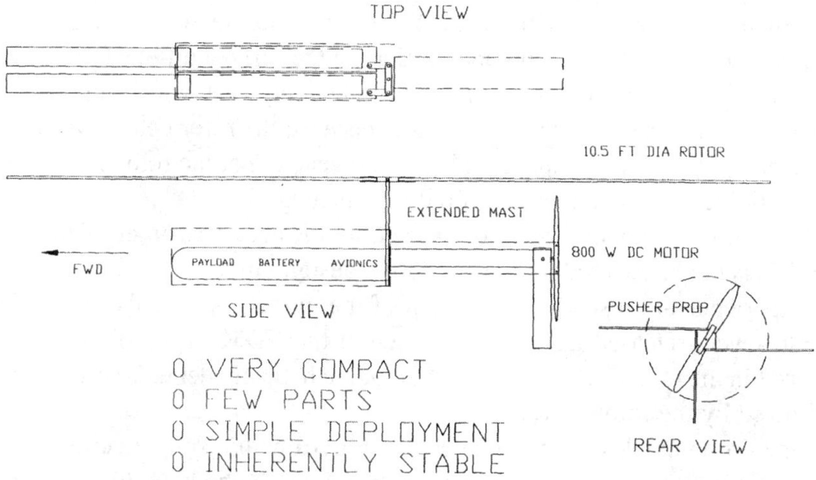

Fig. 8.17 Initial autogyro SSCD concept.

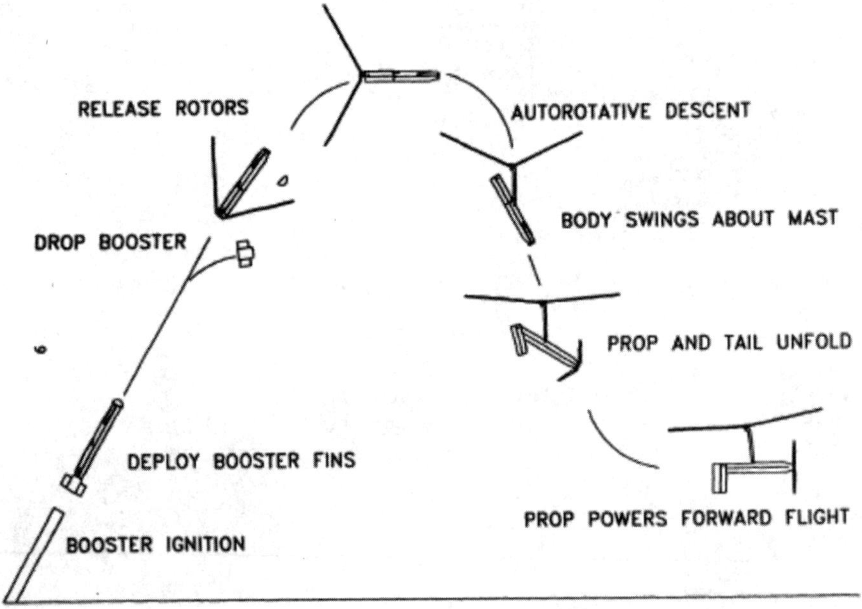

Fig. 8.18 SSCD deployment concept.

The simplest rotary wing design was a teetering hub two-blade autogyro. The rotors had a delta 3 hinge with coupling between coning and feathering. This provided negative pitch with coning angle, which helped the rotor spin up more rapidly. As with Libelle, no rotor head controls were used.

The opening autogyro system was tested in a wind tunnel before the flight demonstrators were built (Fig. 8.19). It was tested in horizontal (perpendicular to the airflow) and vertical (edge-on) conditions to represent opening and forward flight conditions, respectively. The wind tunnel had to operate at low speeds for rotor opening, because the rotor could over-speed and shed parts at higher speeds. The tests revealed that the rotor could open, and little vertical speed was required for opening.

Several drop models were constructed and tested. The order of the drop models is not known today, so the relative design influences and progression of experience are lost. However, each of the models is described here. The models were released from the underside of the NDM I and NDM II utility aircraft in the folded configuration. After launch, the model tails sprung open, followed by the rotors and mast.

One drop model was an unpowered autogyro with a simple vertical body and a rotor on top. The blades were folded along the body at drop, and then opened in descent. The body was constructed of foam. The design did not have a tail (Fig. 8.20). When the blades opened, the body spun in the opposite

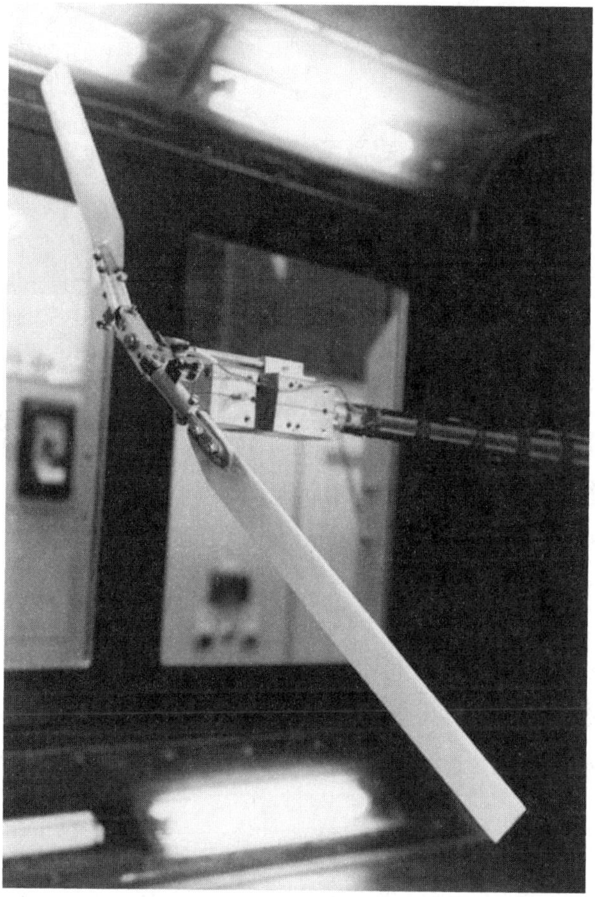

Fig. 8.19 SSCD Autogyro rotor spin-up wind tunnel test.

direction of the rotor. The fuselage had a lower polar moment of inertia than the rotors, so the rotors did not spin as fast as in a version with a tail-equipped horizontal body providing aerodynamic and inertial damping. The rotors did not achieve the required autorotation rpm. The model drifted into Foch's parked car and damaged its roof. Foch had the thrill of informing his auto insurance company that an aircraft crashed into his car and he was the pilot!

Another drop model had a horizontal fuselage with a unique configuration for the tail. The drop model had a folding mast with the rotor blades. The mast was designed to rotate up from the body, to be followed by rotor deployment and spin-up. The rotor and mast deployed before the tail deployed to avoid interference. The large tail resembled the letter "M," with large dihedral for the inboard panel and large anhedral for the outboard panel (Fig. 8.21). Ailerons on the tail provided roll control to counter the large

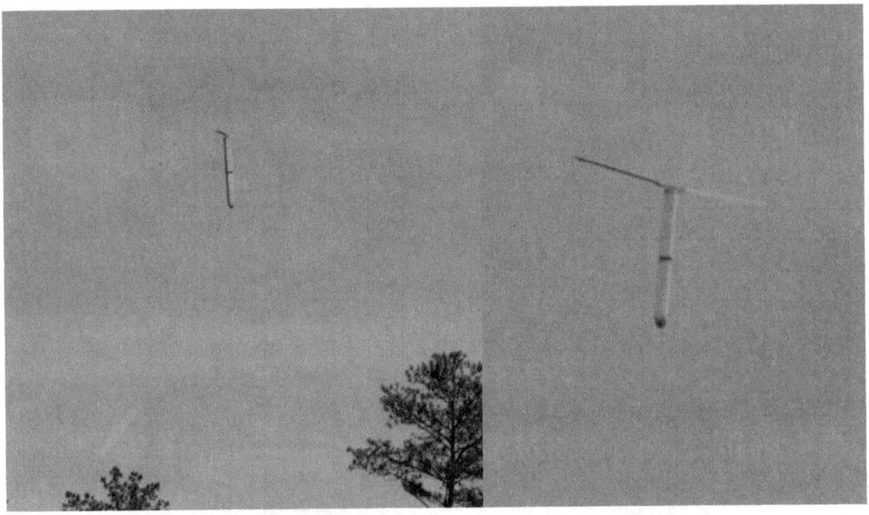

Fig. 8.20 SSCD drop model with foam body and no tail.

gyroscopic rolling moments due to rotor pitch-up. This model was not successful because the rotor never spun up to the desired speed. The fuselage spun in the opposite direction of the rotor while descending.

Another drop model was built with a vertical body and tails (Fig. 8.22). This represented a potential operational configuration with a vertically separated set of antennas. The rotor was mounted to the top of the body. It had a boom-mounted tail that folded down from the fuselage. The tails had a "U" shape, in which the vertical tails projected upward from the tips of the horizontal tail. This system was dropped from the NDM II utility aircraft. As with the M-tail SSCD, the rotor did not spin up to the necessary rpm. This vehicle attained a stable flat spin mode in which the body swung around in an arc. The centripetal acceleration threw the body to the outside of the arc. (Similar behavior was experienced by Pendulus.) Such motion was difficult to control as an RC model, but modern autopilots would have little difficulty in correcting this behavior. This drop model also showed that the rotor mast would need to cant aft more to achieve the necessary rotor pitch angle to operate as an autogyro.

One of the drop models experimented with tip plates to facilitate rotor deployment and spin-up. When the rotors were folded along the fuselage, the tip plates served as small rotors. The plate pitch angle helped facilitate initial rotation, while the drag produced a moment that encouraged blade opening. These tip plates would fling off after the rotors spun up. These devices resembled winglets.

NRL attempted several drops (Fig. 8.23). Several delta hinge angles were also tested, but the blades always stalled. This hinge angle affected the coning

Fig. 8.21 SSCD M-wing drop model.

for the efficient angle of attack during autorotation. Another drop attempted to open the mast, but the rotor would not spin up. Despite several adjustments, the rotor never achieved the desired rpm.

A ground-launched, powered model was built and flown to verify that an autogyro of SSCD size was feasible (Fig. 8.24). Like the air-dropped models, it was equipped with a teetering rotor head. This autogyro did not have rotor

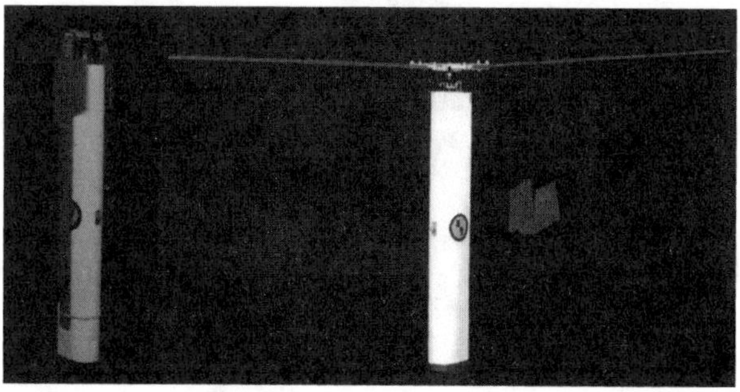

Fig. 8.22 SSCD drop model with vertical body and tails.

Fig. 8.23 Vertical-body SSCD drop vehicle on NDM II. (Test team, left to right: Steve Tayman, Peggy Davidson, Greg Page, Rick Foch, Barry Walden, Mark Haupt, Pete Chaplin, Lew Goldberg, and Chris Bovais.)

head controls, so only rudder and elevator were used. An electric motor and propeller were mounted to the front of the vehicle. The landing gear had a tail wheel configuration. The wheeled vehicle would taxi to generate forward speed to spin up the rotor for initial tests. Later a pull-string prerotator drum was added to facilitate rotor spin-up. The ground handling was challenging due to a forward gear placement, but this was manageable for the RC test pilot. The vehicle would run up and down the runway to get the rotor spun up. Then full throttle was applied to get airborne. The aerodynamic controls did not have sufficient authority. The aircraft rolled to the left as it took off. It continued to roll despite full up elevator and

Fig. 8.24 Ground-launched SSCD Autogyro demonstrator.

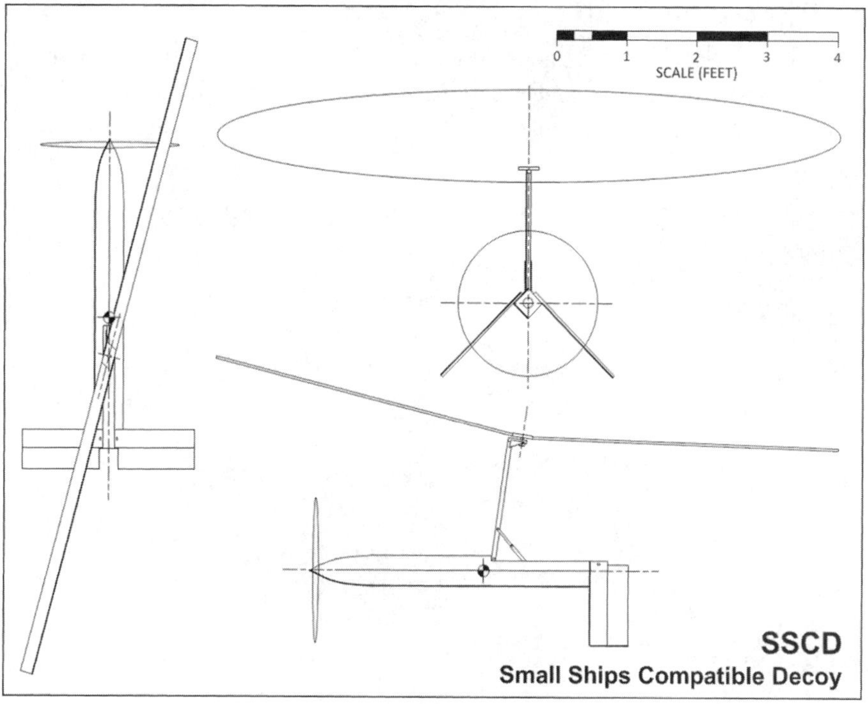

Fig. 8.25 One operational SSCD design three-view drawing.

full right rudder. It could not recover, and it crashed into the ground. The rotor parts flew hundreds of feet, clearly demonstrating the principle of conservation of angular momentum.

Rotor head controls could have prevented the crash by providing the corrective roll control at low speeds. The simple aerodynamic surface controls approach was not adequate for the vehicle dynamics. Like Libelle, it used rudder to turn and elevator for pitch without rotor head controls. Unlike with Libelle, the rotor was unpowered, and forward thrust was generated by a dedicated propeller. Figure 8.25 shows a three-view drawing of an operational SSCD design.

The lessons learned with the SSCD were that a rotary wing design had favorable packaging characteristics and could be developed into a successful, deployable unmanned aircraft so long as rotor controls were incorporated instead of fixed aerodynamic control surfaces. The experience gained here and with NRL's previous rotary wing designs enabled the successful Eager ATD, Dragon Warrior, and SPIDER programs.

SSCD: FURTHER READING

Tayman, S. K., "Wind Tunnel Tests of a 42 Inch Diameter Self-Starting Autogyro Rotor," Naval Research Laboratory, NRL/MR/5710-94-7485, Washington, DC, May 1994.

ROTARY WING DECOY (ROWDY) 1, 1995

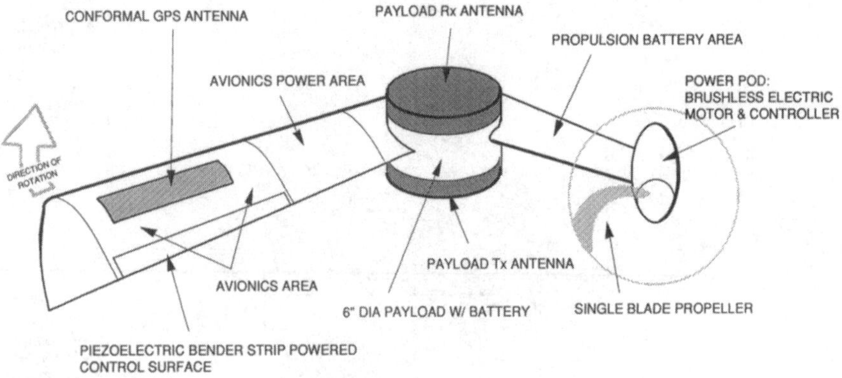

Fig. 8.26 ROWDY 1 concept.

ROWDY 1 was a hovering decoy concept based on a samara (maple) seed (Fig. 8.26). It would serve to protect hovering helicopters against radar-guided missiles. The flight performance goal for this simple, single-blade design was to station-keep within a 100-foot diameter and within 20 to 100 feet of altitude. Its proposed EW payload weighed 12 pounds including the power supply. ROWDY 1 was designed to fly for 3 minutes using electric propulsion. The rotary motion was produced by the propeller. ROWDY 1 was not developed beyond a conceptual design, possibly because as an all-rotating system, the guidance and control were seen as too complex by its potential sponsors. Today, guidance and control for ROWDY 1 would be straightforward because of advances in sensors and the wealth of experience available in automating vehicles with complex dynamic characteristics. In addition to its original EW mission, ROWDY 1 could be readily equipped with chemical or biological agent detectors and would be an ideal platform for mapping the dispersion of such agents as they slowly translate through areas of interest.

The ROWDY 1 concept is the simplest helicopter configuration capable of efficient hovering and forward flight, and it merits further development. Figure 8.27 shows the ROWDY 1 trim equations. Currently, industry and university research projects are developing similar designs as MAVs.

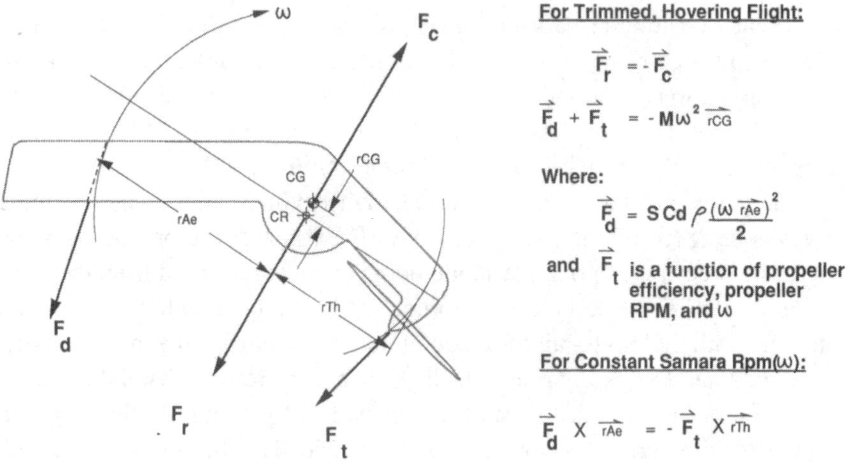

Fig. 8.27 ROWDY trim equations.

EAGER, 1995–1997

Fig. 8.28 Eager in flight.

Eager was a preferential acquisition decoy designed to create an alternative false RF target to spoof radar-guided anti-ship missiles (Fig. 8.28). Unlike the names of most NRL projects, the name "Eager" is not an acronym. The NRL was simply "eager" to develop another VTOL UAS! Regardless, the name was well liked by the sponsors and performers. Eager was a rotorcraft tethered to the ship, with electrical power and signals sent through the tether. Eager was built on the experience gained from developing the earlier STAPL and COVER tethered VTOL programs. It carried a 25-pound EW payload. Eager flew approximately 500 feet away from the ship at a 300-foot altitude with a tether length of approximately 700 feet. The elevated position provided a longer RF horizon due to Earth's curvature, but lower altitude was desired for ship defense. It was designed to fly in all weather conditions continuously for 1000 hours, thanks to the ship-provided power. The system would be recovered after use. Like the FLYRT decoy, Eager was an Advanced Technology Demonstration (ATD) program.

The Eager payload was vertically oriented in a cylindrical fuselage. A single rotor was mounted on the top of the body. The rotor, tail rotor, and airframe were custom designs. Initially, a single boom extended outward from the rotor assembly with a single tail rotor at its end. Subscale model flight tests revealed a problem with the single-tail rotor configuration. To engage the threat missile, the payload had to flexibly orient itself. Under some relative wind directions, the tail rotor blades tended to stall. Eager had to fly alongside the ship and orient in any direction. A second tail boom was added on the opposite side such that at least one tail rotor would be in favorable flow at all times. Now Eager could fly in any direction with no yaw control problems.

Wind tunnel tests revealed an unexpected problem. The fuselage generated a much larger drag and had an oscillatory wake shedding because the fuselage was a cylinder operating at subcritical Reynolds numbers. The solution to the problem was to add vertically oriented boundary trip strips around the fuselage.

The tether system was a major development challenge. It was desirable to make the tether as small as possible. A large tether diameter would create performance problems due to weight and drag effects. The tether formed a catenary in 50-knot winds, and the forces imparted on the aircraft increased with larger diameter. The aircraft would need more power to lift a larger cable, which in turn would require thicker cable to transfer the power. Two tether systems were developed: one with a moderately high voltage and one with a very high voltage. The former had a voltage of a few hundred volts carrying 12 kilowatts and was 3/16 inches in diameter. The latter high-voltage system was beyond the motor and converter technology of the time. The tether was a Kevlar-sheathed multiconductor cable consisting of signal lines and three power cables.

NRL developed an approach to allow Eager to land on a ship deck without tipping over. The aircraft had a high CG with narrow landing gear stance. The contemporary Canadair CL-227 had similar challenges, and it had legs with fittings that locked into a metal surface grid. The fittings functioned as a positive latching device. NRL developed a custom latching system for Eager inspired by that of the CL-227.

NRL took an innovative approach to the flight control system in order to push the state of the art. The autopilot used neural networks. It was taught to fly through simulation, and it improved over time. The control laws were not fully deterministic. Ultimately this approach was not successful because it yielded inconsistent performance.

The next set of control laws had fully deterministic linear inner and outer loops. The demonstration was flown with these traditional control laws. One purpose of an ATD program is to push technologies. NRL attempted the neural network autopilot and very-high-voltage tether system without success. However, much was learned. The fallback technologies satisfied the system needs.

The full-scale Eager rotor and tail rotor system was captive tested at the U.S. Naval Academy rotary wing test lab. These tests helped verify the rotor performance prior to constructing and testing the flying prototype.

NRL developed a capstan reel system. It had a level wind reel that used a figure eight winding pattern that did not introduce twist. By contrast, a traditional drum reel twists the cable once per revolution.

The first flights took place from the ground. Initial flight tests had Eager suspended below a safety cable. Bearings were installed on the top of the rotor head to keep the cable from twisting. A human operator manually maintained tension on the cable. In the event of a controls failure, the cable would snub to prevent the vehicle from falling to the ground. After the initial suspended flights, this safety system was eliminated.

Flight controls issues emerged due to coupling of the tether and the autopilot. The rate and position of the tether extension determined loads on the aircraft. Experience from the STAPL tether system had taught NRL engineers that a more sophisticated system was required for successful operation. The tether tension was actively controlled by the winch. When the vehicle commanded tether pull, the reel system let out more tether. The reel system monitored how much tether was left, and it communicated this status to the aircraft via the tether signal lines. Ultimately, the control system used an algorithm in the winch that had line tension proportional to tether length. This algorithm did not require vehicle feedback. NRL developed a real-time kinematic (RTK) GPS so that Eager could fly precisely relative to the ship. Information from the airborne GPS receiver was sent to the ground via fiber optics, and the RTK processor was on the ground. The system worked quite well, and Eager was able to land with a few centimeters' accuracy. This was NRL's first autonomous hovering vehicle that used GPS.

The Eager program had multiple test flights. The first tests were suspended below a test rig. Next, short tether flight without the test rig was conducted. This flight lasted for 6 hours, and the endurance was limited by test range scheduling rather than the vehicle. Then the system was tested with the full line length at Blossom Point, Maryland.

The final test of the Eager ATD was a flight from a barge anchored in the Chesapeake Bay. The vehicle launched itself successfully, but the winch

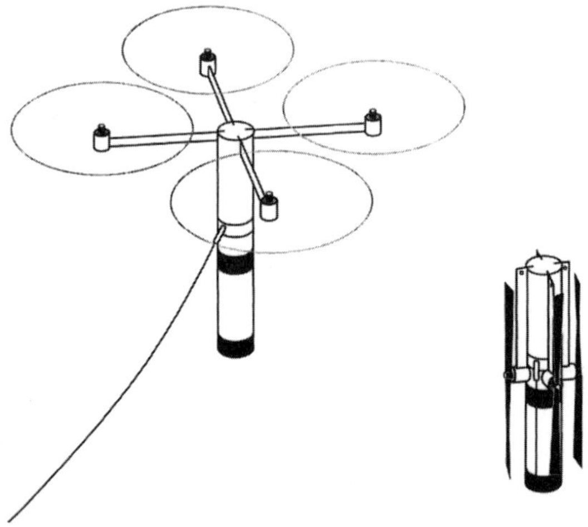

Fig. 8.29 Eager quad rotor concept.

system failed. The winch braking system locked after a power glitch in the control hut. Since the winch system was based on crane technology, it defaulted to braking upon boot-up as a safety feature. This braking system feature was provided by the vendor, though NRL did not ask for it. The winch was intended to provide a line tension that was proportional to the tether length. The system commanded more cable to be extended, but the winch did not comply. The autopilot did not realize that the tether brakes were active, since the line tension hysteresis characteristics produced similar behavior. The aircraft body aligned itself with the taut tether, such that it tilted to the side. Soon the angle increased to the point where there was insufficient lift to maintain flight. The vehicle arced down at the end of the tether until it impacted the water. This first Eager airframe was lost. The second one remains at the Vehicle Research Section (VRS) museum at NRL.

ATD programs last for three years, and the demonstration typically occurs at the end of September on the third year. Like most ATDs, Eager had a single opportunity to demonstrate its technology. NRL would have liked to continue the development and solve the winch control problems. Nevertheless, Eager was considered a successful technology demonstrator by its sponsors, and the effort ended as scheduled in 1997. There was no clear transition path to introduce Eager to the Fleet.

Tethered systems have great advantages, but they have very complicated dynamics. These vehicles are sensitive to relative wind, vibration, and cable motion. The Eager winch system required further development. Eager demonstrated what was possible, and it identified remaining technology challenges. The program helped advance electric motor, electrically driven

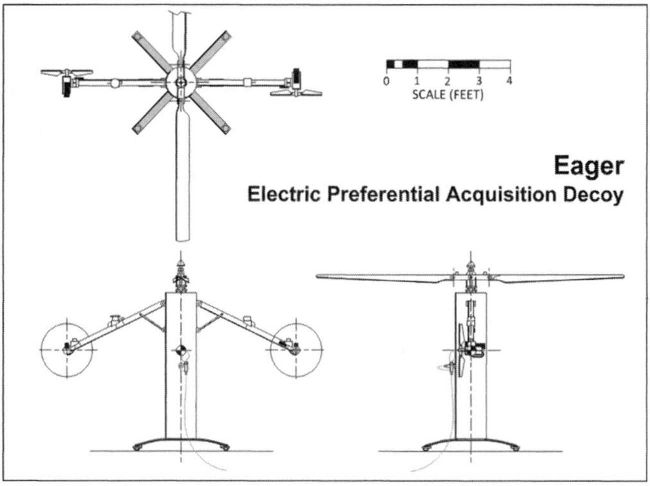

Fig. 8.30 Eager three-view drawing.

tail rotor, RTK GPS, and helicopter autonomy technologies. The electric tail rotor technology was successfully applied to the later Dragon Warrior helicopter program.

One early Eager concept employed an electric quad rotor (Fig. 8.29). The motor-driven rotors were mounted on booms that folded along the fuselage for storage. This configuration was less efficient than the selected single rotor configuration. Electric quad rotors have become quite popular for simple VTOL UASs in recent years. Figure 8.30 shows a three-view drawing of the Eager, and Table 8.4 lists its technical characteristics.

Table 8.4 Eager Technical Characteristics

Parameter	Value
Rotor diameter	10.0 ft
Height	6.0 ft
Takeoff weight	110.0 lb
Empty weight	95.0 lb
Payload weight	15.0 lb
Dash speed	50 KTAS (approx.)
Endurance	1000 hr (design) 6 hr (demonstrated)
Launch method	Vertical takeoff
Recovery method	Vertical landing
Propulsion type	Electric motors
Power	12 kW

HIGH POINT, 1999–2000

Fig. 8.31 High Point.

High Point was a tethered electric ducted rotor unmanned aircraft (Fig. 8.31) intended to provide surveillance capabilities to a special forces unit in an urban warfare environment without putting personnel in harm's way. The vehicle was tethered directly to the operator, who carried the battery pack or power supply. Once the unit reached a corner, High Point would be deployed and would look around the corner.

The electric tether system was first tested with an electric helicopter (Fig. 8.32). An ECO-16 model helicopter was adapted by removing the

Fig. 8.32 High Point proof-of-concept tethered helicopter test vehicle.

battery, adding the power tether, replacing the motor with a high-voltage motor, and adding a video camera.

The design had many notable features. The aircraft had coaxial counter-rotating rotors that were vertically separated. Each rotor had a ring-shaped rotor guard and stators. The large gap between the rotor guard and the rotor tip made the design more of a shrouded propeller than a true ducted fan. The purpose of the rotor guards was to prevent damage to the rotors if the vehicle hit an obstruction such as the side of a building. Each rotor had cyclic and collective control. Yaw was accomplished through differential collective pitch. A single brushless motor drove both rotors via bevel gears. A basket on the bottom of the vehicle had a reloadable spool of tether, which was based on fiber-optic guided missile spools and the towed variants of AAED. The 320-volt battery consisted of four individual 500 amp-hour packs arranged in series. The visible camera, which was left over from COVER, was located on the top of the vehicle. The rotors were 22 inches in diameter. The total vehicle weight was approximately 6 pounds without the tether. Unlike COVER and Eager, the system had relatively low voltage due to the battery pack limitations. The system had rate gyros for a rudimentary stability augmentation system, but it was not fully autonomous.

A prototype system was developed and demonstrated. However, no customer funding was available to fully develop the system. The system showed promise, including low-noise operation. Figure 8.33 shows a three-view drawing, and Table 8.5 lists the High Point's technical characteristics.

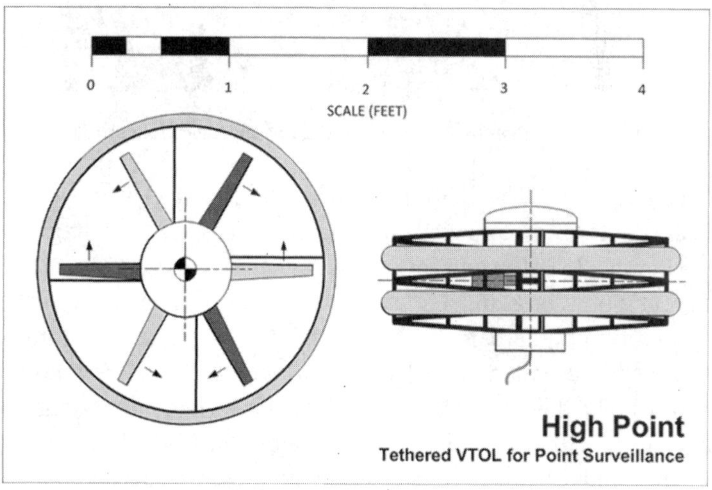

Fig. 8.33 High Point three-view drawing.

Table 8.5 High Point Technical Characteristics

Parameter	Value
Rotor diameter	22.0 in.
Takeoff weight	6.0 lb (no tether)
Empty weight	5.0 lb
Payload weight	1.0 lb
Dash speed	20 KTAS
Endurance	15 min
Launch method	Vertical launch
Recovery method	Vertical landing
Propulsion type	Tethered electric
Power	500 W

DRAGON WARRIOR/VANTAGE, 2001–2004

Fig. 8.34 Dragon Warrior tactical unmanned helicopter.

The Marine Corps Warfighting Laboratory (MCWL) was working on a tactical-class VTOL unmanned aircraft with Sikorsky called Dragon Warrior (Figs. 8.34 and 8.35). This aircraft had a single ducted fan with wings. It had developmental problems and it could not meet the government specifications. The configuration fundamentally limited its growth potential. More fuel was needed to meet the endurance requirement, which had to be housed within the toroidal body that comprised the duct. However, increasing the body internal volume without increasing the outside diameter (limited by USMC requirements) reduced the available rotor diameter. Long endurance favors lighter disk loading, so the reduced diameter fan acted to reduce endurance further. A new solution was required.

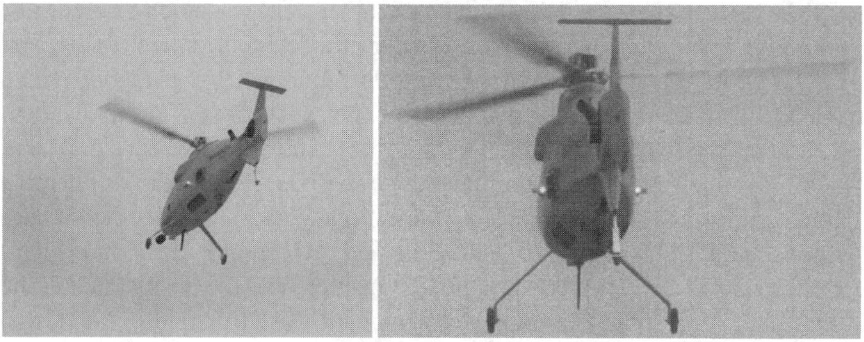

Fig. 8.35 Dragon Warrior in flight.

MCWL tasked NRL to develop a different configuration to meet the mission requirement. NRL and MCWL were already working together on the Dragon Eye program, so the existing working relationship was leveraged. Missions for Dragon Warrior included reconnaissance, surveillance, and target acquisition (RSTA) and tactical communications relay.

MCWL imposed requirements to physically shield the rotor for personnel safety and to protect the rotor from damage. The NRL design looked like a conventional helicopter, but with the addition of a rotor guard that surrounded the rotor (Fig. 8.36). The ring structure had an airfoil cross section to reduce drag in forward flight. Struts connected the rotor guard to the fuselage. NRL engineers discussed the design with helicopter pioneer Frank Piasecki and other rotorcraft designers, but these advisors were not favorably disposed to a shielded rotor design.

Fig. 8.36 Initial rotor guard on the subscale Dragon Warrior demonstrator.

NRL built a subscale RC model helicopter. The complex-shaped, all-composite rotor guard was built in expensive machined female molds. The NRL Dragon Warrior hovered like a conventional helicopter, but it had problems with transition to forward flight. The center of pressure movement on the ring rotor guard shifted forward during transition, creating a large destabilizing nose-up pitching moment. No reasonable amount of collective pitch could allow transition to forward flight. The subscale model was demonstrated to MCWL, and they agreed to remove the rotor guard requirement. The helicopter was able to transition, and the design met all the performance requirements once the rotor guard was eliminated. The design now had the potential safety issues of a standard helicopter, but the trade-off for performance made sense.

Dragon Warrior was designed to carry a 40-pound forward-looking infrared (FLIR) electro-optical infrared (EO/IR) camera ball with a laser range finder. It was designed to be transported in the back of a High Mobility Multipurpose Wheeled Vehicle (HMMWV) and had to be assembled and operated by two nonaviation personnel (Fig. 8.37). The propulsion system operated on heavy fuel, per Navy shipboard requirements. The system had to be simple to operate and maintain, without reliance on aviation facilities and personnel.

The propulsion system was simple (Fig. 8.38). A single belt drive transmitted power from the engine to the main rotor. This eliminated a gearbox and the need for gearbox oil that would be prone to dirt contamination. The tail rotor was electrically driven from power produced by an engine-mounted generator (Fig. 8.39). The tail rotor's digitally controlled electric motor could be modulated in rpm for tail rotor control enabled by a low-inertia tail rotor. The tail rotor was a custom lightweight carbon fiber ducted fan system. The generator also served as a battery-driven starter motor for the engine, which eliminated the need for engine start ground handling equipment. By removing the tail rotor shafts, bevel gears,

Fig. 8.37 Dragon Warrior stowed in a HMMWV (left) and ready for flight (right).

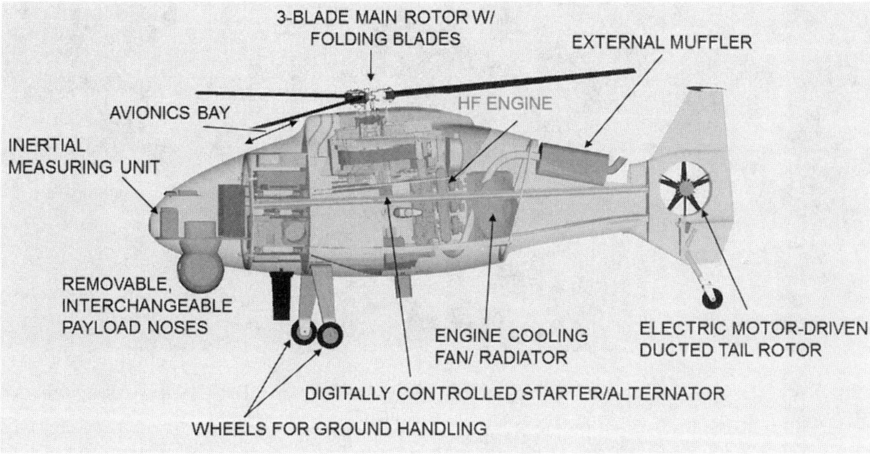

Fig. 8.38 Dragon Warrior internal arrangement.

and pitch mechanisms on the tail, the design eliminated about 100 components. This feature was considered a major breakthrough, and NRL submitted an application for a U.S. patent.

A major change in Navy unmanned aircraft policy took effect during the Dragon Warrior program. The Chief of Naval Operations (CNO) required that all Navy unmanned aircraft, even research aircraft, obtain a Naval Air Systems Command (NAVAIR) flight clearance. The Dragon Warrior now had to have a formalized and more rigorous flight test program. The design documentation, software, and drawings had to adhere to Mil-Spec standards and formats. The avionics and hardware had to adhere to military standards (MIL-STD), which was not a problem since the

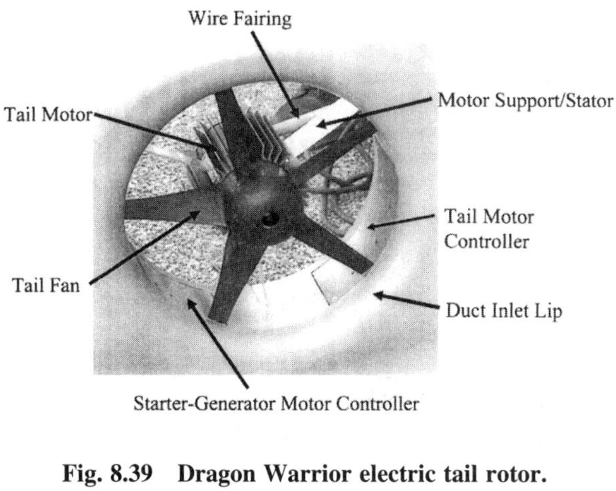

Fig. 8.39 Dragon Warrior electric tail rotor.

Fig. 8.40 Dragon Warrior rotor testing (left) and iron bird testing alongside an airframe (right).

key NRL team members were already FAA-certified airframe and power plant mechanics.

NAVAIR also imposed a Navy requirement to withstand an elevated drop on the landing gear. The Dragon Warrior landing gear had to be redesigned to withstand these additional loads, which added extra weight.

An "iron bird" system was first built to test the various systems (Fig. 8.40). The iron bird consisted of a chassis that could accept all key system components. Dragon Warrior had a complex propulsion-rotor system, power system, payload, and avionics. All wiring and connections were tested by evaluating components singly and then in combination. When the system worked properly, the components were transferred to the flight article. Custom wiring harnesses were built into the aircraft.

A 90-pound-capacity integral fuel tank was built into the fuselage. It was located between the nose avionics bay and the rotor head. This wet fuselage was volumetrically efficient, lightweight, and simple. The engine was located aft of the main rotor.

The UEL AR801 rotary heavy fuel engine initially experienced difficulties with cold engine start. NRL added an air intake heater, which used heating elements to help vaporize the fuel. NRL also solved these problems through extensive testing and modification of the engine map on the engine computer.

NAVAIR required that the Dragon Warrior undergo 25 hours of tethered flights before going to free flight. A swivel rotor head allowed a cable to be suspended from a test rig to the top of the rotor without winding up. The cable length was controlled by a human operator to keep it out of the rotor. If the helicopter lost control, the engine would be shut off and the tether would prevent the helicopter from crashing.

NRL proceeded with free-flight testing following the successful tether testing. The first flight occurred on August 11, 2003. Dragon Warrior ran into

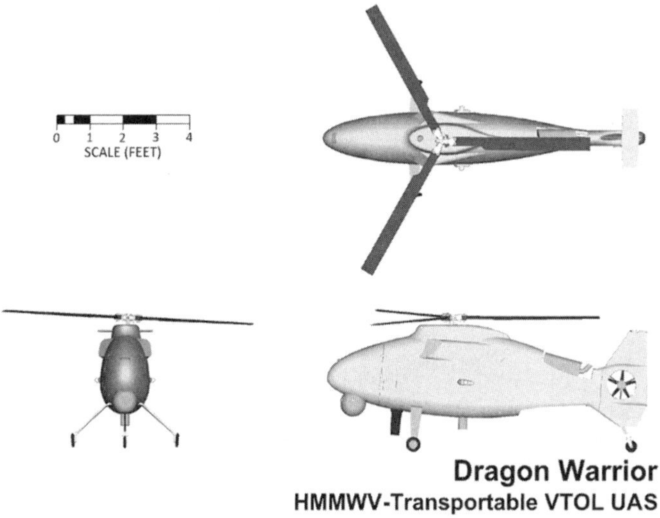

Fig. 8.41 Dragon Warrior three-view drawing.

a programmatic problem during the flight test effort. The Navy decided to adopt a tiered system for unmanned aircraft. Dragon Warrior was successfully meeting the specifications provided by the MCWL sponsor. However, it did not fully satisfy all the criteria for either a Navy Tier I or Tier II UAS. Therefore, Dragon Warrior was not considered a viable and producible unmanned aircraft by the Navy, and USMC interest waned. Figure 8.41 shows a three-view drawing of Dragon Warrior, and Table 8.6 lists its technical characteristics.

Because of the many unique, advanced technology features embodied within the Dragon Warrior program, the Office of Naval Research (ONR) assumed sponsorship in order to complete the flight test program. The vehicle name was changed from Dragon Warrior to Vantage. Under ONR funding, the flight test program demonstrated transition from hover to forward flight and fully autonomous flight. The maximum-speed portion of the flight envelope was not fully opened up at the end of the Vantage program.

Following the ONR-funded effort, the Vantage program was advertised to NATO customers. Unfortunately, while technically successful, Vantage did not fit within perceived NATO requirements.

Vantage has influenced other unmanned helicopter programs. The system was licensed to CybAero of Sweden and developed as the APID system. The design has undergone significant evolution since then. Later variants include the APID 55 and APID 60. The Austrian Schiebel CAMCOPTER tactical unmanned helicopter adopted a very similar appearance to Dragon Warrior

Table 8.6 Dragon Warrior Technical Characteristics

Parameter	Value
Rotor diameter	9.1 ft
Body length	9.3 ft
Takeoff weight	380 lb
Empty weight	212 lb
Payload weight	25–30 lb
Fuel weight	90 lb
Cruise speed	80 KTAS (20 KTAS demonstrated)
Dash speed	100 KTAS
Endurance	3–5 hr
Link range	50 nmi
Launch method	Vertical takeoff
Recovery method	Vertical landing
Propulsion type	Rotary heavy fuel engine
Power	42 hp

after it was made public, and many confuse the two systems. The design was also licensed to American Unmanned Systems LLC, and this company obtained Vantage airframes.

In 2003, NRL proposed an EW variant of Vantage, called Crystal Sun. This program did not go forward.

DRAGON WARRIOR: FURTHER READING

Southwick, J.R., "Vantage™ Unmanned Air Vehicle," *2008 NRL Review*, Naval Research Laboratory, Washington, DC, 2008.

SCIENTIFIC PAYLOAD INSERTION DEVICE ELECTRIC ROTOR (SPIDER), 2004–2005

Fig. 8.42 SPIDER in flight (left) and on the ramp (right).

Fig. 8.43 SPIDER forward fuselage with top fairing removed.

The Scientific Payload Insertion Device Electric Rotor (SPIDER) was a small, autonomous, electric helicopter designed for scientific and other missions in hazardous environments (Fig. 8.42). With the high level of autonomy, the operator did not require extensive training or piloting skills. SPIDER could automatically take off, transit to a waypoint, perform a precision hover, and land autonomously. It could support day or night operations. An Iridium satellite communication link provided beyond-line-of-sight control and payload data return communications to support low-altitude operations.

The electric propulsion system used advanced technologies (Fig. 8.43). Rechargeable lithium polymer batteries were used because of the high specific energy and affordability. Separate brushless DC motors were used for the main rotor and tail rotor. As with Dragon Warrior, no mechanical coupling was needed between the main rotor and tail rotor. This reduced the parts count and made disassembly for transport much easier.

The 40-pound airframe could be disassembled for transport in a case by folding the tail boom and removing the rotor blades. No tools were required to assemble the aircraft. The fuselage structure was mostly constructed of hollow molded carbon fiber. A fiberglass fairing helped reduce drag of the rotor mast and drive mechanisms. The autopilot and other avionics were housed within the top fairing. The batteries fit within the rectangular-cross-section forward fuselage.

A Cloud Cap Piccolo II autopilot was used for control. NRL developed the helicopter neural adaptive control algorithms that were implemented on the Piccolo II autopilot processors. NRL integrated an acoustic altimeter to aid landing, which could be used within 20 feet of the ground (Fig. 8.44).

Fig. 8.44 Acoustic altimeter on the lower fuselage of SPIDER.

SPIDER first flew on May 21, 2004, only 18 months after the program started. It performed more than 179 flights, including 143 autonomous landings. One of the flights included a complete autonomous system demonstration that included flying out 5 nautical miles with satellite communications, hovering over the drop point, emplacing the payload, climbing, and returning to base. This was NRL's most successful rotorcraft to date, and it made large strides in autonomous flight controls. Figure 8.45

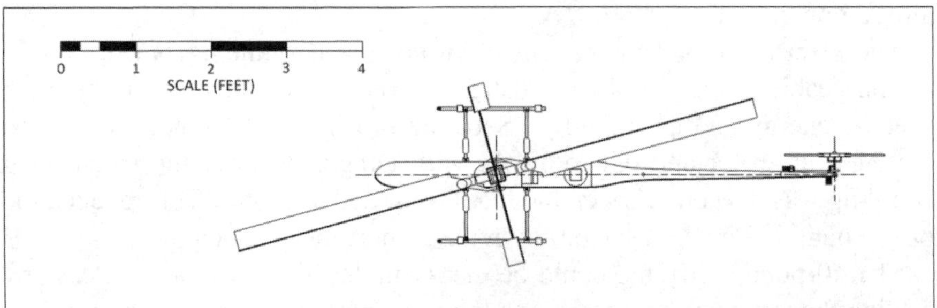

Fig. 8.45 SPIDER three-view drawing.

Table 8.7 SPIDER Technical Characteristics

Parameter	Value
Rotor diameter	7.0 ft
Body length	6.0 ft
Takeoff weight	42.0 lbs
Empty weight	23.0 lbs
Payload weight	4 lbs
Battery weight	15 lbs
Cruise speed	38 mph
Dash speed	50 mph
Endurance	25 min with 20% reserve
Launch method	Vertical takeoff
Recovery method	Vertical landing
Propulsion type	Battery electric
Power	2000 W hover 1500 W cruise

shows a three-view drawing of SPIDER, and Table 8.7 lists its technical characteristics.

Threat Radar Active Countermeasure Expendable (TRACE), 2008

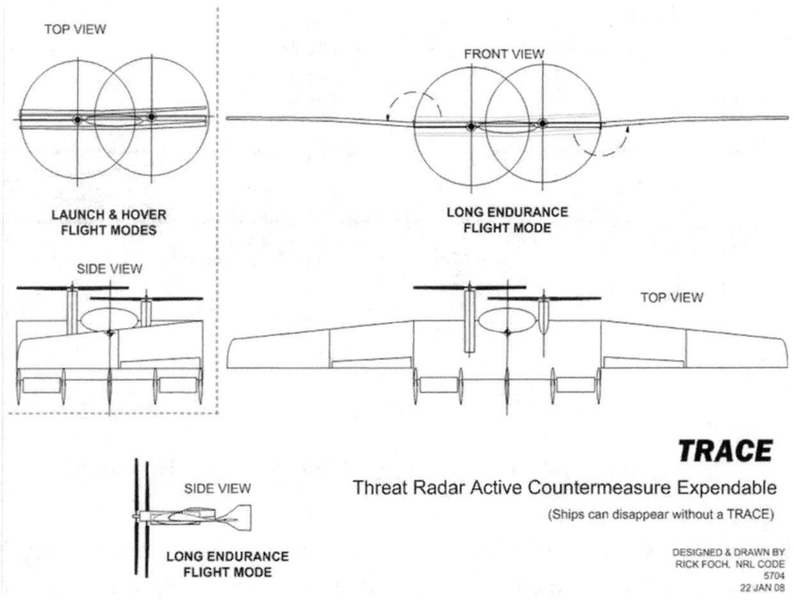

Fig. 8.46 TRACE drawings.

The Threat Radar Active Countermeasure Expendable (TRACE) was a design concept for a long-endurance, fuel cell electric UAS capable of VTOL operations from the deck of a ship (Fig. 8.46). It was inspired by the Boeing Heliwing with two major additions. First, the rotor blades were set in an overlapping configuration to conserve deck space. The use of overlapping rotors for this reason was pioneered by Piasecki in the 1950s beginning with the Piasecki HUP rescue helicopter and continued with the Vertol series of helicopters, today produced by Boeing. The second addition to the Heliwing concept was NRL's in-flight folding and unfolding wing technologies. These allowed a compact vehicle in hover to minimize the impact of wind, and a large-span aircraft in forward cruise or loiter flight to maximize endurance.

Development of TRACE beyond the concept definition phase has not yet occurred, but this concept seems worthy of development for future EW and other military applications.

EXPENDABLE RADAR ACTIVE COUNTERMEASURE ELECTRIC ROTORCRAFT (ERACER), 2008

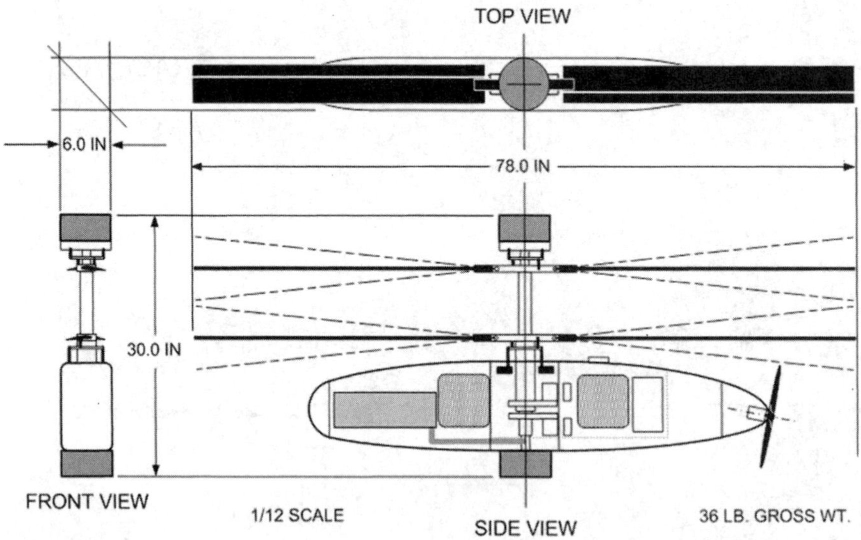

Fig. 8.47 ERACER conceptual design three-view drawing.

Expendable Radar Active Countermeasure Electric Rotorcraft (ERACER) was an NRL internal design study to investigate the potential to develop a simple, high-performance EW decoy for ship defense (Fig. 8.47). It was envisioned to be able to self-launch from a compact container on the ship deck and be expended following its mission. It was required to carry two omnidirectional antennas with the maximum vertical separation for isolating their signals, and to fly these antennas in as near a vertical attitude as possible. In addition, the antenna beam patterns were to be unencumbered by any portions of the airframe. The resulting design that emerged from these notional requirements was a coaxial helicopter with a thrust-compounding, pusher propeller that would provide forward thrust without the need to tilt the main rotor sysytem. ERACER would not require any folding or deployable surfaces, which would allow a rapid and simple deployment.

Although a viable and practical concept for further development, ERACER was supplanted by the Torqueless Stop Rotor project.

Rotary Wing Decoy (ROWDY) 2, 2010

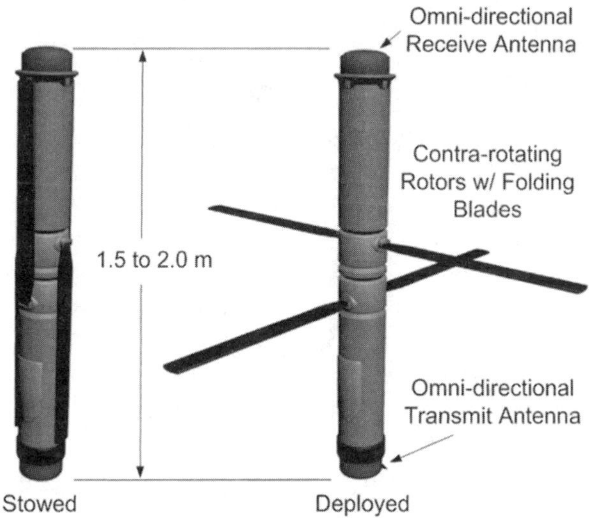

Fig. 8.48 ROWDY 2 design.

The ROWDY 2 was an EW decoy design concept based on the LiteMachines Voyeur small unmanned aircraft configuration (Fig. 8.48). It would essentially be an untethered version of the Eager and was determined to be a feasible, although relatively short endurance, battery electric platform. Like the ERACER, ROWDY 2 development was overtaken by the Torqueless Stop Rotor project.

STOP ROTOR, 2010–PRESENT

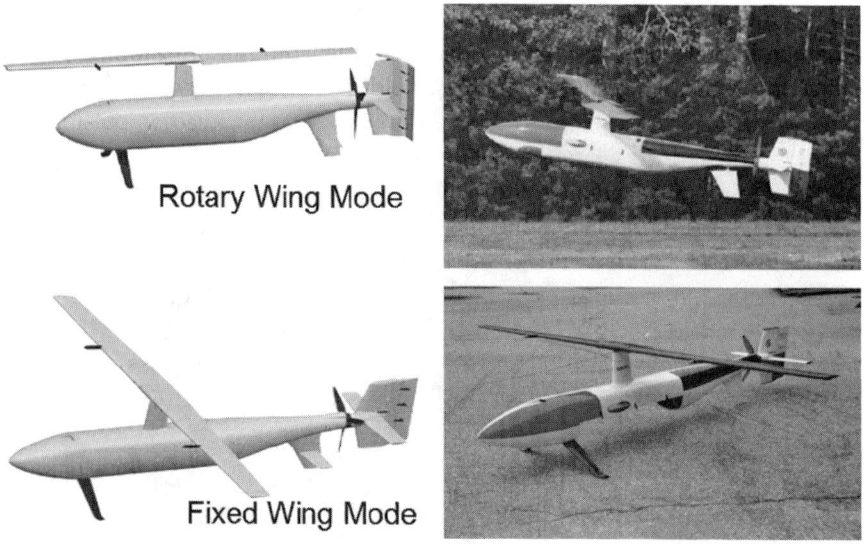

Fig. 8.49 Stop Rotor.

The Stop Rotor was an outgrowth of the idea of a swashplateless rotor (Fig. 8.49). NRL initiated a fundamental research program for an electronically controlled swashplateless rotor system for an unmanned helicopter application. The electromechanical systems provided collective and cyclic rotor control without any mechanical linkages passing through the rotor mast to the rotor head. All the rotor controls were built into the rotor head. This would reduce the width of the rotor mast to enable higher speed potential through lower drag. Another benefit was that the number of parts would be reduced, for improved reliability and lower cost. Static test models and wind tunnel models were constructed and tested under the original swashplateless rotor program.

During this research, it was realized that this technology could enable a vehicle that could act as a fixed wing aircraft and as a helicopter. If the rotors had symmetric airfoil blades, it would be possible to stop the rotor and flip one blade such that both blades served as wings. Similarly, one wing could flip and the rotor could spin up to convert to a helicopter mode. The flexibility and nature of the control system would allow this to happen.

In addition to the blades having direct pitch control by electromechanical actuators built into the rotor hub, the blades featured trailing edge control surfaces. The change in trailing edge control surface deflection angle allows optimal cambering for the forward flight (stopped rotor) mode and the ability to tailor the rotor blade pitching moment. Initially, the Stop Rotor blades were free to rotate in pitch, much like a free wing arrangement. The blades were mass balanced with weights projecting ahead of the blades. The rotor blades had springs to help maintain the pitch angle to improve the blade dynamics. The trailing edge control surfaces were commanded based on rotor position and desired control effect. However, this was found to be inferior to direct rotor blade control, so the more traditional approach of directly controlling the blade angle was implemented. Power and signal for the rotor servos were transferred through slip rings. Power to spin the rotor was produced by an electric motor that spun the main rotor shaft.

The rotor also served as a wing in the fixed wing aircraft mode. The rotor would slow down and lock into a position in which the rotors were orthogonal to the fuselage. One rotor blade would flip over such that the leading edge faced forward. The wing pitch angle was then locked into position. The trailing edge control surfaces then acted as full-span ailerons. These control surfaces had a moderate trailing edge down deflection on both sides to serve as camber flaps. This improved the wing lift-to-drag ratio (L/D) and maximum lift coefficient. Simulations showed that transition from hover mode to fixed wing mode and would result in less than 50 feet of altitude loss. The wear on the rotor system would be minimized by limiting the time in the rotary wing mode.

An electromechanical servo was used for collective and rotor flipping. The servo motor controller allowed a broad angular motion, unlike traditional rotary servos. The servo enabled 180-degree rotor flip in 0.2 seconds, while also allowing finer adjustments for collective pitch while in rotorcraft mode.

The main rotor drive motor provided constant torque to the rotor at all speeds. The motor controller had a trajectory optimization loop. In conversion from aircraft mode to rotorcraft mode, the rotor had to stop in a certain time and smoothly arrive at the proper aircraft wing position. The rotor flip occurred before the rotor crossed over the nose so that the wind did not affect rotor pitching moment. In other words, the wind was along the line of the rotor. After the rotor flipped as it passed over the nose, the wind helped to push the rotor to the wing position. To convert from aircraft mode to helicopter mode, the left blade flipped 180 degrees in pitch as it passed over the tail. The rotor had to pause briefly over the tail while the flip occurred before spinning up as a rotor.

The motor was a four-quadrant motor with a torque controller. The controller closed the loop on motor current, which was equivalent to controlling motor torque. The controller closed the loop on commutation via pulse width modulation (PWM) controls. This approach was common for rotary servos, but not typical for propulsion motors. NRL found it challenging to find a four-quadrant current controlled motor that was lightweight and had the necessary high power rating. Ultimately a custom motor was developed by NeuMotors.

A propeller provided forward thrust in the fixed wing mode and it countered rotor torque while in helicopter mode. The electrically driven propeller rotated around the tail boom, and it was located ahead of the tail surfaces. The multicambered vertical stabilizer was multiply hinged to form slotted flaps. When fully deflected, the propeller flow was deflected 90 degrees to produce a side force that could counter the rotor torque. In fixed-wing flight, the vertical tail was nominally nondeflected. The rudder provided yaw control as with a normal aircraft. An elevator control surface on the horizontal tail provided pitch control.

An electrically powered rotorcraft wind tunnel model was built and tested in the University of Maryland Glenn L. Martin wind tunnel to evaluate the transition from helicopter mode to fixed wing mode. The tests showed correct operations of the electronics, stability and control system, rotor dynamics, and structural features. The advanced motor system provided the appropriate rotor rpm, electronic braking, and stopping at a precise position. The rotor slowly moved to the correct rotational orientation, stopped, and then locked into place. No rotor braking system was required beyond the motor.

Upon successful completion of the wind tunnel testing, an Anti-Submarine Warfare (ASW) Stop Rotor unmanned aircraft design was initiated (Fig. 8.50). The purpose of the program was to investigate the feasibility of a small electric unmanned aircraft for emplacing ASW sensors. This Stop Rotor had unique features. It was capable of vertical takeoff, transition to fixed wing mode for high-speed ingress, hover mode for sensor emplacement, fixed wing egress, and vertical landing. As a result, the system had better range and higher speed than a helicopter while retaining VTOL and hover capabilities.

The Stop Rotor demonstrator was capable of 30 minutes of flight using battery electric propulsion. In fixed wing flight, its projected cruise speed was over 100 knots. It was equipped with a 38-inch-long payload bay that could carry 25 pounds of payload.

The ASW Stop Rotor vehicle began flight testing in 2011. It was first flown in the helicopter mode. The swashplateless rotor operation was verified in flight. The Stop Rotor demonstrated a large helicopter speed range due to the advanced rotor control system. The system completed a fully autonomous flight operation, including takeoff and landing. This version has achieved dozens of flights in the rotary wing mode. At the time of this writing, the Stop Rotor is being prepared for transition to fixed wing flight. The demonstrator aircraft is not equipped with wheeled landing gear, so takeoff and landing in the fixed wing configuration are not possible. Current flight tests are approaching the 70-knot transition speed in rotary wing mode.

Fig. 8.50 ASW Stop Rotor demonstrator in flight.

A parallel effort has been initiated to create a Torqueless Stop Rotor concept demonstrator (Fig. 8.51). Electric motors with tractor propellers are located on the rotor blades. These propulsors spin the rotor. The shaft is no longer mechanically driven, and no torque is imparted to the fuselage. This eliminates the need for a tail rotor or blown rudder. As before, rotor trailing edge control surfaces adjust airfoil camber and become flight controls in the fixed wing flight mode. Power is also transferred to the rotor via slip rings. The slip rings have six additional conductors to allow three phases for two motors. The motor controller and batteries are housed within the fuselage so that the spinning mass is minimized. The rotor propulsors provide forward thrust in the fixed wing mode. The primary advantage of this arrangement is that the tail motor can be eliminated, which reduces complexity and mass. The fuselage can be made more compact since no tail rotor is required. The torqueless rotor arrangement is compatible with lighter disk loading designs, which enable longer-endurance aircraft. The rotor-mounted propellers can slow the rotor once flipped to aid conversion. The current torqueless rotor design uses the ASW Stop Rotor fuselage to minimize development cost, so the full benefits are not yet realized. This design has a telescoping outer rotor to aid packaging

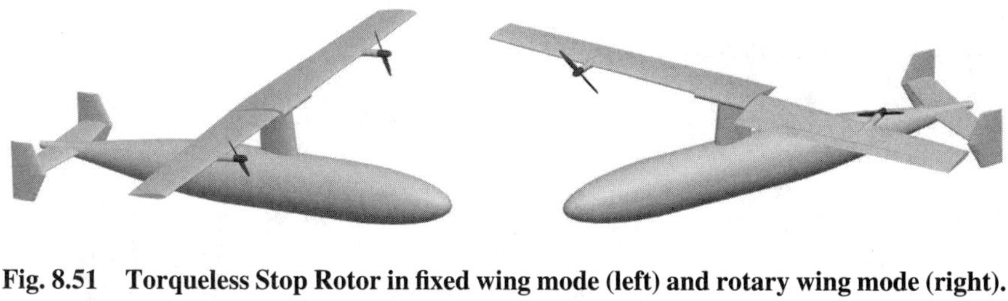

Fig. 8.51 Torqueless Stop Rotor in fixed wing mode (left) and rotary wing mode (right).

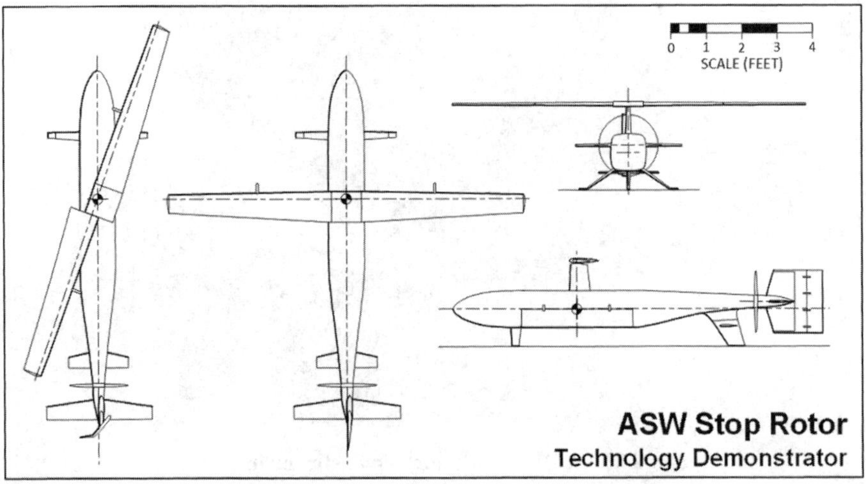

Fig. 8.52 ASW Stop Rotor three-view drawing.

for storage and transportation. This new design is under development at the time of this writing. The rotor has been successfully ground tested.

These systems are capable of approximately 30 to 60 minutes of flight using battery electric propulsion. Future variants could use hybrid electric

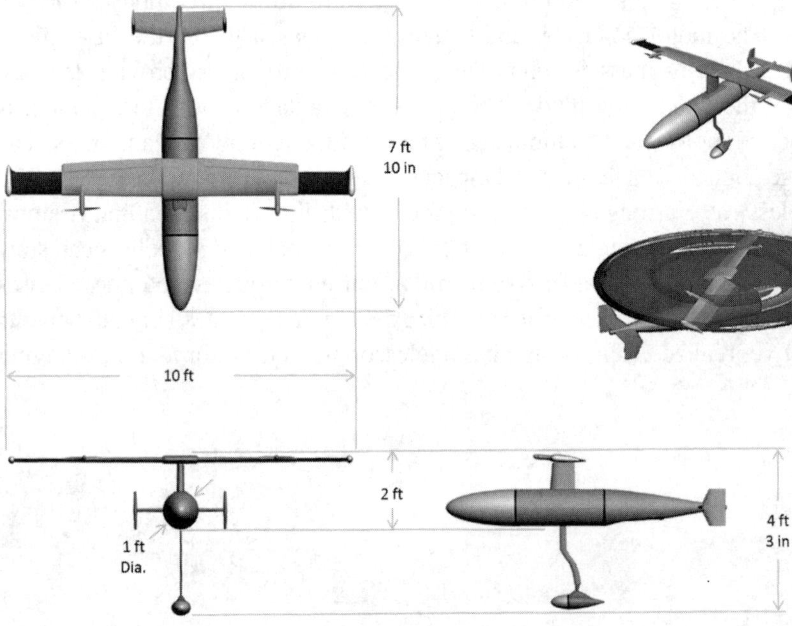

Fig. 8.53 Torqueless Stop Rotor views.

Table 8.8 Stop Rotor Technical Characteristics

Parameter	ASW Stop Rotor Value	Torqueless Stop Rotor Propulsion Value
Wing span	10.0 ft (same as rotor diameter)	10.0 ft (same as rotor diameter)
Wing area	8.4 ft^2	8.75 ft^2
Length	10.33 ft	7.0 ft
Takeoff weight	120.0 lb	90.0 lb
Empty weight	60.0 lb	50.0 lb
Payload weight	25.0 lb	0 lb (testing)
Battery weight	38.0 lb	38.0 lb
Cruise speed	90 KTAS	80 KTAS
Dash speed	110 KTAS	90 KTAS
Endurance	30 min	30 min
Max L/D	12 (fixed wing mode)	12 (fixed wing mode)
Airfoils	Eppler 474	Eppler 474
Launch method	Vertical takeoff	Vertical takeoff
Recovery method	Vertical landing	Vertical landing
Propulsion type	Battery electric	Battery electric
Power	3000 W cruise 6000 W hover	2500 W cruise 5000 W hover

propulsion or fuel cells to provide electric power, which should yield much longer endurance.

Figure 8.52 shows a three-view drawing of the ASW Stop Rotor. Figure 8.53 shows views of the Torqueless Stop Rotor. Table 8.8 lists the technical characteristics.

STOP ROTOR: FURTHER READING

Tayman, S.K., "Stop-Rotor Rotary Wing Aircraft," U.S. Patent No. 8,070,090, Dec. 6, 2011.

Chapter 9

PLANETARY FLIGHT

There is considerable confusion and lack of consensus on what to call vehicles that fly through the atmosphere of other planets or moons. It would be erroneous to call them "aircraft" because they do not fly through the air. Air is uniquely associated with the Earth's atmosphere. One common name is "flyer" because the vehicles fly through an atmosphere. Others have suggested "planetary aircraft" or "planetary flyers." However, it is feasible to fly in the atmosphere of Titan, which is a moon and not a planet. Foch suggests "extraterrestrial aircraft" because they are aircraft that fly in atmospheres not on Earth. However, to most people, this name implies that they are aircraft flown by extraterrestrials. Here we will call them "flyers" or "vehicles."

NRL has designed vehicles that can fly in the atmospheres of Mars and Saturn's moon Titan. While NRL is the Navy's corporate laboratory, it often collaborates on significant projects with other U.S. government organizations. In the case of planetary flyers, all the NRL studies were collaborations with the NASA Ames Research Center. NRL's expertise in, and technologies for, fully autonomous, deployable unmanned aircraft directly support the development of these intriguing systems. The Vehicle Research Section (VRS) has conducted five major studies for Mars flyers and one for a Titan flyer.

KITTY HAWK, 1997–1998

Fig. 9.1 Kitty Hawk over Mars, artist's concept.

Foch has had a sustained interest in Mars (Fig. 9.1) flight since he was a junior engineering student in 1978. He learned that the Viking lander used a parachute as a decelerator, so there must be sufficient atmospheric density to support flight. At that time, Foch wrote to the Viking program office at NASA Langley to request atmospheric data. He received three pages of accelerometer data along with the parachute ballistic coefficient data. Using instructions from NASA, he was able to back out the density as a function of altitude and thereby began a lifelong quest to design Mars flyers.

Foch was introduced to Dale Reed through his correspondence with NASA Langley. Reed was working on Mars aircraft designs at NASA Dryden Flight Research Center at the time. Reed developed an advanced Ackerman hydrazine expansion engine as an enabling technology for flight on Mars. This engine catalytically reformed hydrazine fuel into a gas to drive a reciprocating engine. It operated much like a steam engine. Reed developed the NASA Sniffer and Mini-Sniffer unmanned aircraft to fly these engines to high altitude on Earth. This important propulsion technology would be used by Foch and other Mars flyer design teams.

Many years elapsed between Foch's college interest in Mars flyers and the time when he had an opportunity to work on a system in a professional capacity.

Fig. 9.2 MAGE proposal cover.

He followed NASA Mars flyer concepts and studies with interest, but he did not have the opportunity to participate. Foch developed deployable aircraft, autonomous flight controls, and advanced propulsion systems at NRL, with the hope that these technologies could eventually be applied to planetary flight.

Finally, in the mid-1990s, Foch was contacted by Larry Lemke of NASA Ames. Lemke was leading a team to propose a Mars flyer for a NASA Discovery Program mission. He contacted industry, universities, and government labs to find an engineer possessing experience designing unmanned aircraft and aircraft that could unfold from packages. Eventually, Lemke reached Foch and explained the Mars flyer concept. He was surprised to learn Foch had had a strong interest in Mars aircraft ever since the Viking program. Foch eagerly agreed to support Ames. This was NRL's first aircraft collaboration with NASA.

The program was named the Mars Airborne Geophysical Explorer (MAGE). The flyer was named "Kitty Hawk" (Fig. 9.2). The launch window that would provide minimum launch energy and maximum payload would result in an operational mission in the Martian atmosphere on or around December 17, 2003. This was the 100th anniversary of the Wright Brothers' historic flight in North Carolina. What better way to commemorate this historic anniversary than a flight on Mars with a flyer named Kitty Hawk?

The problem was how to design a high-performance, sailplane-like aircraft that could fold into an atmospheric entry vehicle. Any hinges reduce aircraft performance through the addition of drag-generating gaps and steps, as well as weight. The folding mechanisms also reduce the reliability for deployment. Foch was able to convince NASA Ames that it would be best to optimize an aircraft that would fit inside the entry vehicle rather than design an optimum aircraft for performance and then find a way to fold it. The final design was simple and offered high performance.

Kitty Hawk was a plank flying wing configuration with a blunt fuselage. The wings folded into the reentry vehicle with two hinges on each side and folding control surfaces. The fuselage rested on the bottom of the entry vehicle. The inboard wing panel folded up, and the outboard panel folded down on the outside. This arrangement was simple for deployment.

The deployment occurred in multiple steps (Fig. 9.3). The entry vehicle would descend under parachute to the lower atmosphere, and the thermal

EVENT	TIME (Seconds)	ALTITUDE (km)
E/V encounters atmosphere	0	125
Drogue chute deploys	120	17
Main parachute deploys	122	16.5
Heat shield released	152	13
A/C drops from E/V	162	12
A/C parafoil deployed	167	11.5
A/C wings deployed	180	10.5
A/C level flight	187	9.5

Fig. 9.3 Kitty Hawk deployment sequence.

protection heat shield would be jettisoned. The flyer would then drop away from the back cover of the entry vehicle. The descending flyer would deploy a parafoil to slow the descent, stabilize the flyer, and provide time for the wing deployment. Under parafoil, the wings would unfold and the engine would start. Then the parafoil would be released and the flyer could begin the science mission.

The flyer was designed for range rather than endurance. The design range for Kitty Hawk was 2000 kilometers at an altitude of 9.5 kilometers. The flying wing configuration is well suited to range missions despite operating at a lower lift coefficient, because of its inherently high lift-to-drag ratio (L/D) compared to conventional configurations. It should be noted that conventional configurations tend to come out ahead for endurance missions where the endurance parameter ($C_L^{3/2}/C_D$) is more important. Flying wings and conventional configurations are generally comparable for range missions. Flying wings offer the advantage of simpler packaging for the Mars mission, thanks to the smaller number of major airframe sections.

The flyer's propulsion system consisted of a reciprocating hydrazine monopropellant engine and a large-diameter, low Reynolds number (LRN) propeller. The engine was a modern version of the Ackerman engine that Dale Reed advanced in the 1970s. The engine produced a maximum continuous rated power of 10.5 kilowatts and a maximum power rating of 11.2 kilowatts.

The NASA-NRL team proposed the system, but NASA Headquarters elected not to fund Mars exploration missions at that time. While waiting for the proposal evaluation, NRL continued to develop the technology and built and flew an air-dropped gliding version of Kitty Hawk. Figure 9.4 shows an air-drop Kitty Hawk test model. Figure 9.5 shows testing modified with displaced tail for experimentation, which supported the Kitty Hawk 2 program. Figure 9.6 is a drawing showing the features of the Kitty Hawk.

Fig. 9.4 Air-drop Kitty Hawk test model.

Fig. 9.5 Kitty Hawk flight testing modified with displaced tail for experimentation. This supported the Kitty Hawk 2 program.

The 1/6-scale model was successfully flown in 1998 (Fig. 9.7). It was also used to support early development of the Kitty Hawk 2 Mars flyer design (discussed later in this chapter). In this role, it was tested with a taller fin, designed to house a tall gravity gradiometer instrument, and displaced tail surfaces. Table 9.1 lists Kitty Hawk's technical characteristics.

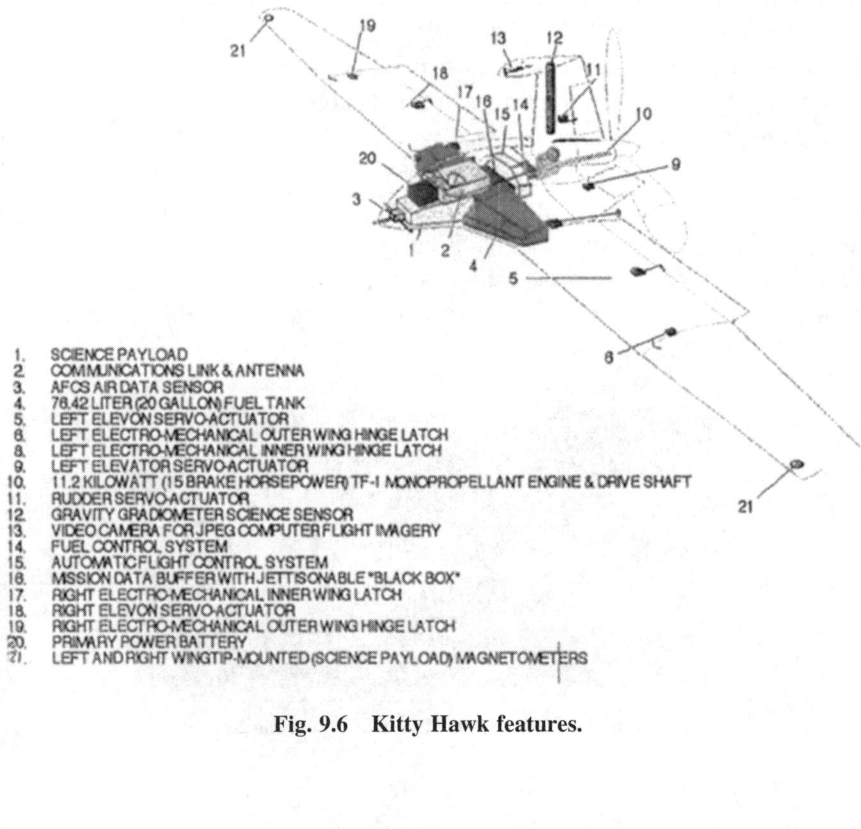

Fig. 9.6 Kitty Hawk features.

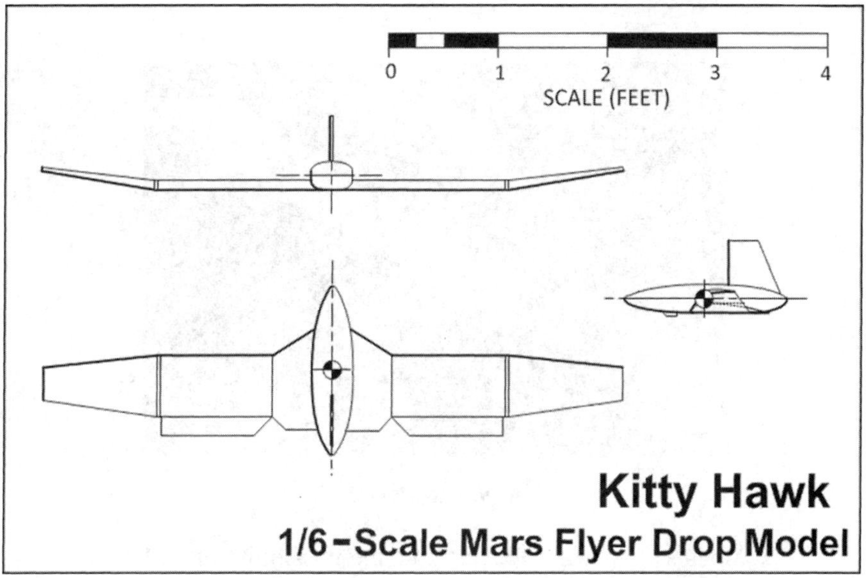

Fig. 9.7 Kitty Hawk 1/6-scale drop-model three-view drawing.

KITTY HAWK: FURTHER READING

Malin, M.C., "Mars Airborne Geophysical Explorer (MAGE)," a proposal to NASA AO No. 98-OSS-04 (Discovery Program), Malin Space Science Systems, NASA Ames Research Center, Orbital Sciences Corporation, and the Naval Research Laboratory, June 1998.

Table 9.1 Kitty Hawk Technical Characteristics

Parameter	Value
Wing span	9.72 m
Wing area	11.09 m^2
Length	2.60 m
Takeoff mass	133.8 kg
Fuel mass	54.0 kg
Payload mass	20.0 kg
Cruise speed	157 m/s (0.68 Mach)
Cruise altitude	9.5 km
Max L/D	16.3
Launch method	Entry vehicle deployment
Recovery method	Expendable
Propulsion type	Ackerman hydrazine engine
Power	11,200 W (maximum)

NASA 723, 1999

Fig. 9.8 NASA 723.

NASA Ames asked NRL to investigate the possibility of a Mars flyer that could be operated by astronauts on the surface of the planet. Unlike other concepts, this flyer would be configured and launched on the surface rather than while descending in an entry vehicle. This would have lower risk, because the vehicle would not reconfigure from the folded configuration in a dynamic environment, and there would be more time for system checkouts. The unfolding would be assisted by robotic systems contained within the box. The box would then serve as a rail launcher, which would be powered by mechanical springs or compressed gas. The flyer was referred to as NASA 723 (Fig. 9.8).

The design was to be as simple as possible. The flyer had a conventional configuration with a V-tail. The outboard panels folded under the inboard panels, and the tail boom folded over the fuselage. Only three hinges and three latches were needed. The system could be unfolded and prepared for launch either by an astronaut or autonomously on an unmanned lander. In the latter case, the flyer would be considered an auxiliary payload, and the mission success would not depend solely on the flyer mission performance.

The payload would include a camera to provide reconnaissance for the astronauts. After landing, the flyer could serve as a ground sensor and navigation aid.

NRL built and flew a dynamically scaled demonstrator aircraft. It flew slower in Earth's atmosphere than it would in the Martian atmosphere, but its airfoil operated at similar Reynolds numbers. Battery electric propulsion was used for the demonstrator. A full-scale mock-up was also fabricated for display. Figure 9.9 shows a three-view drawing of the NASA 723, and Table 9.2 lists its technical characteristics.

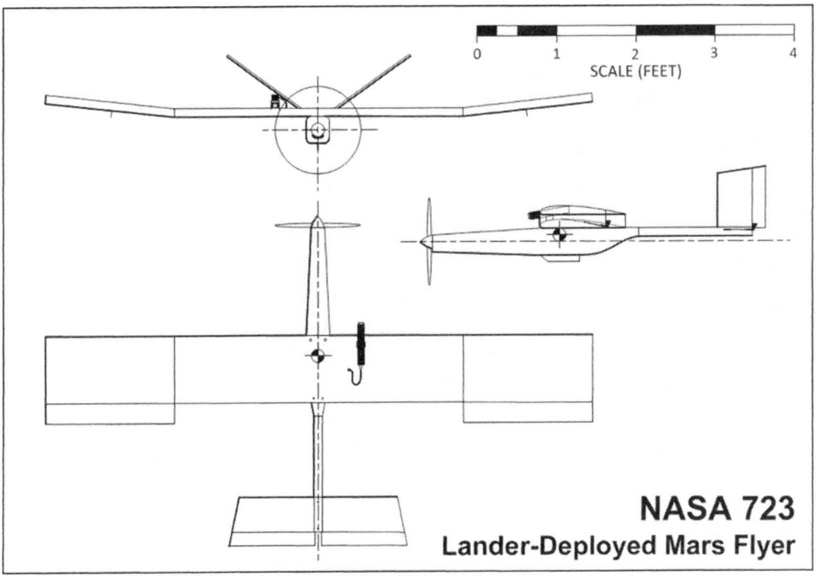

Fig. 9.9 NASA 723 three-view drawing.

As with the other Mars flyers, this program was not selected for a mission. The concept was the forerunner for the smaller MAC-1 (discussed later in this chapter). NRL's NASA 723 design was displayed in a *National Geographic* article, where an image showed an astronaut holding the flyer.

Table 9.2 NASA 723 Technical Characteristics

Parameter	Value
Wing span	2.14 m
Wing area	0.56 m^2
Length	1.33 m
Takeoff weight	2.2 kg
Payload weight	0.4 kg
Cruise speed	75 m/s
Endurance	1.0 hr
Max L/D	15
Launch method	Rail launcher from lander
Recovery method	Belly landing
Propulsion type	Battery electric
Power	250 W

KITTY HAWK 2, 2000–2001

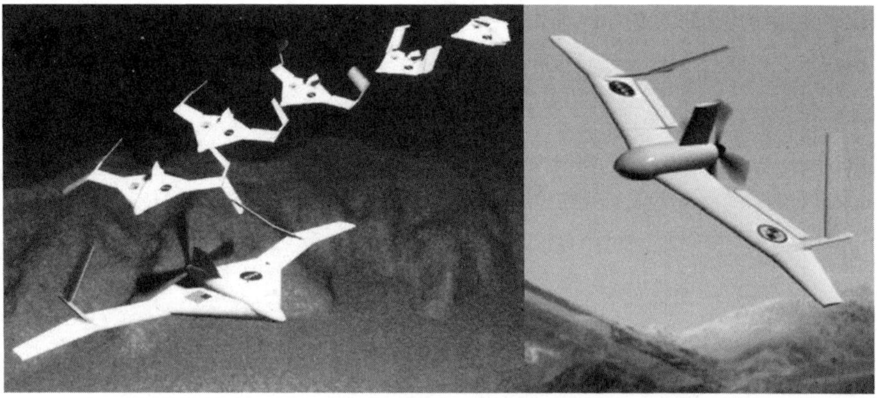

Fig. 9.10 Kitty Hawk 2 initial design (left) and final design (right).

Following the Kitty Hawk study and proposal, NRL and NASA Ames developed the Kitty Hawk 2 (also called the Airplane for Mars Exploration, AME) for another Discovery Program mission proposal. The first Kitty Hawk's entry vehicle was negatively evaluated by NASA Headquarters because it had an elongated vertical shape like a Mercury capsule from the early manned spaceflight program, rather than a proven Viking Mars entry vehicle geometry. Kitty Hawk 2 corrected this by using the Viking Mars entry vehicle geometry, and it was also a smaller, lower-cost Mars flyer than previous designs in order to use a smaller entry vehicle and lower-cost booster.

The configuration (Fig. 9.10) that emerged was based on NRL's Extender aircraft. Extender was designed to deploy from a P-3 life raft module, which had a similar internal volume form factor as that of a Viking Mars entry vehicle. The Kitty Hawk 2 flyer had a cranked wing with a displaced tail. As the design matured, it was so similar to Extender that the airframe could be reused with only minor adjustments. This allowed the Extender program flight test data, air-drop data from the vertical wind tunnel, mechanical design, and geometry to apply directly to the Kitty Hawk 2 program. Unlike with the first Kitty Hawk, no parachute was required for atmospheric entry, because the aircraft was stable through the entire deployment sequence. The Kitty Hawk 2 would unfold for flight while descending toward the planet surface.

Kitty Hawk 2 was designed to use either battery electric propulsion or a hydrazine Ackerman reciprocating engine. A large-diameter propeller was housed at the back of the fuselage. The propeller was a key risk for the program, because the flight performance was heavily dependent upon the

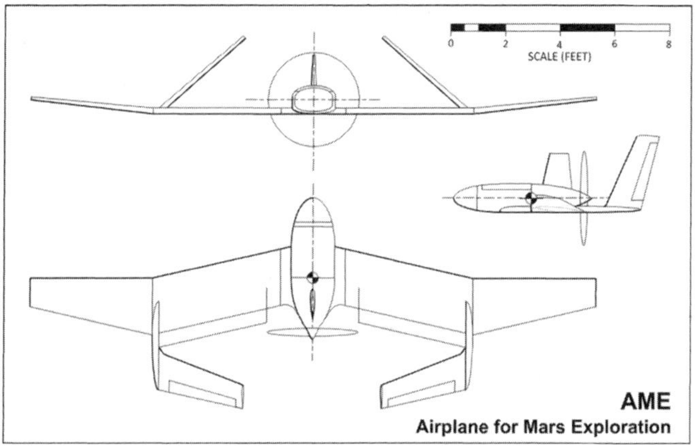

Fig. 9.11 Kitty Hawk 2 three-view drawing.

propeller performance. If the propeller efficiency was off by a few percent, the flyer might not be able to sustain level flight. The deployment in the Martian atmosphere would be the first time the propeller was ever tested in a relevant environment. Testing on Earth could not replicate the temperature, pressure, speed, and gravity simultaneously. This concern led to the rocket propulsion system on the later Matador program.

The Kitty Hawk 2 would conduct a surface geology mission. Data collected by the flyer would be relayed back to Earth via a satellite orbiting Mars. Figure 9.11 shows a three-view drawing of Kitty Hawk 2, and Table 9.3 lists its technical characteristics.

Table 9.3 Kitty Hawk 2 Technical Characteristics

Parameter	Value
Wing span	6.33 m
Wing area	3.67 m^2
Length	2.29 m
Takeoff mass	65 kg
Fuel mass	26.0 kg
Payload mass	10.0 kg
Cruise speed	140 m/s (0.61 Mach)
Cruise altitude	6.5 km
Max L/D	15.5
Launch method	Entry vehicle deployment
Recovery method	Expendable
Propulsion type	Ackerman hydrazine engine or battery electric
Power	7500 W (maximum)

Like every NASA Mars flyer program proposed to date, Kitty Hawk 2 was not selected for award. NRL built and flew several demonstrator aircraft. One Extender aircraft had the large-diameter Mars propeller installed and delivered to the NASA Ames Research Center for display.

MARS ASTRONAUT COMPANION (MAC)-1, 2003–2004

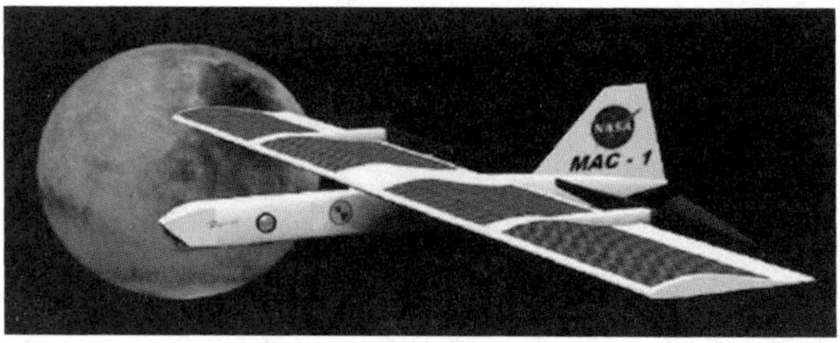

Fig. 9.12 MAC-1 configured for flight.

Most of the NRL efforts to fly aircraft on Mars involved deployment from an atmospheric entry vehicle. This approach was frequently viewed as high risk, due to the complex unfolding while descending through the atmosphere. An alternative approach was to develop as a tool for astronauts an unmanned flyer that could be deployed by them from the planet surface (Fig. 9.12).

The flyer would perform local reconnaissance, geological surveys, and atmospheric sciences missions while flying (Fig. 9.13). After the system was worn out, it would fly to a remote location to serve as a navigation aid and ground sensor. The MAC-1 design had interchangeable payloads including video and infrared imagers, atmospheric sensors, magnetometers, communications relay systems, and navigation beacons.

Fig. 9.13 Graphic of MAC-1 shown to scale with astronauts (left) and unfolding from case (right).

The MAC-1 was based on the Dragon Eye configuration. It was larger and had a lighter wing loading to operate in the thin Martian atmosphere. It would be launched by a simple compressed gas catapult and fly at approximately 170 mph. Solar panels recharged the batteries between flights, and the panels would also power the aircraft when it served as a ground sensor. Large-diameter propellers were located behind the wings due to the stability benefits. It should be noted that propellers are generally not placed behind the trailing edge due to the high noise of the blades passing through the wake, but noise was not a consideration on Mars because sound waves are extremely weak in the thin Martian atmosphere.

The system would be handled by a gloved astronaut, so the system had to be easy to set up. No plug-in components or tools were permitted. The single-piece flyer fit within a case with multiple folds. Two skewed hinges enabled the fuselage to be stowed between the folded wings, and each wing had two hinged sections. The astronaut's gloved hand would rotate and unfold the wings into place with a few simple motions, and they would snap into place. The large-diameter propellers had a fold that allowed the propeller blades to align with the trailing edge while stored. The case also served as a ground control station.

NRL performed a detailed system design study. MAC-1 has not yet been selected for development. The simulation and analysis showed that this class of system is feasible. Figure 9.14 shows a three-view drawing of MAC-1, and Table 9.4 lists its technical characteristics.

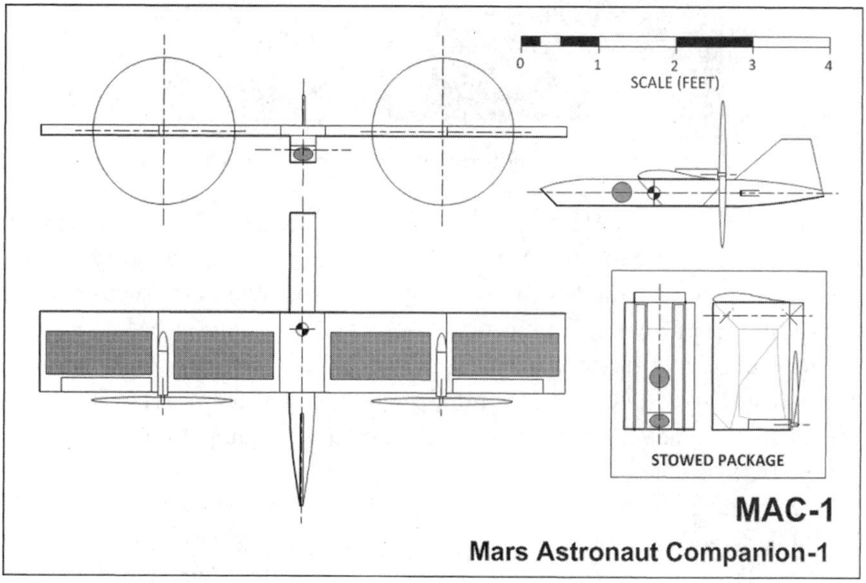

Fig. 9.14 MAC-1 multiple-view drawing.

Table 9.4 MAC-1 Technical Characteristics

Parameter	Value
Wing span	2.1 m
Wing area	0.53 m^2
Length	1.1 m
Takeoff mass	3.2 kg
Payload weight	0.5 kg
Cruise speed	76 m/s
Endurance	30 min
Max L/D	9
Launch method	Compressed gas catapult
Recovery method	Belly landing
Propulsion type	Battery electric
Power	500 W

MATADOR, 2004–2006

Fig. 9.15 Matador with rocket propulsion, artist's concept.

Following the Kitty Hawk and Kitty Hawk 2 studies, NRL developed the ARES/Matador system for NASA Ames as a preproposal design study (Fig. 9.15). NRL conducted a trade study between a hydrazine reciprocating engine propeller vehicle and a rocket vehicle. The vehicles had to carry a 5-kilogram science payload and fit within a 2-meter-diameter entry vehicle. The payload would consist of a gravity gradiometer and a public outreach camera. The name ARES changed to Matador after a competing NASA team adopted the original name.

The hydrazine reciprocating engine used an advanced Ackerman engine. The hydrazine was converted from a liquid to a gas through a catalytic reaction. The volume expanded through the phase change, which powered the engine. The hydrazine engine operated in the same fashion as a steam engine.

The propeller-driven flyer had impressive calculated performance. It was capable of 3 hours' endurance and 1250-kilometer range. The flyer cruised at 106 to 140 meters per second (Mach 0.45 to 0.6) at altitudes of 1 to 1.5 kilometers above the mean surface level. The flight speed and performance were constrained by propeller tip Mach limitations. It had a 5-meter wingspan when fully deployed. The deployment from the entry vehicle required 10 hinges and 16 latches. The disadvantages of this design were the dependency upon a parachute and the large number of mechanisms for deployment. In addition, significant engine development would be required.

The propeller-driven flyer was considered to be a risky design for a number of reasons. Aerodynamic design of vehicles in the LRN transonic regime was challenging. The hydrazine reciprocating engine was a new design. There was little experience with transonic LRN propellers. The number of wing folds would create a high-risk deployment.

The rocket-powered flyer used a hypergolic bipropellant consisting of hydrazine and nitrogen tetroxide. The two fuels would react in a combustion chamber to power the rocket motor. This motor was mature, fully characterized, and qualified for spaceflight. The thrust versus fuel burn characteristics were known.

The rocket flyer was a subsonic cranked delta wing with a deployed span of 4.1 meters (Figs. 9.16 and 9.17). It required only two hinges and four latches for deployment. The outboard wings had less wing sweep than the inboard section, which increased the aspect ratio and improved the aerodynamic efficiency. It would operate at an altitude range of 1 to 3 kilometers above the mean surface level. The cruising speed was 103 to 170 meters per second (Mach 0.45 to 0.75). The range was 545 kilometers with a flight duration of 1 hour.

The delta wing had predictable high angle of attack characteristics. The vehicle would fall from the entry vehicle at 90 degrees angle of attack with no parachute. The rocket would fire horizontally to accelerate the flyer to forward flight while the wing drag provided aerodynamic deceleration. As

Fig. 9.16 Artist's concept of Matador in flight on Mars.

Fig. 9.17 Concept showing folded Matador after release from the entry vehicle.

the forward speed increased, the angle of attack decreased. The flyer would attain level flight at an altitude of 4 kilometers and Mach 0.9. The team performed numerous entry simulations. Even under the worst-case scenario of a late (low-altitude) entry vehicle parachute deployment, the rocket-powered flyer still pulled out 2 kilometers higher than the propeller solution. A rocket attitude control system would stabilize the flyer until the aerodynamic control surfaces became effective. The delta wing had well-understood aerodynamics. Thanks to the large chord, it operated at a higher Reynolds number than the propeller design. Also, the mature rocket motor was insensitive to Reynolds number effects, and its performance was fully established. Overall, the rocket flyer was considered a lower-risk design solution.

The rocket-powered flyer was selected for continued evaluation. The solution went through a detailed design phase (Figs. 9.18, 9.19, and 9.20).

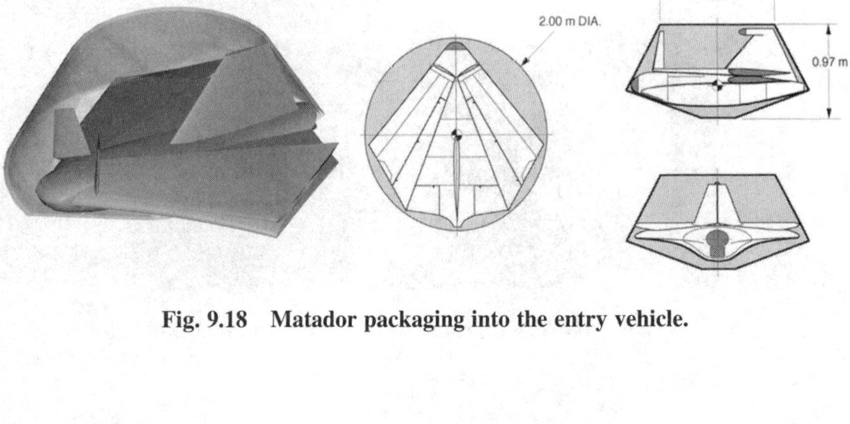

Fig. 9.18 Matador packaging into the entry vehicle.

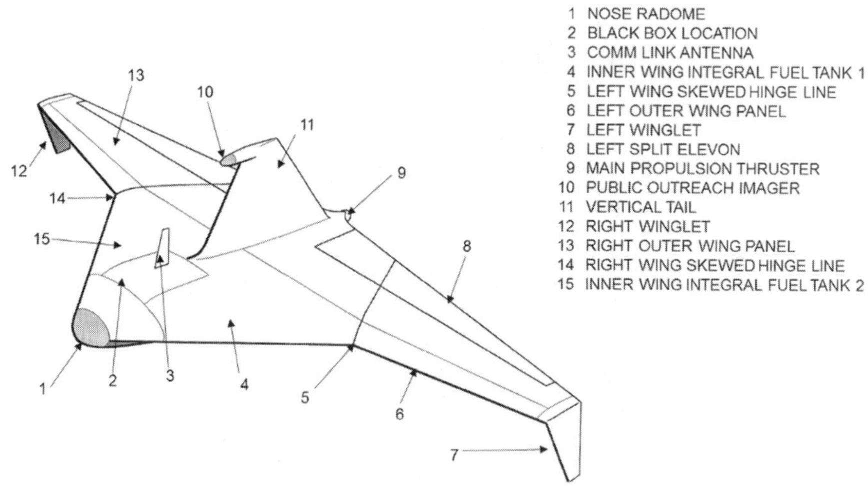

Fig. 9.19 Matador external features.

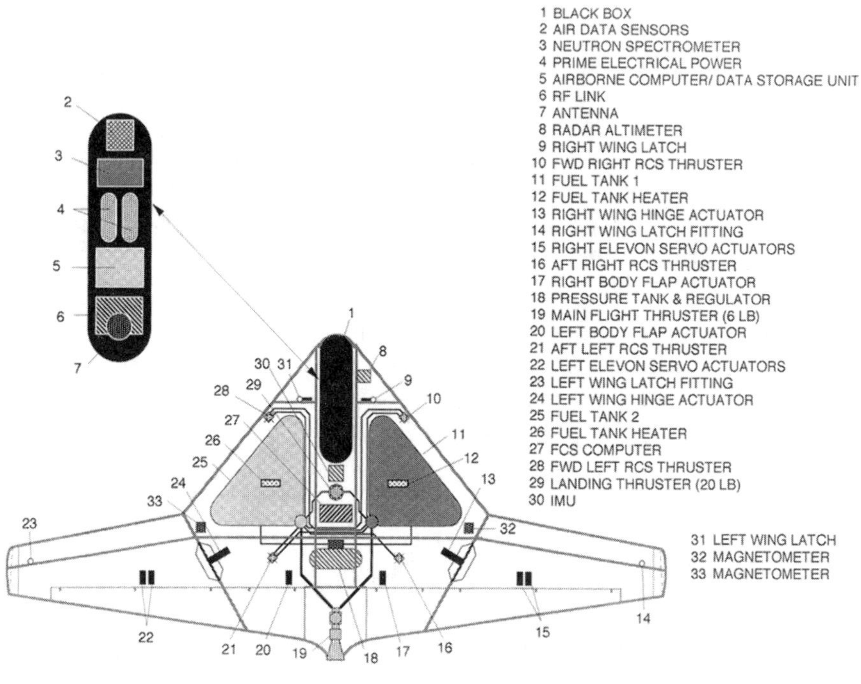

Fig. 9.20 Matador internal features.

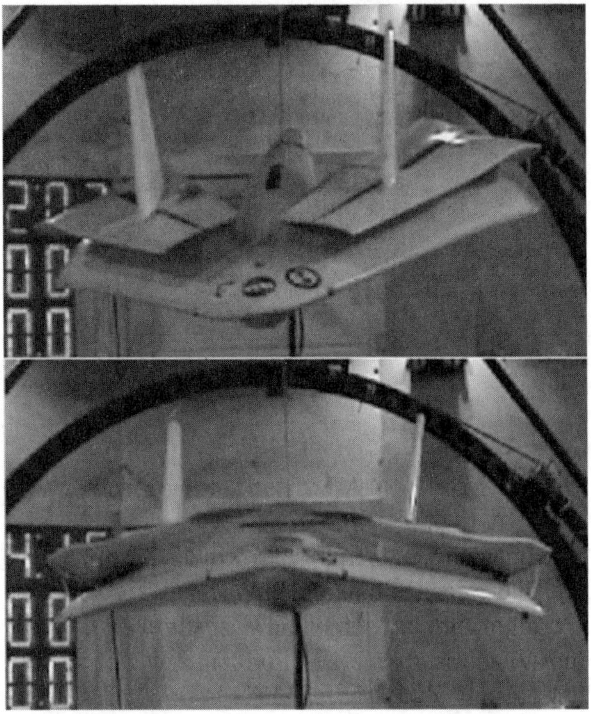

Fig. 9.21 Matador scale model in vertical wind tunnel testing.

Boeing Phantom Works conducted computational modeling using advanced paneling methods on the aircraft. Antony Jameson of Stanford University used computational fluid dynamics (CFD) to computationally tailor the wings along the span. The aerodynamics were particularly challenging due to the LRN transonic flight geometry and the cranked delta wing planform. The team built and tested a half-scale wind tunnel model to verify the aerodynamic characteristics and the deployability. The model was tested in the Wright Patterson Air Force Base vertical wind tunnel to characterize the pull-out (Fig. 9.21). Considerable risk reduction efforts were completed to support the proposal, and the team had high confidence in thc system feasibility.

Matador was proposed as a NASA Discovery Program mission. As with Kitty Hawk, this proposal was not selected. There were other competing Mars flyer proposals as well, but none were selected for the Discovery mission. A three-view drawing and planform details of Matador can be found in Fig. 9.22 and its technical characteristics in Table 9.5.

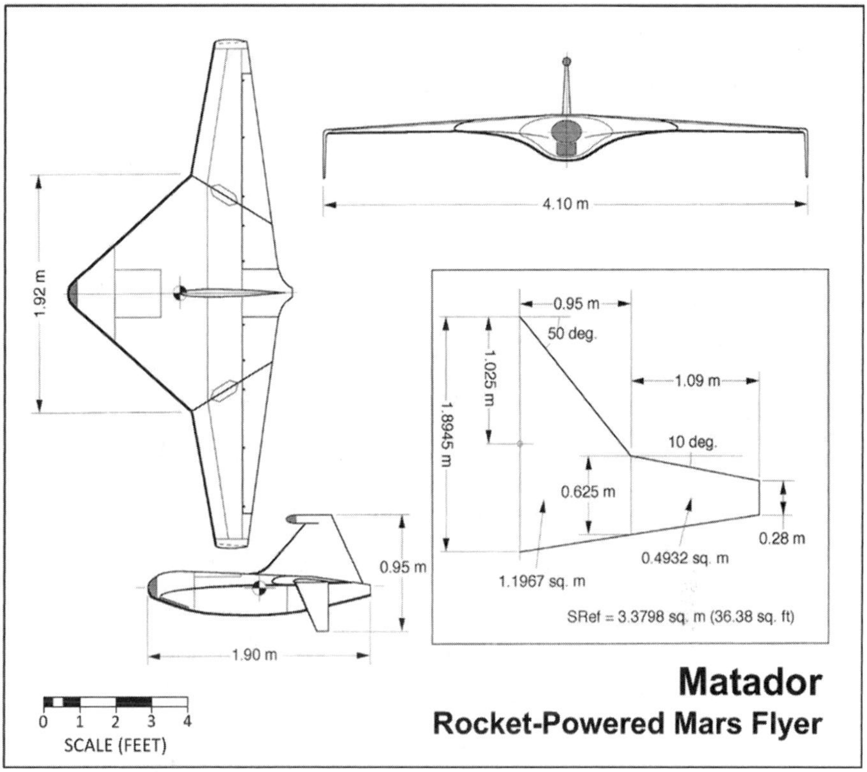

Fig. 9.22 Matador three-view drawing and planform details.

Table 9.5 Matador Technical Characteristics

Parameter	Value
Wing span	4.1 m
Wing area	3.38 m^2
Length	1.9 m
Payload weight	5 kg
Cruise speed	103–170 m/s
Endurance	1 hr
Range	545 km
Max L/D	14.5
Launch method	Entry vehicle, rocket pull-out
Recovery method	Expendable
Propulsion type	Bipropellant rocket

MATADOR: FURTHER READING

Boothe, K., Page, G., MacKrell, J., and Carruthers, S., "MATADOR Technical Report," Naval Research Laboratory, Nov. 2006.

Vassberg, J.C., and Jameson, A., "Aerodynamic Shape Optimization Part II: Sample Applications," *Von Karman Institute Lecture*, Brussels, Belgium, March 9, 2006.

TITAN FLYER (AVIATR), 2011

Fig. 9.23 Titan Flyer.

Flight in the dense atmosphere of Saturn's moon Titan is more benign than sea level flight on Earth, from an aerodynamic perspective. The density at a representative altitude of 8 kilometers is 3.91 kilograms per cubic meter, which is more than three times the density at Earth's sea level. The low gravity of 1.372 meters per second squared is roughly 14% that of Earth. Flyers designed to fly within Titan's atmosphere can be compact and low power and can fit within atmospheric entry vehicles with few joints. The Reynolds number, which is proportional to the ratio of density to viscosity, is a governing aerodynamic parameter. Fortunately, the Reynolds number for the Titan Flyer is in the same regime as for manned terrestrial aircraft, eliminating the challenging LRN regime typical of flight on Mars and enabling a high L/D. Flyer's airspeed of 6 to 7 meters per second is well within the subsonic regime, eliminating complexities and penalties of transonic flight typical of Mars flyers. The final design is shown in Fig. 9.23.

Foch designed a Titan flyer that could fit within a 3.75-meter-diameter entry vehicle without any folds. The lack of folding would improve the

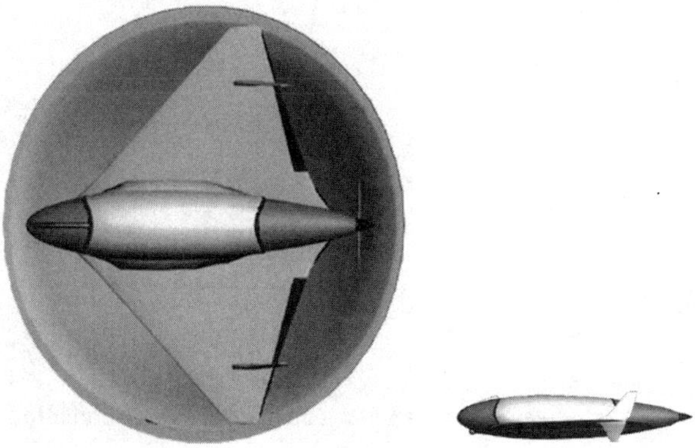

Fig. 9.24 Initial Titan Flyer design.

vehicle's reliability by eliminating failure mechanisms. The flyer was to be powered by two Advanced Stirling Radioisotope Generators (ASRGs), which convert nuclear energy to electrical power for very long-endurance flight. The low solar radiation at Titan plus the atmospheric attenuation would not make solar flight practical. The design featured a delta wing, vertical tails at mid-span on each wing, and a pusher propeller. An elliptical array antenna measuring 2 meters by 0.64 meters fit in the middle of the fuselage.

The concept became part of a NASA Discovery Program mission proposal. The principal investigator was Jason Barnes of the University of Idaho. Larry Lemke of NASA Ames was a key contributor. The vehicle design was led by Foch of NRL, with collaborative design support from Jay Gundlach and Francesco Giannini of Aurora Flight Sciences. The vehicle was given the name AVIATR, Aerial Vehicle for In-situ and Airborne Titan Reconnaissance.

More detailed analysis of the initial design revealed several problems (Fig. 9.24). A more refined weight estimate revealed that the weight would grow by at least 30%. The ASRGs were also found to be 30% less powerful than originally assumed. The antenna grew to 0.8 meters long to provide the required data rate back to Earth. The fuselage packaging would grow beyond the permissible entry vehicle size.

A revised design was necessary to fulfill the mission (Fig. 9.25). The reduced ASRG power output required reduced power consumption by the flyer, which was accomplished through a higher L/D and a lighter wing loading. The aerodynamic efficiency was improved with a span increase, which also increased the aspect ratio. The large span required two skewed wing joints to package within the entry vehicle.

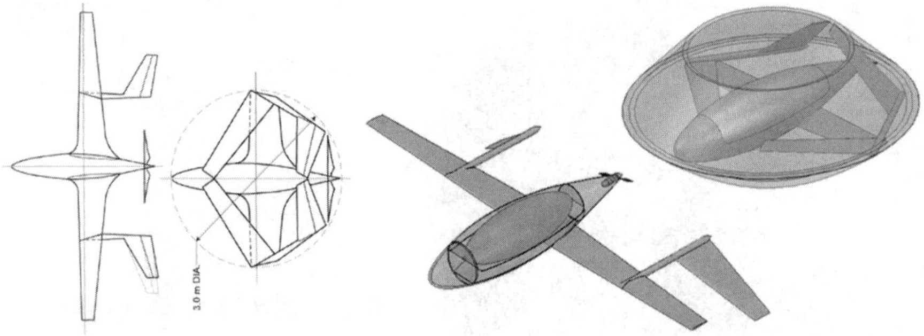

Fig. 9.25 Second Titan Flyer conceptual design iteration.

The new configuration had a displaced tail. The tail booms were attached to the outboard wing panels. The tails consisted of a single surface on each side, and these surfaces were oriented down and outboard. This tail arrangement was equivalent to an inverted V-tail, except the halves were not connected. The tail anhedral required wing dihedral to provide suitable lateral-directional handling characteristics. The tail configuration permitted folding within the entry vehicle without interference.

As the antenna grew and the ASRG geometry definition improved, it was discovered that the vehicle could not be packaged to fit within the entry vehicle (Fig. 9.26). The ASRGs were significantly larger than initially anticipated. The elliptical antenna occupied too much fuselage length. This packaging problem nearly caused the program to end.

Yet another design iteration was made (Fig. 9.27). The antenna changed from an elliptical antenna in the middle of the fuselage to a circular phased array antenna in the nose. The antenna area was 0.6 square meters, which was greater than the 0.4-square-meter threshold. The ASRGs were located in the middle of the fuselage. The fuselage cross sections were tailored along

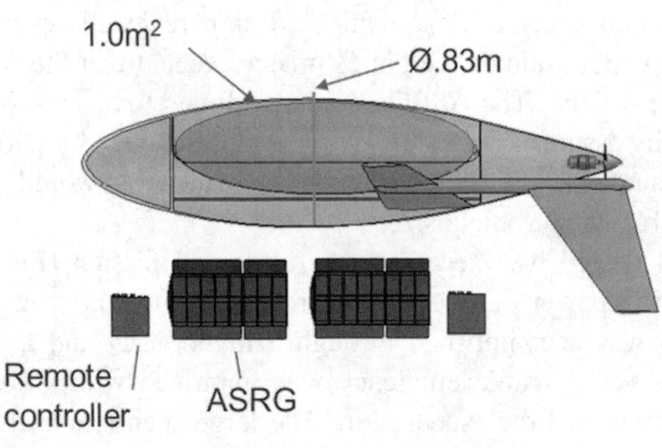

Fig. 9.26 Titan Flyer packaging challenges.

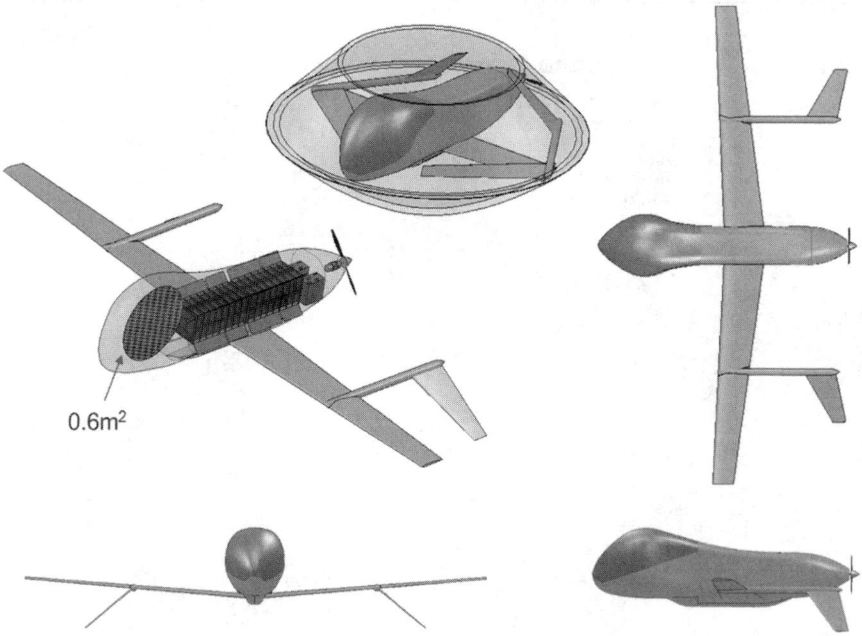

Fig. 9.27 Third Titan Flyer conceptual design iteration.

the length to accommodate a science payload bay below the wings. All the components could fit within the flyer thanks to the new fuselage arrangement.

Foch constructed small balsa wood free-flight models of the configuration. It was readily apparent that the configuration had stability problems. The models were directionally unstable, as large yaw rates built up during the gliding flight. The model tail geometry was modified to include separate horizontal and vertical stabilizers. With this change, the flight characteristics were acceptable. The Titan Flyer baseline configuration was updated to include this revised tail geometry (Fig. 9.28).

The Titan Flyer was required to carry 10 kilograms of payload plus 5 kilograms of communications equipment. The atmospheric flight mission segment duration was to be one Earth year. Achieving this performance required miniaturized electronics, refined aerodynamics, and lightweight composite structures.

The fixed power output of the ASRGs required power management throughout cyclical phases of the mission. The flyer adopted a climb-collect-glide profile to improve effective airspeed. During climb, much of the science payload power and all of the communications system power were inactive, thus providing more power for propulsion. The flyer would climb to an altitude of 14 kilometers, where collection operations occur. The

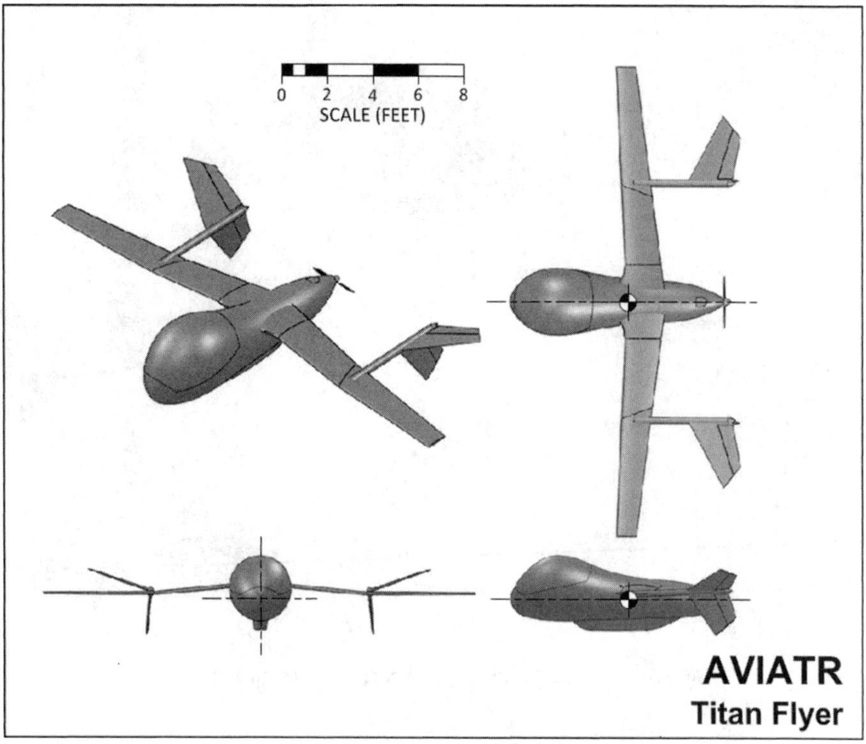

Fig. 9.28 Titan Flyer final configuration multiple-view drawing.

communications system was active during descent, where non-flight-critical power was supplied for communications purposes while the flyer was in a glide.

The Titan Flyer's projected performance was obtained with a vehicle L/D of 15 and 80% propeller efficiency. The cruise lift coefficient was 0.7, which provided a 2-*g* stall margin to contend with an uncertain gust environment. Any foreseeable performance margin degradation due to decrease in propulsion power available, drag increase, or weight increase would result in a longer climb segment.

The atmospheric entry and vehicle deployment were lower risk than for Mars missions. The entry vehicle consisted of two major components, the lower shell/heat shield and the upper shell (Figs. 9.29 and 9.30). During entry, atmospheric heating would be relatively benign due to the low entry speeds. The dense atmosphere would yield long descent times upon parachute deployment, which would permit more time for each deployment event with simpler technologies.

Upon decelerating to a threshold velocity, the lower shell/heat shield would jettison using separation nuts. The flyer was secured to the upper back

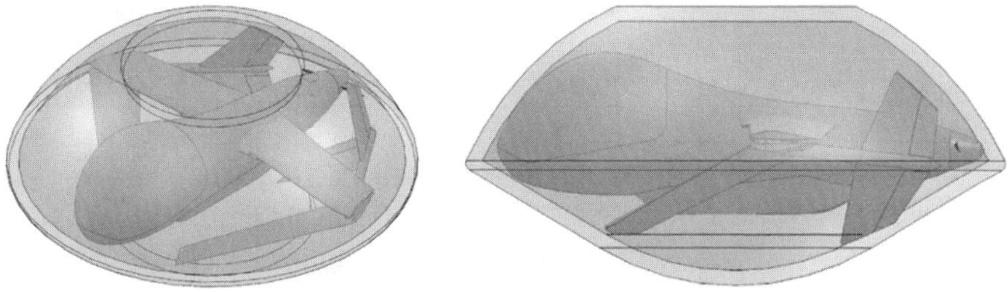

Fig. 9.29 The Titan Flyer folds within the entry vehicle with only two articulation mechanisms, resulting in a low deployment risk.

shell of the entry vehicle with brackets at the nose, at the wing joints, mid-fuselage, and using a sliding rail at the aft end. Explosive separation nuts in the brackets would fire, and the sliding rail would pivot such that the flyer faced nose down. The flyer would proceed to slide down the rail and to clear the upper back shell with the desired nose-down, low angle of attack orientation relative to the initial flight path. Once it was fully clear of the back shell, the outer wings with integral tail assembly would passively unfold and lock into the flight configuration. Finally, the flyer would execute a pull-out maneuver, and the electric motor would start, enabling level powered flight. The entire deployment sequence was controlled via onboard timers

Fig. 9.30 Titan Flyer deployment and operations concept.

and flight data sensor output, so no commands or round-trip communications with mission control were required.

The key to packaging a large aircraft into the entry vehicle was the ability to reliably unfold wings in flight. Wing unfolding was fully passive and used spring-driven mechanical actuation assisted by aerodynamic forces. The aerodynamic forces acting upon the wing provided the correct moment, effectively "flying" the outboard panels into position. This wing unfolding was similar to the technology demonstrated on the NRL LAURA, Venture, and FLYRT ATD programs. The low airspeed and high density provided damping and minimal wing opening shock loads. The autopilot could correct for any roll rate caused by asymmetric wing deployment. Centripetal acceleration induced by the roll further provided additional favorable wing deployment moment. The mechanical actuation to rotate the outer wing panels about their hinges used torsion springs to provide a moment that supplemented the aero forces and ensured reliable wing deployment regardless of wind gusts or unplanned acceleration loads.

The dense Titan atmosphere required a two-blade propeller of 1 meter diameter. No hinges were required, which greatly simplified the deployment. The propeller would be directly driven by a brushless DC motor. Sizing the propeller for maximum efficiency required motors that provided high torque at low rpm. Well-developed "outrunner" motor configurations provided the necessary performance with high reliability and low weight. They also improved propulsion system reliability and lifespan by eliminating the need for mechanical gear boxes that would have to operate at low temperatures.

Advanced composite structures similar to the type required for the Titan Flyer were demonstrated in the NRL Swallow, FLYRT, and Eager programs as well as in Aurora's production composite structures used on Global Hawk and other aircraft. These vehicles had among the lowest ratios of airframe mass to total mass of any aircraft in their category. Largely mass constrained, the Titan Flyer capitalized on this technology to achieve superior performance. The Titan Flyer employed a preimpregnated carbon fiber/epoxy or fiberglass sandwich structure with composite or metallic honeycomb core and low-temperature-qualified matrix and fiber materials. Core selection would be optimized to the requirements of each component or region of the structure. The high stiffness and structural rigidity of this construction were necessary to withstand the deployment loads at the light structural weight required for this high-performance flyer.

The Titan Flyer fuselage configuration was inspired by terrestrial unmanned aircraft designs. It enabled efficient payload packaging, enabled ample fields of regard for science payloads and the communications system, and was volumetrically efficient. The pusher propeller arrangement eliminated forward-facing payload view obstruction and provided additional

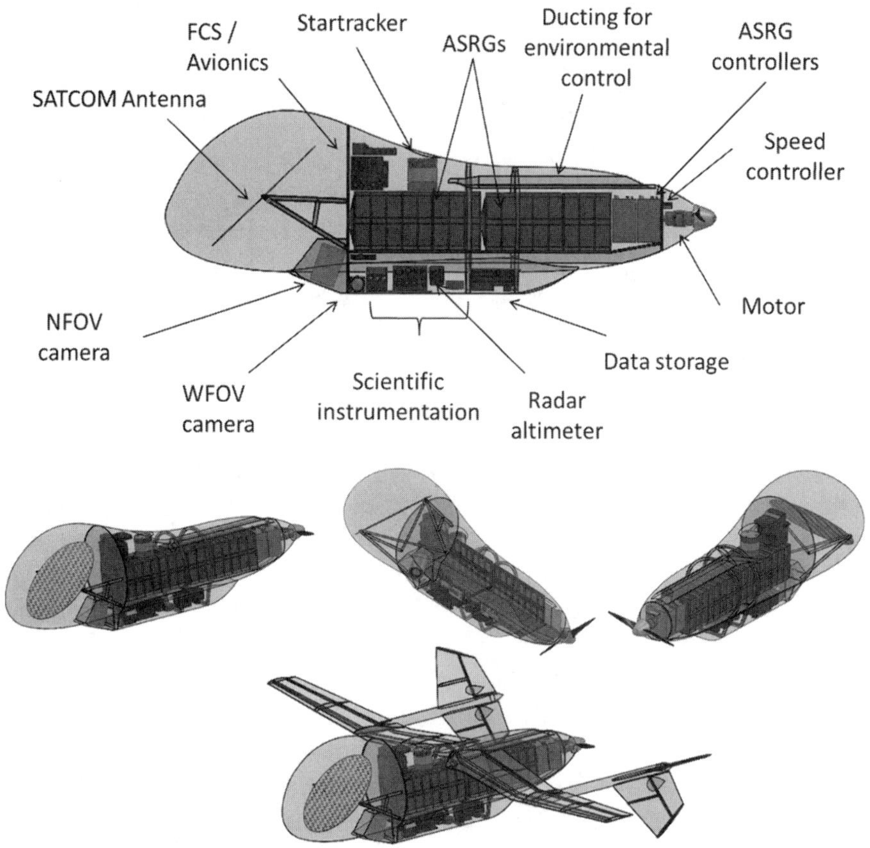

Fig. 9.31 Titan Flyer internal layout details (FCS = flight control system; SATCOM = Satellite Communications; NFOV = narrow field of view; WFOV = wide field of view).

stability. All the primary flyer systems resided inside the central fuselage. Only the flight control actuators were located in the deployable outer wing panels and stabilizer assemblies. The majority of the science payload was housed in a lower fuselage fairing that provided forward, nadir, and side field of regard tailored to the mission. A gimbaled circular phased array antenna with 0.6-square-meter area was located in the nose and was surrounded by a Radio Frequency (RF)–transparent radome structure (Fig. 9.31). The ASRGs were located near the center of gravity (CG) to reduce the mass moments of inertia and to provide heating for the surrounding systems.

The Titan Flyer automatic flight control system (AFCS) consisted of an autopilot, inertial navigator, and digital controller. The autopilot used deterministic flight control algorithms that would be extensively tested in a

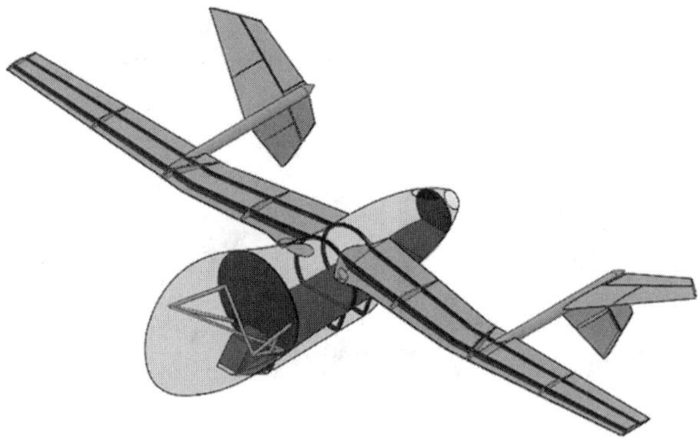

Fig. 9.32 Titan Flyer structural arrangement.

simulation environment and in terrestrial flight test beds. The flyer also employed a space-qualified inertial reference system. The vehicle state solution used a star tracker, laser altimeter, and air data system (pitot-static and barometric pressure) combined in a Kalman filter.

The Titan Flyer (Fig. 9.32) employed servo-actuated aerodynamic surfaces for control about all three axes. The torque-proportional electromagnetic actuators would have higher inherent reliability and longer life than electromechanical actuators. Controls included two rudders and two elevons. The elevons were the horizontal stabilizer control surfaces. The flyer was designed to be stick-fixed, statically stable about all axes, with dynamic stability augmentation provided by the flight control system. The adaptive capabilities of the autopilot enabled it to adjust control rates and feedback gains to continuously optimize aircraft behavior and stability. This allowed the flyer to minimize the effects of atmospheric turbulence.

The inertial measurement unit would integrate attitudes and accelerations during deployment, from release by the entry vehicle until reaching trim flight. Inertial data was sufficient to indicate the heading quadrant, thus enabling the sun tracker to fine tune the heading for the initial turn to the flight course. At the beginning of powered flight, the flyer would obtain an accurate initial heading to follow the desired ground track. The lightweight inertial measurement unit and star tracker would provide the navigation solution.

The air vehicle was designed to operate in a safe mode when one of the two flight computers was down for software updates or diagnostics. In the event of two inoperable flight computers, the autopilot-generated lateral damping would no longer be present, so the air vehicle would fly in its natural stability state, which was a spirally stable mode with lightly damped, low-amplitude, roll oscillations (Dutch roll, described in further detail in Appendix A). The aircraft would be trimmed to execute a low-rate, large-radius turn and to maintain its current altitude via an outer autopilot-independent control loop that used a barometric altimeter. Upon reset of an autopilot, the lateral damping, laser altimeter–based altitude, and inertial navigation/antenna pointing modes would be restored.

The Titan Flyer electrical system was comprised of the ASRG power source, a power distribution unit, and the power buses. The electrical system provided conditioned 28 volts of DC power to the AFCS, science payload, communications system, data buffers, and flight control actuators as required.

The ASRGs are the primary power source to the spacecraft during transit to Titan. The avionics, subsystems, and science payload may be powered in standby mode during this phase, and these would be kept within permissible temperature limits by this heat source. Any excess power would either be (1) absorbed by the propulsion system and converted to useful work in the form of climb or higher velocity or (2) applied as additional transmit power for communications.

The communications antenna was a phased array gimbaled along the flyer's pitch axis. The antenna could operate without interference through 100 degrees in elevation from horizontal forward, upward to 10 degrees past vertical in the vehicle body axes. This permitted the flyer to fly in the direction of Earth while transmitting, thus providing the greatest communication window availability. The azimuth pointing was controlled by adjusting the flyer heading.

Figure 9.33 shows the Titan Flyer mission profile. The flyer would traverse Titan's atmosphere on a 24-hour cycle. Table 9.6 lists its technical characteristics.

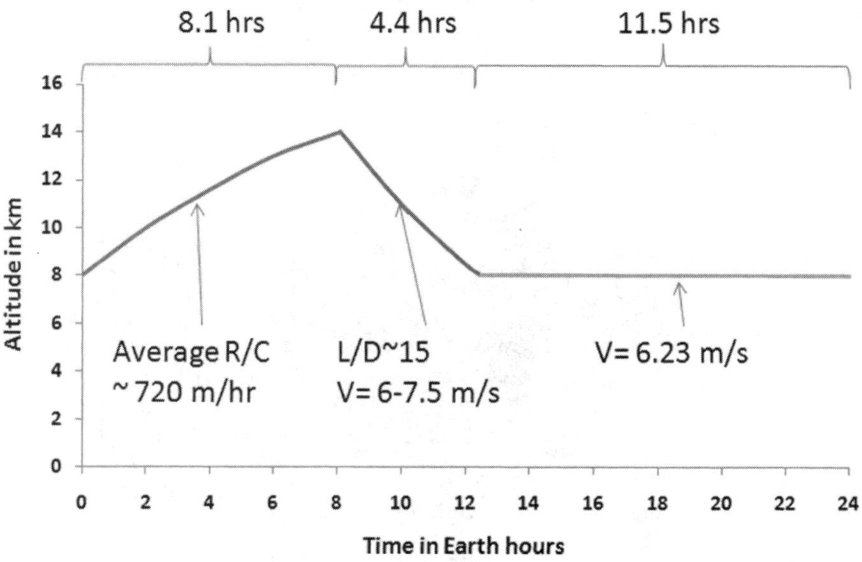

Fig. 9.33 Titan Flyer mission profile.

Table 9.6 Titan Flyer Technical Characteristics

Parameter	Value
Wing span	6.0 m
Wing area	3.0 m^2
Takeoff weight	115.0 kg
Payload weight	10.0 kg
Loiter speed	6–7.5 m/s
Endurance	1 year
Max L/D	15
Launch method	Atmospheric entry vehicle
Recovery method	Expendable
Propulsion type	Two ASRGs

VENUS FLYERS, 2010–2011

NRL conducted preliminary studies of flight on Venus. It would be possible to fly in the planet's upper atmosphere above the clouds. It would be very difficult to land or fly in the lower atmosphere due to the high temperature, high pressure, and corrosive atmospheric chemistry. It would be possible to briefly fly within the clouds to conduct a science mission and then rise above the cloud tops to cool off. The Venus flyer design studies have not been published, and NRL has not yet produced any Venus flyer proposals.

MARS HELICOPTER, 2012

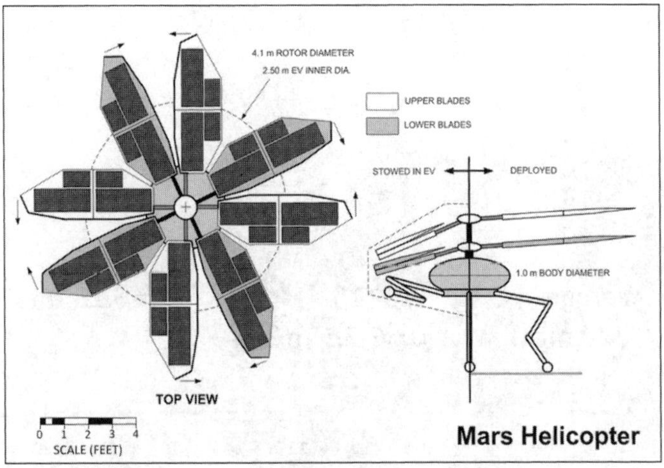

Fig. 9.34 Mars Helicopter concept.

Recently, Mars orbital imaging has indicated the strong possibility of brine water flows appearing in real time near the edges of certain Martian craters. Lemke suggested that an interesting and useful mission would be to deploy a Mars lander near the center of a candidate crater and to land a Mars flyer nearby. Then, when cued by an orbital sensor, the flyer would be sent to investigate the newly appearing brine flow gullies. Ideally the flyer would land, take a physical sample, and then fly to the lander, where an onboard robotic laboratory would analyze the sample. Such a mission requires a flyer with vertical takeoff and landing (VTOL) capability and a landing gear capable of coping with a wide range of terrain features and ground angles. The obvious solution would be a Mars helicopter.

A few years earlier, students at the University of Maryland (UMD) performed conceptual design studies for a Mars helicopter and won a design competition for their concept. In addition, NASA Ames had performed rotor test measurements in their Mars atmosphere chamber and determined baseline rotor performance characteristics under Martian conditions. These UMD and NASA studies indicated that a Mars helicopter was challenging, but feasible, so NRL undertook a design study to meet Lemke's requirements.

The resulting design concept was a coaxial helicopter with surface-adaptable, four-leg landing gear (Fig. 9.34). It featured battery electric propulsion. The batteries would be secondary lithium cells recharged via solar panels integrated into the upper skin of the rotor blades. The range would be relatively short, on the order of several kilometers, with a recharge time of several hours between hops. This level of performance was adequate for the envisioned application. The number of flights would depend on the degradation of the solar panels, but a useful life of dozens of flights was envisioned. In addition to the crater search mission, a Mars helicopter could conceivably explore many sites across the Martian surface in multiple short flights.

At the time of this writing, the NRL/Ames team is awaiting the opportunity to further develop this interesting concept for a NASA proposal opportunity.

Chapter 10

Fuel Cell Aircraft

The period from the early 1990s to about 2004 was the heyday for the development of battery electric powered unmanned aircraft systems (UASs), both at NRL and throughout the world. Major breakthroughs were occurring frequently in electric motors, motor controllers, and new battery chemistries. Flight endurances grew from a few minutes to a few hours during this period, enabling useful military missions. UAS operators liked the reliability, the instant starting under all weather conditions, the lack of noise, the ability to stop and restart while in flight, and the minimal maintenance required for these electric aircraft. Unfortunately, there was one major limitation to battery electric flight: short flight endurance in comparison to liquid fuel systems powered by internal combustion engines. NRL researchers realized that next-generation military UASs should feature all the advantages of battery electric aircraft, but have flight endurance equal to or better than that of existing liquid fuel UASs. The approach to achieve this goal was to develop small, efficient, and affordable fuel cells to generate electrical power.

Spider Lion, 2005

Fig. 10.1 Spider Lion in flight.

The purpose of Spider Lion (Fig. 10.1) was to see if it was possible to use emerging field-portable fuel cell technology as a baseline for a small unmanned aircraft propulsion system. It was hoped that the new fuel cells could be used on a Dragon Eye–like small UAS to extend the endurance. Operators liked Dragon Eye's silent operation and runway-independent launch and recovery. The major system limitation was the 30- to 60-minute flight duration. Battery technology was making important advances in specific energy, but multihour endurance with battery power was still many years away, if even achievable. NRL's Director of Research asked the Vehicle Research Section (VRS) and NRL's Chemistry Division to investigate fuel cell propulsion. The name Spider Lion was inspired by the name of the NRL electrochemist who rose to the challenge to become one of the leading authorities on fuel cells for unmanned systems, Dr. Karen Swider-Lyons.

After surveying the field, NRL found that Protonex Technology Corp. had the best available fuel cell for NRL's purposes. Protonex was developing man-portable hydrogen proton exchange membrane (PEM) fuel cells that could replace battery packs (these fuel cells were for soldiers). One system was capable of producing 95 watts.

NRL conducted a system design study around this Protonex PEM fuel cell. There was a question about whether the air vehicle could operate on 95 watts with the all-up system weight. The entire fuel cell system included far more components than just the fuel cell stack; other items included the humidifier, hydrogen tank, regulators, cooling systems, fluid pumps, and air pumps. These components beyond the core fuel cell are known as "the balance of plant." NRL evaluated the full balance of plant to identify weight-reduction opportunities.

NRL found that at this stage of technology development, small unmanned aircraft that used this class of fuel cell technology had advantages for power ranges from 90 watts to 1 kilowatt. Below 90 watts, battery-powered aircraft had an advantage due to the weight of the fuel cell balance of plant. Above 1 kilowatt, reciprocating engines offered better endurance. Within the identified power range, fuel cells offered about four to five times better endurance than lithium batteries.

Based on the results of these studies, the NRL Director of Research initiated a modest project to build and test a fuel-cell-powered small UAS (Fig. 10.2). To minimize costs, on-hand materials were used whenever possible. The fuel cell was procured from Protonex, and to minimize aircraft construction costs, NRL proceeded by modifying a commercial off-the-shelf, 2-meter-span competition radio-controlled (RC) sailplane. The wings and tails were reused, and a custom balsa wood fuselage was built to house the fuel cell system.

Many of the fuel cell balance-of-plant items had to be custom developed. A 5000-psi paint ball gun CO_2 tank was used to store the compressed

Fig. 10.2 Spider Lion atop NDM III for air launch.

hydrogen gas. A model aircraft field box fuel pump was used to pump water for the humidifier, and Protonex developed a custom humidifier. Aquarium pumps were used to pump air for the oxygen source. NRL developed a custom cooling system that consisted of aluminum tubing with flattened thin-sheet aluminum fins from beer cans.

NRL first ran the system on the bench for a few months. The system functioned within expected performance parameters at the conclusion of the testing. The balance of plant was not excessively heavy, but it occupied considerable volume. The Spider Lion fuselage was therefore relatively large to accommodate all the systems.

The initial Spider Lion flight tests used batteries so that the basic flying qualities could be evaluated. Then the fuel cell was installed (Fig. 10.3). An initial problem was that the paint ball gun pressure tank could only be filled to pressures of 3000 psi rather than the desired 5000 psi. This was due to the storage pressure of NRL's hydrogen source and not any technical limitations of the power plant system. Therefore, Spider Lion could not carry the intended hydrogen fuel quantity. Despite this problem, the aircraft achieved approximately 1 hour's endurance on its first endurance test. This was not

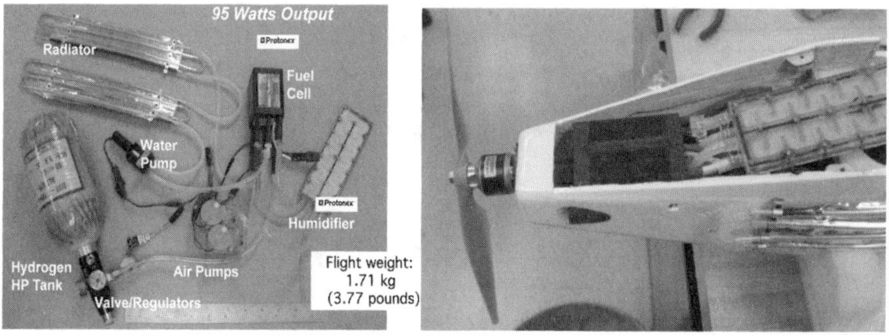

Fig. 10.3 Fuel cell system (left) and fuel cell installed on Spider Lion (right).

Table 10.1 Spider Lion Technical Characteristics

Parameter	Value
Wing span	6.5 ft
Takeoff weight	5.6 lb
Hydrogen weight	15–29 g
Loiter speed	35 KTAS
Stall speed	25 KTAS
Endurance	3.3 hr
Launch method	Air launch
Recovery method	Belly landing
Propulsion type	Compressed hydrogen fuel PEM fuel cell
Power	95 W

superior to lithium battery performance. Recall that the SARA achieved 2 hours and 45 minutes' endurance in the mid-1980s. This benchmark had to be exceeded to prove the potential of fuel cell technology.

On November 9, 2005, Spider Lion achieved 3.3 hours' endurance, flying until the hydrogen fuel was depleted. It consumed 15 grams of hydrogen, which was compressed to 4500 psi. This flight exceeded NRL's best battery flight record. The Spider Lion technology formed the foundation for the XFC and Ion Tiger programs, discussed next.

NRL further developed the compressed gas hydrogen storage tank, resulting in a new 6000-psi tank that could store 29 grams of hydrogen. The final fuel cell system and hydrogen storage weighed 3.8 pounds. Spider Lion's technical characteristics are listed in Table 10.1.

Experimental Fuel Cell (XFC), 2006–Present

Fig. 10.4 XFC in flight.

Spider Lion and the later Ion Tiger were fuel cell propulsion research test beds. They proved many of the technologies that were necessary for an operational aircraft. The Experimental Fuel Cell (XFC), on the other hand, was designed from the outset as a tactical, fuel-cell-powered unmanned aircraft that could be placed into production for military use (Fig. 10.4). The XFC and Ion Tiger programs progressed in parallel as the fuel cell technology was developed and improved. NRL sought and received external sponsorship from the Office of the Secretary of Defense (OSD) Rapid Reaction Technology Office (RRTO) and the Office of Naval Research (ONR) SwampWorks.

NRL had flexibility to establish its own technical requirements for the system. The system should have all the advantages of a small, battery electric, tactical unmanned aircraft. Some key capabilities included fully autonomous flight and reliable self-deployment at launch under a broad range of weather conditions. These benefits should be combined with long endurance enabled by a fuel cell. NRL adopted a requirement for the aircraft to autonomously deploy from a container without a booster or pyrotechnics. XFC had to have visual, acoustic, and radar low observable characteristics. The system also had to be affordably expendable for tactical missions, which was challenging due to the high cost of fuel cells, high-pressure tanks, and regulators.

The aircraft was designed to fit within a tube measuring 19 inches in diameter and no more than 8 feet long. This would readily fit on a small boat or High Mobility Multipurpose Wheeled Vehicle (HMMWV) and would even have the potential for launch from a submarine. The stowed system included the aircraft and launch devices.

The XFC configuration took advantage of many lessons learned from past programs. Reliability is inversely proportional to the number of deployment mechanisms, so hinges, latches, and other folding devices were minimized. Many past self-deploying unmanned aircraft would take the best aerodynamic design, simply make it fold, and then live with the performance loss. XFC was approached from the opposite perspective. Foch evaluated vehicle geometries that would be easy to unfold, and then he evaluated the achievable performance. This ultimately led to a reliable, higher-performance design.

The final configuration was unconventional and asymmetric (Fig. 10.5), but it offered high performance, high packaging efficiency, and high launch reliability. A key feature was a scissoring X-wing, in which an upper and lower wing lay along the top and bottom of the fuselage, respectively, when stowed (Fig. 10.6). A single torsional spring deployed the upper and lower wings together, so this wing deployment arrangement was effectively a single mechanism. There was insufficient span and wing area to obtain the desired performance with only the X-wing, so wing tip extensions were added. These

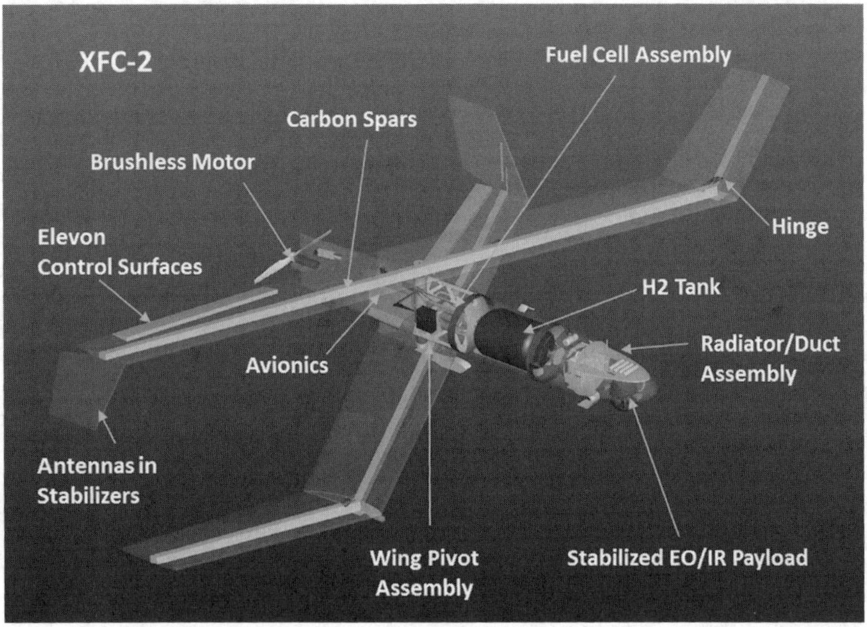

Fig. 10.5 XFC configuration and internal layout.

outboard panels unfolded from underneath the upper and lower wings. The addition of two more wing hinges and latches would reduce the launch reliability, but the performance advantages were significant. The wing tip extensions were attached to the forward swept side of the upper and lower wings, and these panels were swept aft to maintain the desired aerodynamic center position. The aft swept side of the lower wing had an upward-facing winglet, and the aft swept side of the upper wing had a downward-facing winglet. These winglets provided directional stability, though they also slightly reduced the induced drag.

The highly unconventional configuration initially generated significant controversy. Ultimately, it was determined that the configuration could be made to perform well and that there were many merits. The combination of upper and lower wing systems and fuselage could be tailored to simultaneously generate an elliptic lift distribution and trim. The airfoil

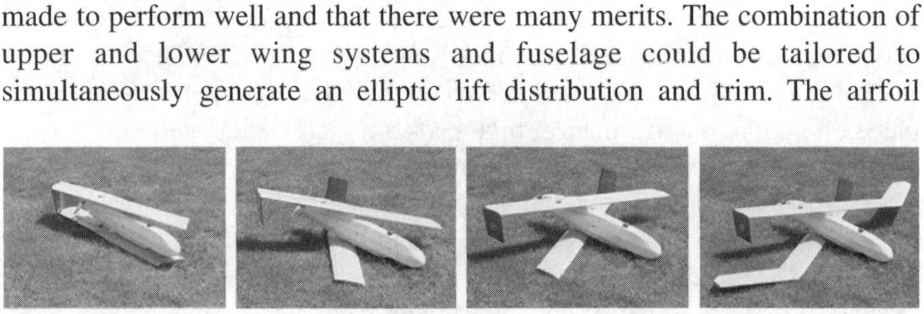

Fig. 10.6 XFC deployment sequence.

and twist distributions of the upper and lower wings could be independently optimized using computer tools. The wings are not simply mirror images of each other. Vortex lattice modeling and computational fluid dynamics (CFD) were used to design the wings. Initially, the main wings had no taper, and the tips had taper. Later versions had main wing taper. The final, best lift-to-drag ratio (L/D) for the baseline XFC design was approximately 15.

The system underwent wind tunnel testing. The team was concerned that the high sweep on the wing would produce a drag penalty. The wings were swept at 25.29 degrees, which has the arctangent of 0.5. This sweep allows designers to quickly keep track of dimensions in the ratio of 1 unit back for every 2 units of span. It is well known among aerodynamicists that wing sweep introduces cross flow instabilities and Tollman-Schlichting waves that can cause a laminar boundary layer to transition to a turbulent boundary layer earlier than would be seen in an unswept wing. In high Reynolds number applications, the turbulent boundary layer produces more profile drag than a laminar boundary layer.

The wind tunnel tests yielded surprising and encouraging results. There appeared to be no L/D penalty due to the sweep. In comparison with an unswept wing tested at the same Reynolds number, flow visualization tests revealed that the wing sweep provided beneficial characteristics. As expected, sweep did promote early transition. However, this was beneficial at chord Reynolds numbers between 150,000 and 250,000 evaluated in the tests. A turbulent boundary layer has more energy, and it is capable of remaining attached through adverse pressure gradients on the upper side of the airfoils. A challenging low Reynolds number (LRN) phenomenon is laminar separation bubbles, in which a laminar boundary layer separates in an adverse pressure gradient. The laminar separation bubbles greatly increase the profile drag and can lead to undesirable stall characteristics. The usual location for laminar separation bubbles is the aft upper surface of the lifting airfoil. Since the XFC wing sweep caused transition to turbulent flow before reaching the adverse pressure gradient, laminar separation bubbles were not experienced. The profile drag was lower than for the equivalent unswept wing. As expected, the wing sweep reduced the lift coefficient for a given angle of attack. The combined effects of reduced lift coefficient and reduced profile drag produced an approximately equivalent L/D as an equivalent unswept wing. Additional tests revealed that the benefits of sweep are valid from sweep angles between 15 and 30 degrees. By chance, the XFC's sweep of 25.29 degrees showed about the maximum benefit. The benefits of sweep for aerodynamic efficiency on LRN vehicles have large implications for small unmanned aircraft design. The conventional wisdom that sweep should be avoided for this class of aircraft is invalid.

Flight testing began after the wind tunnel test campaign. The initial prototypes had fixed surfaces that did not unfold, and used batteries rather than fuel cells for power. The purpose of the early tests was to evaluate the handling qualities of the design. The first flights were made under radio control.

Based on the outcome of the Shearwing flight tests (circa 1980; see Chapter 4), Foch was concerned that XFC would be too difficult for an external radio-controlled (RC) pilot to control with visual feedback due to the highly asymmetric configuration. He recommended a pilot camera in the front of the aircraft to allow the pilot to fly the aircraft through a video monitor on the ground. The development team did not use the camera. The assigned test pilot was not available, so Foch had to conduct the first flight. To his pleasant surprise, Foch was able to fly the aircraft with minimal difficulty. An additional 20 years of flight test experience after the Shearwing definitely helped. In later flights, another NRL test pilot commented that XFC was "the straightest flying crooked airplane I ever flew."

The first prototype was constructed of foam and fiberglass without any tooling. The tolerances were not good, resulting in some performance shortfalls. The handling qualities were evaluated with the first prototype.

Later prototypes used Computer Numerically Controlled (CNC) milled female tools and hollow molded construction. The tolerances were greatly improved. The molded aircraft were built much like SENDER, using "lobster claw" construction. The wing and fuselage skins used a fiberglass and Rohacell foam sandwich. The wing spar had a carbon fiber cap strip, and the shear web was made from dense foam. The structure was very lightweight due to light, minimum-gage materials.

The second batch of prototypes had some design modifications from the first prototype. The first vehicle had an annular ring around the motor and spinner to provide cooling flow exhaust into the propeller disk. The new design had a vertical edge ("knife edge") aft closure to reduce drag. The size of the winglet vertical stabilizers was increased, and the shape changed. The initial design had an inward cant on the upper winglet and an outward cant on the lower winglet. The new aircraft had vertical winglets. Previously, the wings were of the same length, but the new winglets without the cant required wings of different length to accommodate the wing folding. The wing twist distribution was modified to improve efficiency. The wingtips had to be hinged with torsional springs to fold along the wing in order to fit within the container. The new prototypes were designed to be folded and deployed to support the full flight test program.

NRL developed the fuel cell in parallel with the aircraft. The cruise power was around 155 watts. The payload consumed 25 watts, and the avionics required 17 watts. NRL wrote the fuel cell control algorithms to manage the cooling system, regulator, and fuel flow. This control software was housed in a Cloud Cap autopilot, which also served as the flight control computer.

A key challenge was providing low-drag cooling for the fuel cell system. The first approach was to use a skin radiator. This consisted of connected rows of aluminum tubing with aluminum tape on either side. The skin radiator ran along the length of the fuselage, with fairings on either end to reduce drag. An advantage of this arrangement was that it consumed no internal volume. However, it was discovered that the system provided insufficient cooling on a hot day. The designers determined that the best system solution was to use a small motorcycle radiator with a controlled cooling fan, mounted inside the fuselage. This internal radiator system produced more drag but was more robust than the skin radiators.

The fuel cell system was flown on the XFC prototype. An early flight encountered a critical problem with the autopilot control of the fuel cell. These electrical system design problems were resolved and the system worked as intended. The flying aircraft was found to be inaudible at 250 feet from the observer. The loudest components were the propeller and the water pumps for the humidifier, but these could be easily improved in future designs.

The test program incrementally proved the flight performance. The threshold endurance was 4 hours, and the goal performance was 6 hours. The first fuel cell flight lasted approximately 2 hours. The next flight was 4 hours, which met the threshold requirement. Later, XFC achieved more than 6 hours' endurance, which exceeded the goal performance. The climb rate with the initial fuel cell was low but acceptable. NRL found that a climb rate of over 300 feet per minute was needed for a satisfactory military unmanned aircraft. The Ion Tiger fuel cell system development was progressing in parallel (Figs. 10.7 and 10.8). It produced 550 watts of power with no weight increase over the XFC fuel cell system. When it was later installed on XFC, the climb rate performance requirements were achieved.

A key system development was a single-stage pressure regulator, which enabled variable power output from the fuel cell (Fig. 10.9). The final design could regulate pressure between 10 psig and 5000 psig. It had electronic monitoring of the high- and low-pressure sides. The low-pressure side included a safety valve. This system was a vast improvement over commercially available pressure regulators. Small, single-stage pressure regulators that cover this pressure range were not available in this weight class. When the tanks were stored with pressurized hydrogen for over a month, no pressure leaks were detected from the regulators.

A later phase of the test program aimed to demonstrate system deployment from a tube. Up to this point, all the flight tests had used a nondeploying, fixed configuration. Very early in the XFC concept design phase, one

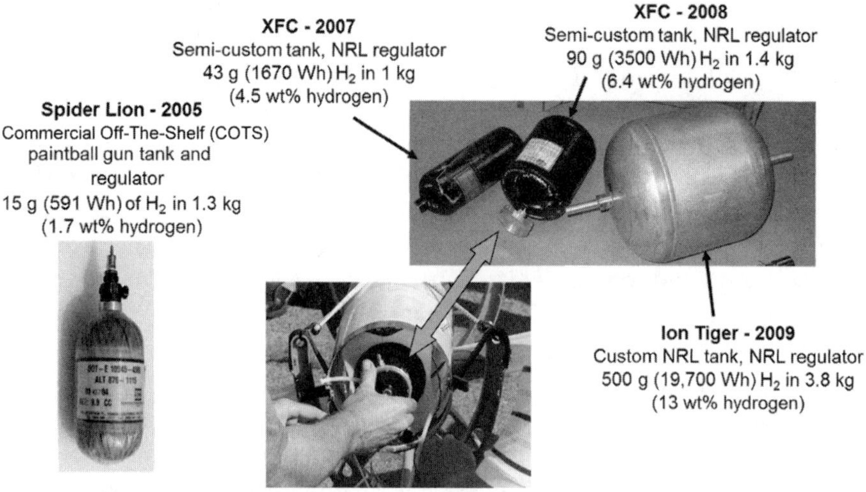

Fig. 10.7 Compressed hydrogen storage tank development.

possibility of a nonpyrotechnic launch was envisioned by using an electrical tether to provide sufficient supplemental power to enable the aircraft to self-propel out of the tube using its motor and propeller. However, this umbilical approach was infeasible due to the pusher propeller and geometric interference with the folded wings of the selected configuration.

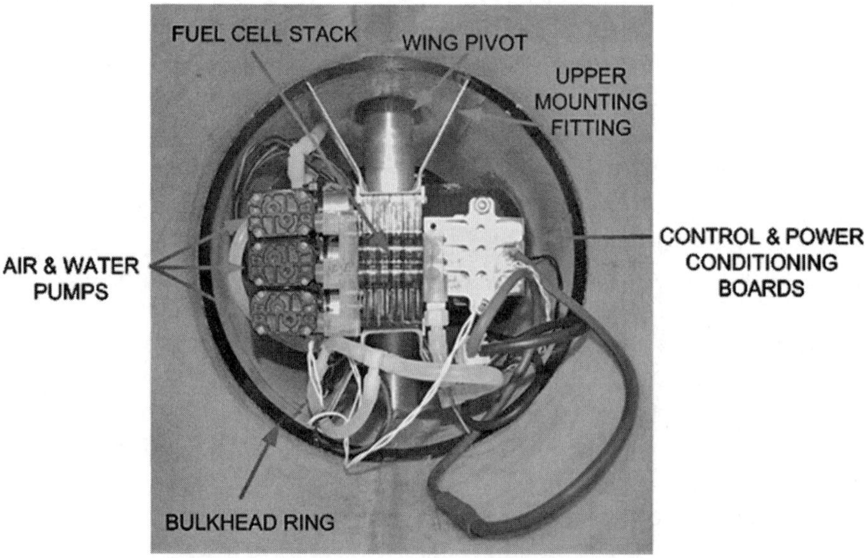

Fig. 10.8 XFC integrated fuel cell system.

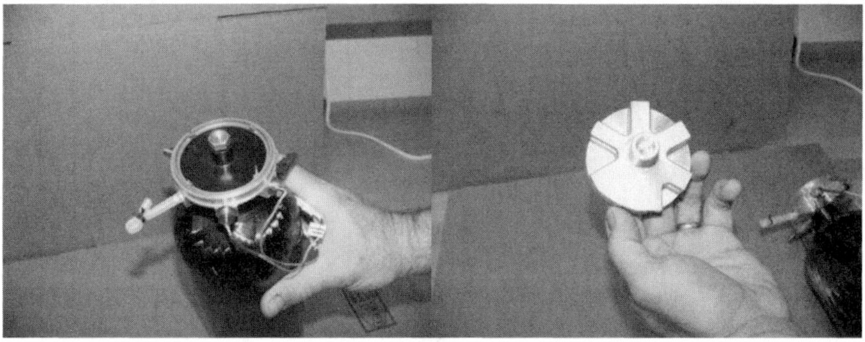

Fig. 10.9 XFC custom pressure regulator.

The selected launch approach used a novel, electric propulsion booster device called the Electric Assisted Take-Off (EATO), shown in Fig. 10.10. ITT designed and patented an expendable booster system comprised of batteries, a motor system, and a tractor propeller, which was adapted for the XFC by NRL and ITT. The unit was mounted at the front of the aircraft. The propeller diameter was smaller than the tube diameter, requiring a folding EATO propeller. The EATO thrust would pull the XFC from the tube, climb away, and provide the aircraft sufficient time to unfold. The EATO would fall away after the aircraft was configured for controlled flight with sufficient altitude and airspeed.

Initial EATO tests were conducted with old SENDER airframes. Initially the EATO had a single propeller, but this was found to produce excessive torque. The device was modified to have contrarotating propellers driven by two motors with bevel gears.

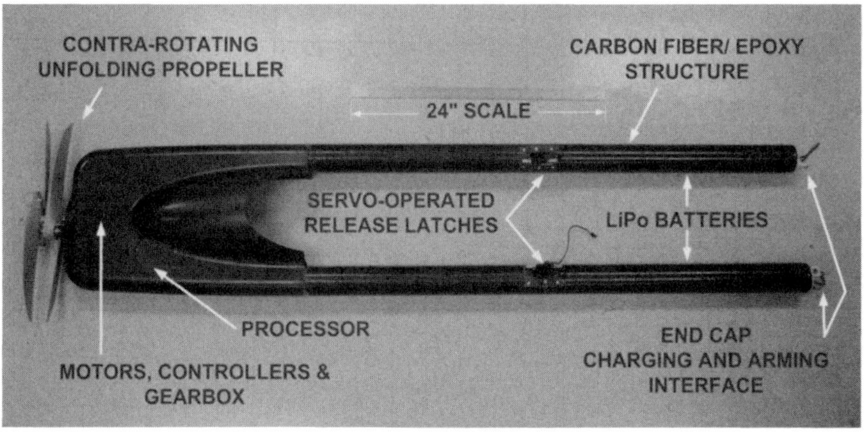

Fig. 10.10 EATO system.

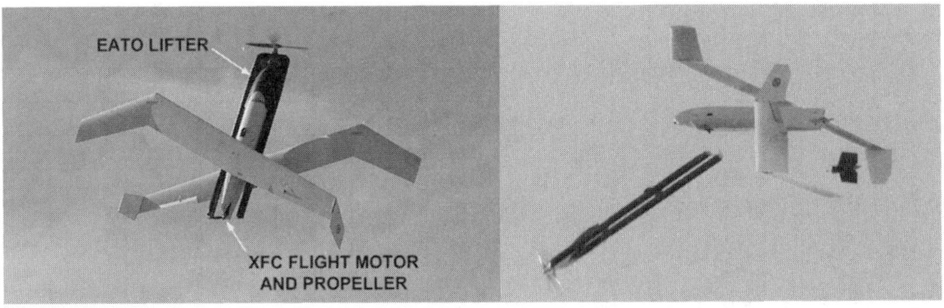

Fig. 10.11 EATO deployment.

Once the initial EATO development tests were completed, the full XFC deployment development began. New XFC prototypes were configured with deployment mechanisms. The main wing scissor mechanism was developed, which included an internal spring and a cam mechanism. The wings were held closed by the container. The mechanism had a soft stop and lock mechanism, which permitted some oscillation to prevent excessive shock loads on the structure.

An open frame tube structure was built to house the stowed XFC. The frame allowed the test team to see the aircraft at the early stages of deployment. The frame included rollers to guide the aircraft out of the tube with minimal friction.

Initial deployments occurred in calm, controlled conditions (Fig. 10.11). The tube was oriented at an 80-degree angle. The first launch was not successful because the aircraft was too nose-heavy. The aircraft started to unfold prior to hitting the ground. Wind tunnel testing showed that XFC had suitable ballistic characteristics when folded, but the EATO shifted the center of gravity (CG) forward more than the flight control system could compensate. The system launched in a ballistic trajectory, which generated a large pitch rate at the end of the launch. XFC could not recover from the combination of high nose-down pitch rate and a low nose attitude with respect to the horizon after the EATO separated. The solution was to run tubes containing the batteries along the fuselage length to shift the CG farther aft.

The next test, after the EATO modification, was successful. The wings scissored symmetrically, and the wing tips deployed nearly simultaneously. The wings were not exactly the same length due to packaging constraints. The moments of inertia due to the asymmetry introduced yaw rate under acceleration. After the first deployment test, springs were added to the outer wing panels and winglets to ensure proper unfolding in sideslip conditions.

The deployment tests continued. Most tests were successful, but there were some failures. The failures occurred when an outer wing panel or winglet would not unfold. If an outer wing panel failed to unfold, the aircraft had a roll rate higher than could be compensated for by the elevon control surfaces, resulting in a crash.

The initial deployment solution had a passive EATO, and the XFC would provide all flight control through the deployment sequence. The wings scissored open immediately after clearing the tube. The elevons on the wings were the only control surfaces, and they combined both roll and pitch functions. The elevons were scheduled to hold the nose up at launch, so there was limited roll authority available. This favored steeper launches, where less pitch authority was necessary. However, the launch attitude could not be effectively limited on a small ship. A more robust launch control system was necessary. The solution was to put control vanes in the slipstream of the EATO propeller. These vanes provided pitch and yaw commands to hold the

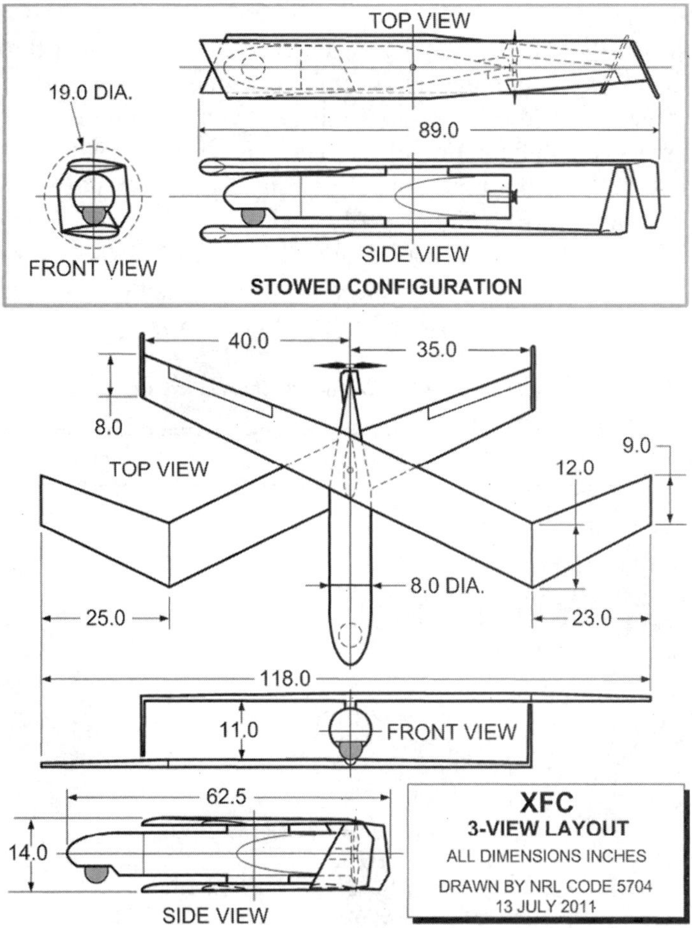

Fig. 10.12 XFC three-view drawing and stowed geometry.

yaw rate to zero. Roll was controlled through the XFC elevons. This control solution yielded multiple successful launches.

Problems persisted with the wing tips not deploying properly. The problem was acute in certain combinations of crosswind and container attitude. The failures resulted when one wing tip would not unfold, and the aircraft rolled over. Full differential elevon deflection could not counter the roll moment. The final design solution was to add actuators to the outer wing hinges. This was more complex and heavier. However, the wing deployment was now a fully closed loop. All requisite technologies for reliable deployment were in place.

The XFC has demonstrated multiple payloads. It has a 2.5-pound payload capacity. The baseline electro-optical infrared (EO/IR) surveillance payload is a Cloud Cap TASE ball. NRL has also flown a quantum well modulating laser retroreflector. A ground-based laser is directed at the retroreflector, the payload data signal modulates the reflector, and the laser beam is reflected back at the ground. This system provided covert communications through the laser without any radio frequency emissions and demonstrated minimal power required aboard the XFC for high-bandwidth data transmission.

The innovative configuration of the XFC has been awarded a U.S. patent (No. 8,444,082). At the time of this writing, the system is ready for full mission demonstrations. Figure 10.12 shows a three-view drawing and stowed geometry for the XFC, and Table 10.2 lists its technical characteristics. The appendix at the end of this chapter contains a press release regarding the U.S. Navy's launch of this UAS.

Table 10.2 XFC Technical Characteristics

Parameter	Value
Wing span	9.8 ft
Wing area	14.7 ft^2
Length	5.2 ft
Takeoff weight	20.0 lb
Payload weight	2.2 lb
Fuel weight	90 g
Cruise speed	30 KTAS
Dash speed	35 KTAS
Stall speed	21 KTAS
Endurance	7 hr
Max L/D	15
Airfoils	Progressively blended NACA 4-digit series
Launch method	EATO launch
Recovery method	Belly landing or expendable
Propulsion type	PEM fuel cell
Power	550 W

XFC: FURTHER READING

Foch, R., MacKrell, J., and Schultz, W., "XFC–Fuel Cell Powered Long Endurance Tactical UAS," *Proceedings of AUVSI*, 2011.

NAVY LAUNCHES UAS FROM SUBMERGED SUBMARINE

Our discussion of the XFC ends with a copy of a press release from the U.S. Navy about the launch of an all-electric, fuel-cell-powered UAS from a submerged submarine.

Naval Research Laboratory Public Affairs Office News Release

Washington, DC • Stennis Space Center, MS • Monterey, CA
http://www.nrl.navy.mil
202-767-2541

96 -13r
Editor contact: Daniel Parry
daniel.parry@nrl.navy.mil
202-767-2326

(WASHINGTON • Dec. 5, 2013) – The U.S. Naval Research Laboratory (NRL) with funding from SwampWorks at the Office of Naval Research (ONR) and the Department of Defense Rapid Reaction Technology Office (DoD/RRTO) demonstrated the launch of an all-electric, fuel cell-powered, unmanned aerial system (UAS) from a submerged submarine.

From concept to fleet demonstration, this idea took less than six years to produce results at significant cost savings when compared to traditional programs often taking decades to produce results.

"Developing disruptive technologies and quickly getting them into the hands of our sailors is what our SwampWorks program is all about," said Craig A. Hughes, Acting Director of Innovation at ONR. "This demonstration really underpins ONR's dedication and ability to address emerging fleet priorities."

The successful submerged launch of a remotely deployed UAS offers a pathway to providing mission critical intelligence, surveillance and reconnaissance (ISR) capabilities to the U.S. Navy's submarine force.

XFC UAS launch in time-lapsed photography. Deployed from the submerged submarine USS Providence, the NRL developed XFC unmanned aircraft is vertically launched from a 'Sea Robin' launch vehicle (bottom right). The folding wing UAS autonomously deploys its X-wing airfoil and after achieving a marginal altitude, assumes horizontal flight configuration. (*NAVSEA-AUTEC*)

Operating under support of the Los Angeles class submarine USS Providence (SSN 719) and the Naval Undersea Warfare Center-Newport Division (NUWC-NPT), the NRL developed XFC UAS – eXperimental Fuel Cell Unmanned Aerial System – was fired from the submarine's torpedo tube using a 'Sea Robin' launch vehicle system. The Sea Robin launch system was designed to fit within an empty Tomahawk launch canister (TLC) used for launching Tomahawk cruise missiles already familiar to submarine sailors.

Once deployed from the TLC, the Sea Robin launch vehicle with integrated XFC rose to the ocean surface where it appeared as a spar buoy. Upon command of Providence Commanding Officer, the XFC then vertically launched from Sea Robin and flew a successful several hour mission demonstrating live video capabilities streamed back to Providence, surface support vessels and Norfolk before landing at the Naval Sea Systems Command Atlantic Undersea Test and Evaluation Center (AUTEC), Andros, Bahamas.

"This six-year effort represents the best in collaboration of a Navy laboratory and industry to produce a technology that meets the needs of the special operations community," said Dr. Warren Schultz, program developer and manager, NRL. "The creativity and resourcefulness brought to this project by a unique team of scientists and engineers represents an unprecedented paradigm shift in UAV propulsion and launch systems."

The NRL Chemistry and Tactical Electronic Warfare Divisions team includes the design-builder of the Sea Robin, Oceaneering International Inc., Hanover, Md.; the fuel cell developer Protonex Technology Corp., Southborough, Mass.; and NUWC-NPT's Autonomous and Defensive Systems Department for Temporary Alteration (TEMPALT) and test demonstration support.

The XFC is a fully autonomous, all-electric fuel cell-powered folding wing UAS with an endurance of greater than six hours. The non-hybridized power plant supports the propulsion system and payload for a flight endurance that enables relatively low cost, low altitude, ISR missions. The XFC UAS uses an electrically assisted take off system which lifts the plane vertically out of its container and therefore, enables a very small footprint launch such as from a pickup truck or small surface vessel.

The Naval Research Laboratory is the Navy's full-spectrum corporate laboratory, conducting a broadly based multidisciplinary program of scientific research and advanced technological development. The Laboratory, with a total complement of nearly 2,500 personnel, is located in southwest Washington, D.C., with other major sites at the Stennis Space Center, Miss., and Monterey, Calif. NRL has served the Navy and the nation for 90 years and continues to meet the complex technological challenges of today's world. For more information, visit the NRL homepage or join the conversation on Twitter, Facebook, and YouTube.

Ion Tiger, 2008–2013

Fig. 10.13 Ion Tiger in flight.

After the initial success of the Spider Lion, NRL was ready to support the development of a mission-capable, tactical UAS that used fuel cell propulsion (Fig. 10.13). ONR sponsored NRL to develop a fuel-cell-

Fig. 10.14 Ion Tiger fuel cell system.

powered technology demonstrator aircraft with performance that was competitive with traditional fueled aircraft such as Insitu Inc.'s ScanEagle UAS. ONR set a goal to exceed ScanEagle's range and endurance performance with quiet electric propulsion.

NRL determined that a fuel cell with continuous power of 200 to 250 watts was required. The system had to be fully throttled for thrust control. The fuel cell system efficiency had to be 30% to 40% to achieve the endurance objectives. The design goal was to demonstrate 24 hours' endurance, or eight times that of the Spider Lion, through a significant increase in system efficiency and a reduction in the mass fraction of the balance of plant. This included improving the mass fraction storage of hydrogen with a lighter pressure tank, more efficient cooling systems, and a fully integrated power plant design. The final system developed under the Ion Tiger project was capable of producing 550 watts of peak power at an efficiency of 42%.

NRL started the fuel cell power plant development in the first year of the program (Fig. 10.14). The aircraft design progressed in parallel. The aircraft sizing showed that a fuel tank with a volume of approximately 2 liters was required. The pressure vessel was a short cylinder with hemispherical end caps. The fuselage would need to have a large diameter to house the tank.

NRL simultaneously improved the fuel cell performance while making it more robust. The fuel cell membrane was protected from contamination that would be typical of nonsterile field conditions. Other enhancements

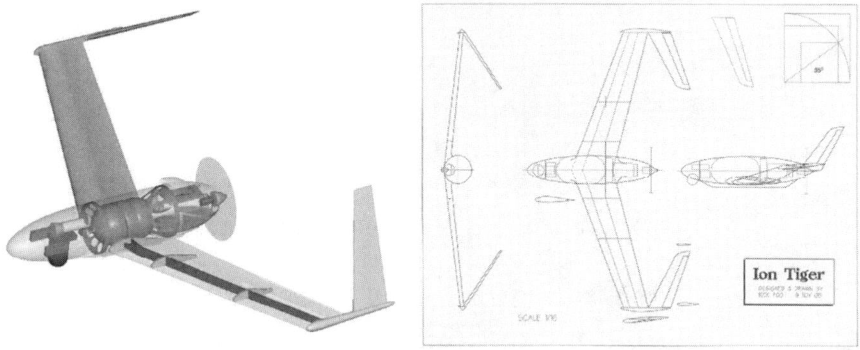

Fig. 10.15 Initial displaced tail Ion Tiger configuration.

included better longevity, storage, and starting capabilities. NRL developed a digital controller to manage the complex fuel cell system. NRL also developed a custom single-stage regulator that could step down from 5000 psi to 7 psi and that was easily controlled. This yielded a major weight savings. NRL worked with Protonex to develop an integrated fuel cell system that included all the balance-of-plant components such as pumps, valves, and tubing. Now, the integrated fuel cell system could simply be hooked up to the gas sources and electrical leads for installation. This user-friendly arrangement was almost as simple as hooking up a battery.

The aircraft had to be lightweight and aerodynamically efficient. The airframe was constructed of hollow molded composites. These molded structures could be produced with high-precision surface tolerances to enable low-drag aerodynamics. The fuselage was a semi-monocoque carbon fiber shell. The wings had carbon fiber spars and fiberglass-foam shell skins. Airframe fabrication was contracted to industry.

The configuration was conventional, with proportions like those of a powered sailplane. The initial designs had a displaced tail configuration (Fig. 10.15), but this was abandoned early on to minimize the perception of development risk for an unconventional configuration. The wing was mounted as a high wing on the fuselage, and it had moderate dihedral. The tail was a cruciform configuration. The tractor propeller was mounted at the nose of the fuselage (Figs. 10.16 and 10.17). The nose was removable, and it contained the fuel cell system, motor, and motor controller. The hydrogen tank slid into the main fuselage. Avionics were housed in the fuselage aft of the wing.

Ion Tiger was winch launched and landed conventionally. The aircraft had a monowheel landing gear and tip skids. The aircraft would be pulled by a model sailplane winch as it rolled down the runway. It would lift off and climb under winch power. After the aircraft reached an altitude of 20 to 30

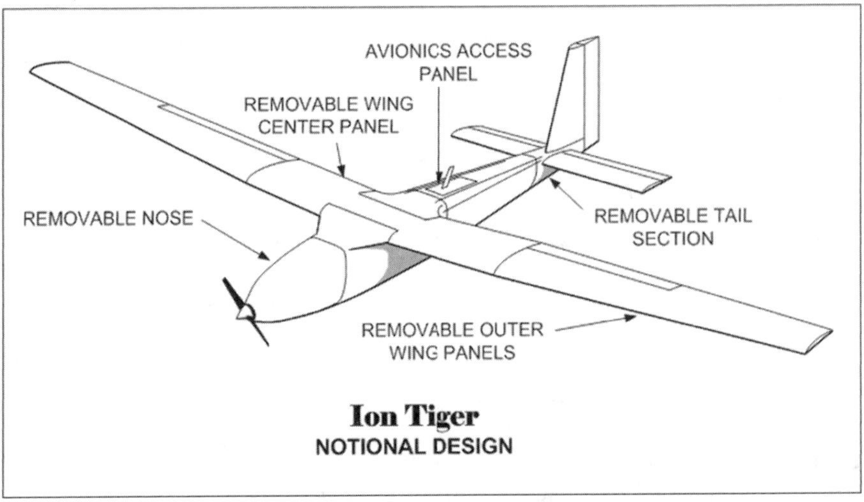

Fig. 10.16 Low-risk Ion Tiger airframe configuration concept.

feet, the motor would start and the aircraft would climb away. At the conclusion of the flight, it would land and roll out on the monowheel. Once it slowed to a near stop, one wing would safely drop to the runway.

Ion Tiger had autonomy and mission systems representative of an operational aircraft. It had a Cloud Cap autopilot to permit GPS waypoint navigation with minimal operator interaction. This autonomy was important for long-duration flights. A Cloud Cap TASE gimbaled camera payload

Fig. 10.17 Final Ion Tiger aircraft.

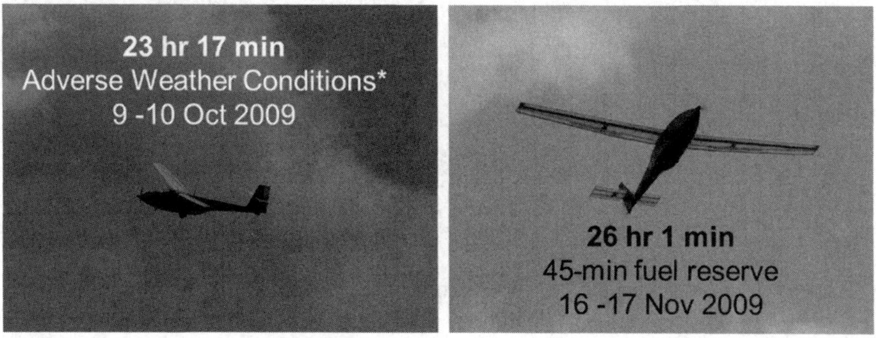

Fig. 10.18 Key Ion Tiger flights.

provided video imagery. The TASE ball was mounted on the lower fuselage aft of the wing where the fuselage cross section transitioned, to minimize drag of the installation. The combination of autonomy and payload capability enabled a realistic mission demonstration.

The Ion Tiger flight test program progressively tested the system capabilities. The first flight took place in 2008. Initial flights were under radio control. Next, the autopilot control laws were tested and refined. The system had good handling qualities and was able to operate in a wide range of weather conditions. The rate of climb was adequate for the mission but not high relative to other NRL unmanned aircraft. The low power relative to the vehicle weight was a design compromise to maximize endurance. The best endurance reached in the initial flight test series was more than 7 hours.

The Ion Tiger program was running out of funding and schedule, but the 24-hour endurance goal had not been demonstrated. To stay on schedule, NRL was forced to fly the full system demonstration under adverse weather conditions. Ion Tiger was launched early in the morning. A weather front passed through during the flight, producing strong winds and heavy rain. The system flew autonomously through the night, and landed after 23.2 hours. The poor weather conditions forced the aircraft to fly for over 3 hours with 100% throttle to maintain position in winds that reached 25 knots (Fig. 10.18). The normal cruise throttle setting was 60%.

The NRL team was disappointed that Ion Tiger fell short of the 24-hour goal. However, the sponsor was ecstatic about the accomplishment. By flying through the bad weather, which was representative of operational conditions, the system achieved Technology Readiness Level (TRL) 7, surpassing the program goal of TRL 6. With this success, ONR provided additional resources to continue the program for another year.

The Ion Tiger design was able to produce all its propulsion power from the fuel cell. Other programs have taken a hybrid propulsion approach, in which the fuel cell provides the sustained power but a battery provides peak power. Such a system would not have worked for this flight in bad weather, during which the 100% throttle was required for over 3 hours. The battery would have run out of charge. NRL has concluded that fuel cell aircraft require a fuel cell system designed to provide the full rated power for the entire flight when necessary to fly the mission.

Ion Tiger was prepared for additional demonstration flights a few months later. On November 16–17, 2009, the aircraft flew continuously for 26 hours and 1 minute under good weather conditions. The problem of fuel cell membrane contamination, which degraded the fuel cell's chemistry and performance, was solved prior to this flight, resulting in a U.S. patent. The temperature ranges of the fuel cell could be adjusted through the control system. This required a few seconds of power down, but then the system could return to the full performance. The endurance of Ion Tiger was more than eight times greater than could be achieved with lithium ion battery technology.

The Ion Tiger was seen as a valuable research aircraft for fuel cell research. It had significant residual research value after the initial objectives were met. The next big leap in fuel cell performance was to use liquid hydrogen (LH_2) rather than compressed hydrogen gas. With an insulated Dewar tank, it would be possible to achieve 72 hours of endurance for the equivalent volume and mass as with the compressed hydrogen gas tank system.

NRL embarked on a major project for LH_2 fuel systems (Fig. 10.19). Working with industry, the Lab was able to develop greatly improved hydrogen storage systems. Pressurized-gas hydrogen tanks could achieve a hydrogen mass fraction of 13%, while the new Dewar system could achieve a hydrogen mass fraction of nearly 40%. The Dewar tank was constructed of welded spun aluminum. The two hemispheres were made by spinning sheet aluminum on a mandrel to achieve uniform thickness to uniformly distribute the stresses. The two hemispheres were welded together. There were two nested tank spheres separated by Kevlar thread catenaries. The Dewar tank was insulated by creating a vacuum between the inner and outer spheres. The fuel line from the inner tank to the outside of the tank and its fittings were heavily insulated. The inner sphere contained the liquid hydrogen. The system operated effectively on the bench under static conditions. However, the fuel sloshing under dynamic conditions generated mechanical energy in the tank, and this led to a higher boil-off rate. This problem was eventually resolved through bench tests and flight experiments. On April 16, 2013, the Ion Tiger took off on its initial long-endurance LH_2 flight test and was landed on April 18, after

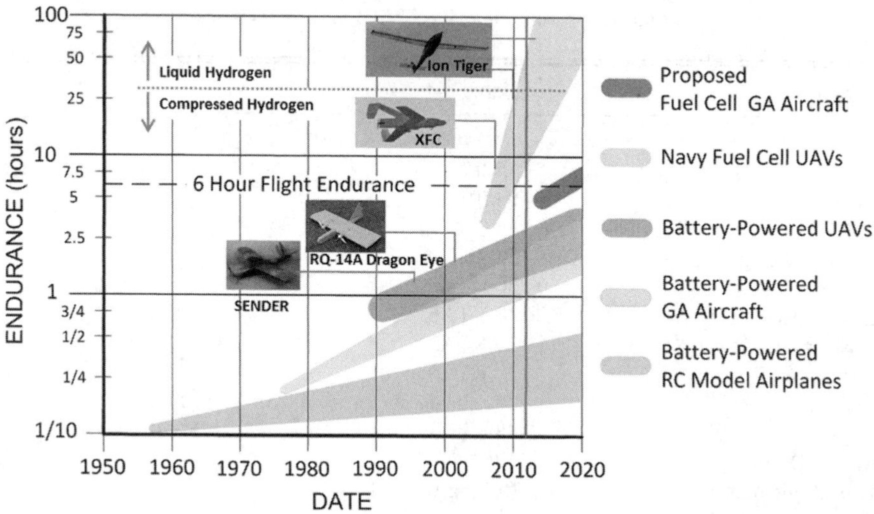

Fig. 10.19 Fuel cell technology improvement.

flying for 48 hours and 1 minute. This flight is possibly the world's record for the longest-duration fuel cell flight. Ion Tiger is being prepared at the time of this writing to demonstrate a flight endurance of 72 hours. A three-view drawing of Ion Tiger is shown in Fig. 10.20 and its technical characteristics are listed in Table 10.3.

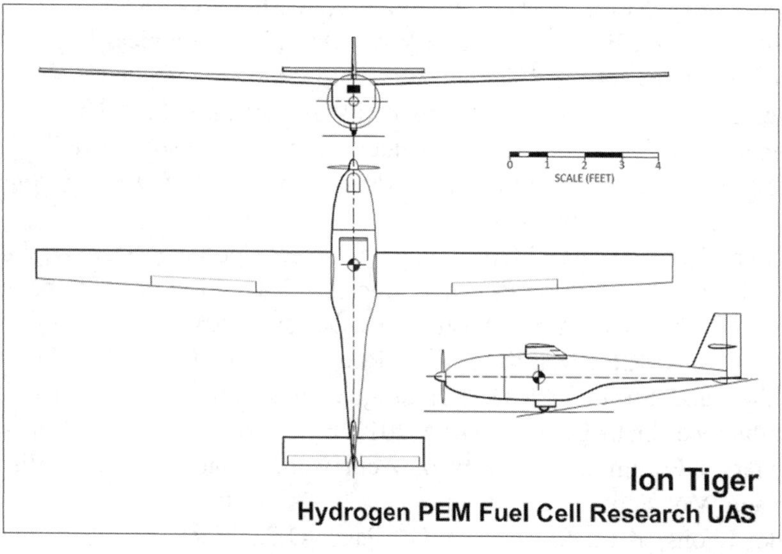

Fig. 10.20 Ion Tiger three-view drawing.

Table 10.3 Ion Tiger Technical Characteristics

Parameter	Value
Wing span	17.0 ft
Wing area	15 ft^2
Length	8.0 ft
Takeoff weight	35.0 lbs
Payload weight	5.0 lbs
Fuel weight	500 grams
Loiter speed	30 KTAS
Dash speed	35 KTAS
Stall speed	20 KTAS
Endurance	26+ hr on compressed hydrogen (2009) 48+ hr on liquid hydrogen (2013)
Max L/D	17
Launch method	Rolling winch launch
Recovery method	Wheeled landing
Propulsion type	PEM fuel cell
Power	550 W

ION TIGER: FURTHER READING

Anon., "The Ion Tiger Fuel Cell Unmanned Air Vehicle," *2010 NRL Review*, Naval Research Laboratory, Washington, DC, 2010, pp. 16–18.

McKinney, D., Thompson, D., and Parry, D., "Surveillance Vehicles Soar on Fuel Cell Power," *SPECTRA, The Magazine of the Navy's Corporate Laboratory*, Naval Research Laboratory, Washington, DC, Fall 2010.

Rocheleau, R.E., Bethune, K., and Virji, M., "Final Technical Report, Fuel Cell Stack Testing and Durability in Support of Ion Tiger UAV," Grant Number N00173-08-2-C008, University of Hawaii, Feb. 2010.

Stroman, R.O., Kellogg, J.C., and Swider-Lyons, K., "Testing of a PEM Fuel Cell System for Small UAV Propulsion," *42nd Power Sources Conference,* 2006, pp. 487–490.

Stroman, R.O., Schuette, M.W., and Page, G.S., "Cooling System Design for PEM Fuel Cell Powered Air Vehicles," Naval Research Laboratory, NRL/MR/6110-10-9253, Washington, DC, June 2010.

Swider-Lyons, K.E., MacKrell, J.A., Rodgers, J. A., Page, G.S., Schuette, M.W., and Stroman, R.O., "Hydrogen Fuel Cell Propulsion for Long Endurance Small UAVs," AIAA 2011–6975, AIAA Centennial of Naval Aviation Forum, "100 Years of Achievement and Progress," Virginia Beach, VA, Sept. 21–22, 2011.

Swider-Lyons, K.E., Stroman, R.O., Page, G.S., Mackrell, J.S., Rodgers, J.A., and Schuette, M.W., "The Ion Tiger Fuel Cell Unmanned Air

Vehicle," *Proceedings of the 44th Power Sources Conference*, 2010, pp. 561–563.

JAKE SPIRIT, 2011

Fig. 10.21 Jake Spirit.

The late James Kellogg was one of NRL's pioneers of fuel cell electric flight. The Jake Spirit was named in his honor, combining the first two letters of his first and last names.

The purpose of the Jake Spirit design concept was to fly a very long-range mission using fuel cell propulsion (Fig. 10.21). The Director of Innovation at ONR challenged NRL with the mission of flying from San Diego to Honolulu (2226 nautical miles) with an unfolding mission aircraft. The potential of fuel

Fig. 10.22 Jake Spirit, Ion Tiger, and XFC shown for size comparison.

cell propulsion was becoming apparent from the successes with XFC and Ion Tiger. A bolder mission seemed appropriate as a next step. The Jake Spirit could accomplish this flight in 64 hours in still-air conditions while flying at 35 knots. The aircraft was capable of 72 hours' endurance, so there was some margin for winds. The system required liquid hydrogen fuel technology. By comparison, the Ion Tiger could fly for 24 hours and cover 1200 nautical miles using compressed hydrogen gas fuel.

The real-world operational constraints were challenging compared to the idealized mission. The prevailing winds are from west to east for most of the year, so the Jake Spirit's performance would be inadequate for most of the year. However, with a tailwind, an aircraft could travel from Hawaii all the way to Kansas. The geographic mission goal was not a fully defined requirement, just an idea for a demonstration.

NRL developed the design and proposed it to ONR. Jake Spirit was not funded in the proposed form. Rather, the Ion Tiger program received additional funding to develop a liquid hydrogen fuel system. For the same tank weight and volume as the current compressed gas hydrogen tank system, the Ion Tiger can reach 72 hours' endurance. Figure 10.22 shows a comparison of the Jake Spirit, Ion Tiger, and XFC. Figure 10.23 shows a three-view drawing of the Jake Spirit, and its technical characteristics are listed in Table 10.4.

In 2012, the Jake Spirit proposal was supplanted by an operationally improved version named Ion Eagle, which should be capable of reaching all

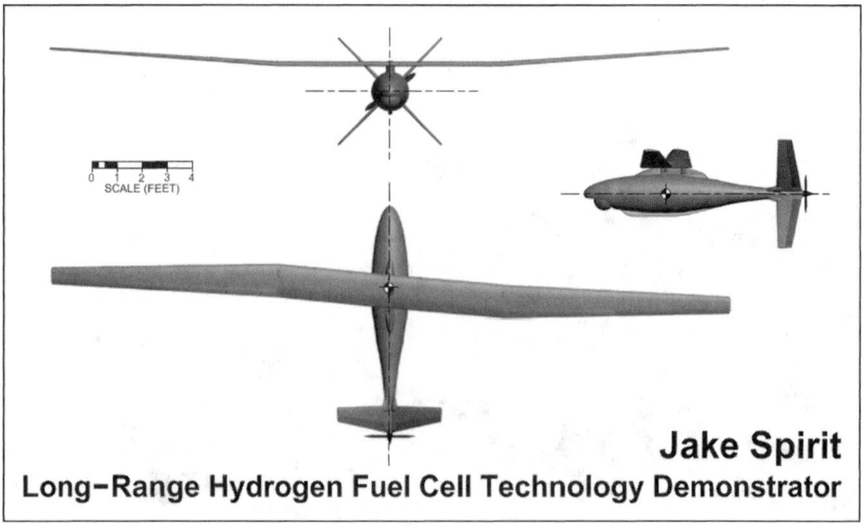

Fig. 10.23 Jake Spirit three-view drawing.

Table 10.4 Jake Spirit Technical Characteristics

Parameter	Value
Wing span	28.3 ft
Wing area	36.5 ft^2
Length	10.0 ft
Takeoff weight	60.0 lb
Empty weight	45.0 lb
Payload weight	10.0 lb
Hydrogen weight	5 lb
Cruise speed	35 KTAS
Dash speed	45 KTAS
Stall speed	22 KTAS
Endurance	72 hr
Range	3500 nmi
Max L/D	25
Launch method	Rolling winch launch
Recovery method	Belly landing
Propulsion type	PEM fuel cell
Power	1200 W

Jake Spirit performance goals and be capable of autonomous launch from the ship-based Vertical Launch System (VLS).

Chapter 11

Manned Aircraft

While NRL is best known for its unmanned aircraft programs, there have also been manned aircraft efforts. The vehicle configuration, autonomy, and propulsion technologies originally developed for unmanned systems are transferred to the manned domain. These manned aircraft are mostly autonomous or highly automated so that the operators require minimal training.

Fly Back Ejection Seat, 1991

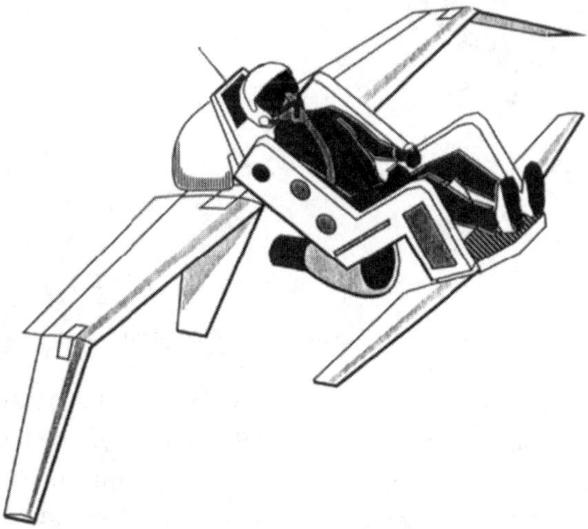

Fig. 11.1 Fly Back Ejection Seat concept.

The Fly Back Ejection Seat came about when John Montgomery asked Rick Foch and Frank Klemm how small turbine engine technology could be applied outside of the unmanned aircraft domain. The resulting ejection seat concept (Fig. 11.1) would deploy wings and start a small turbojet engine after ejection to allow a pilot to get back to friendly territory. The idea was first explored by Fairchild using sailwing technology at the end of the Vietnam

War. The NRL concept would use rigid folding wings and newly developed small turbojets. The rigid wing would provide more efficiency. However, the low achievable life-to-drag ratio (L/D) and high fuel consumption would yield short range. The ejection seat design was not constrained to fit within existing aircraft; rather, it was assumed that next-generation tactical aircraft could be co-developed with this device. The design was not taken further than an early conceptual design.

VEENESS, 1992

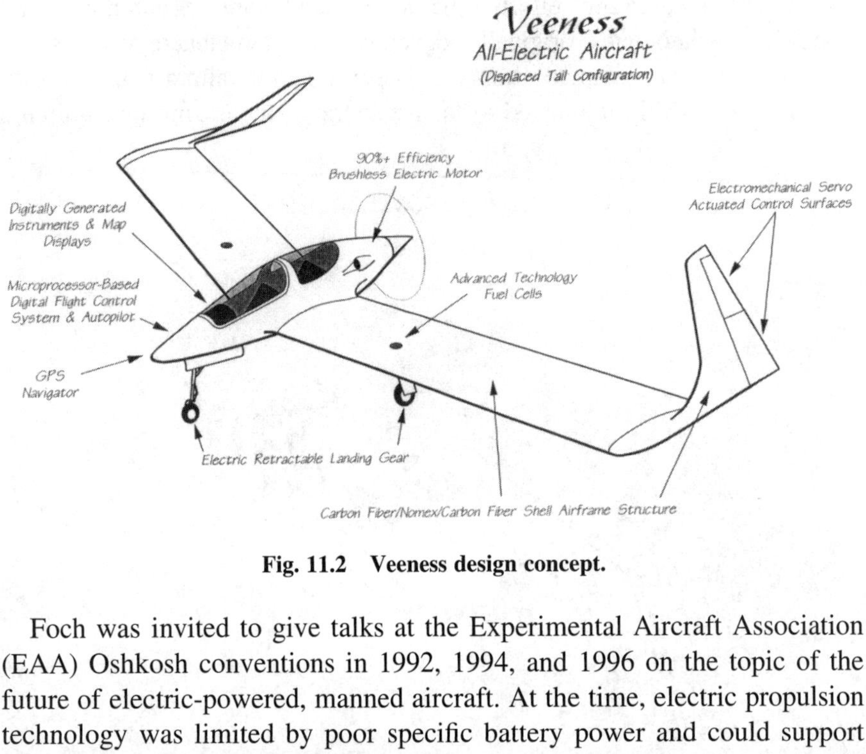

Fig. 11.2 Veeness design concept.

Foch was invited to give talks at the Experimental Aircraft Association (EAA) Oshkosh conventions in 1992, 1994, and 1996 on the topic of the future of electric-powered, manned aircraft. At the time, electric propulsion technology was limited by poor specific battery power and could support only electric-assisted sailplanes. Foch presented the Veeness all-electric aircraft concept in 1992 (Fig. 11.2).

The Veeness concept looked further into the future. It used a lithium hydroxide fuel cell that was then under development by Lockheed for military systems and could provide a range of a few thousand miles. Veeness would use 90%-efficient brushless motor technology. Since it had an electric power source, the design used advanced fly-by-wire avionics, electromechanical actuators, and electric subsystems. The structure was made of hollow molded carbon fiber–Nomex sandwich skins. It turned out that the lithium hydroxide fuel cell was technically but not commercially feasible, due to the regulatory impact of handling the highly alkaline byproducts.

Many of the features of Veeness were incorporated into the Ion Dasch design two decades later (discussed later in this chapter). Ion Dasch used a hydrogen proton exchange membrane (PEM) fuel cell. The pressurized hydrogen gas storage required a high-volume tank, which required a configuration change. The all-electric system concept remained the same.

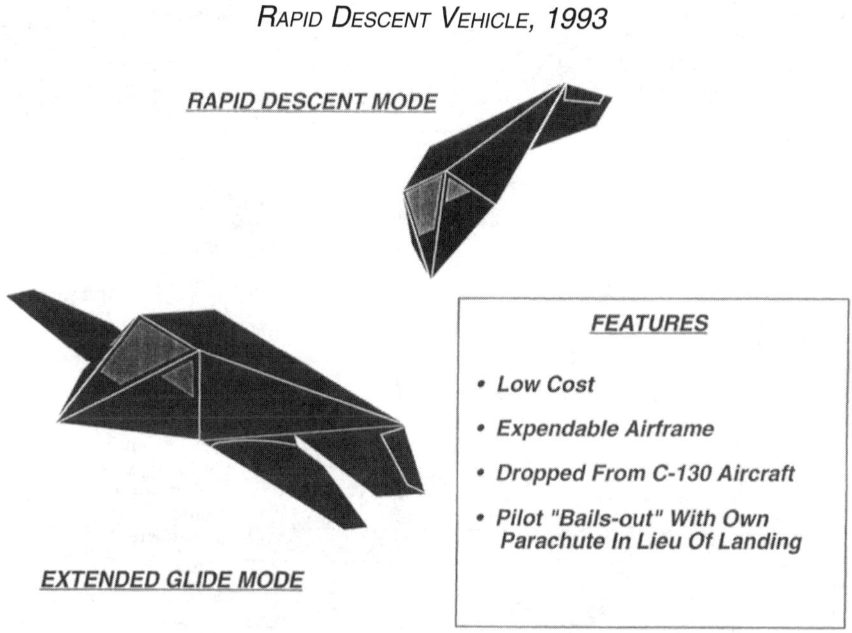

Fig. 11.3 Rapid Descent Vehicle concept.

The Vehicle Research Section (VRS) evaluated a homebuilt aircraft called the Facetmobile, designed by aerospace engineer Barnaby Wainfan, and observed wind tunnel testing. This unique aircraft had a low aspect ratio, tailless configuration and was formed of a series of flat surfaces. The Facetmobile offered potential for very affordable airframes constructed of premolded flat panels.

Foch considered various military missions for this technology. The mission with the greatest potential was deep penetration of special forces agents in high-low insertion. High-low means that the personnel jump out of the host aircraft at high altitude and open their parachutes at low altitude to minimize the time under parachute. The Rapid Descent Vehicle would enable the agents to launch from a C-130 or C-17 at high altitude, fly at high speed initially,

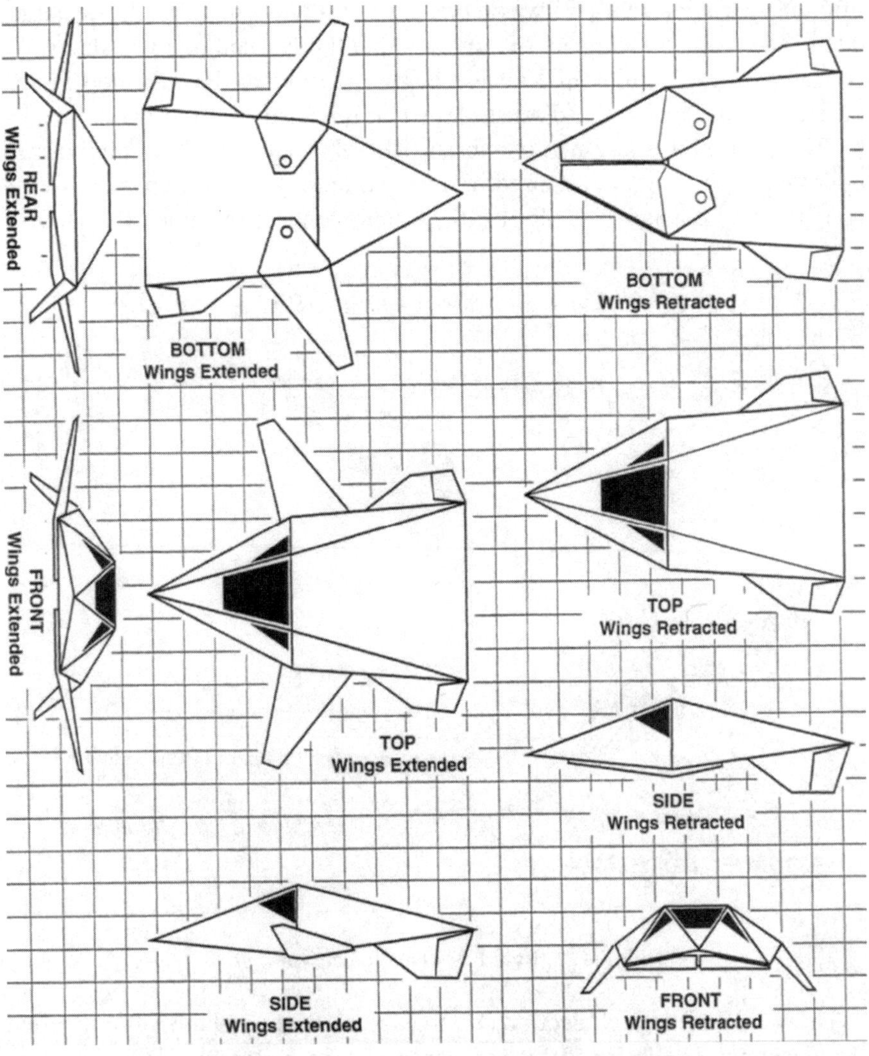

Fig. 11.4 Rapid Descent Vehicle configurations.

deploy wings for better range, and then bail out of the vehicle. The faceted shape could employ radar-absorbing materials or other low-Radar Cross Section (RCS) treatments to minimize detection (Figs. 11.3 and 11.4).

The basic Facetmobile design had a poor L/D that was not much of an improvement over that for parafoils. To help enhance range, the Rapid Descent Vehicle design had wings that would swing out from the lower fuselage. No wind tunnel tests were conducted on this configuration, but the intersection of the wing and fuselage chine were expected to generate large drag due to separation. The fuselage had ample volume to stow the wings.

The design was briefed to potential customers, but the MOLE (discussed next) was considered a better solution.

MAN-OPERATED LIGHT EXPENDABLE (MOLE), 1993

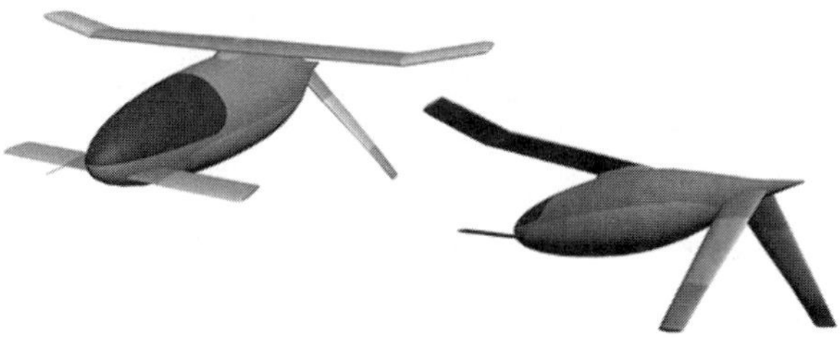

Fig. 11.5 MOLE conceptual design.

The Man-Operated Light Expendable (MOLE) aircraft was intended to insert a special operations person into enemy territory (Fig. 11.5). A single individual fit within each aircraft. The MOLE would roll out the back of a C-130 or C-141 at high altitude. After the MOLE cleared the host aircraft, the wings deployed, and gliding flight would begin. The person would bail out of the craft with his/her equipment at an altitude of a few thousand feet, and would parachute to the ground. The aircraft had an L/D of 15, which is considerably higher than that of a parafoil. Also, the high wing loading enabled high penetrating airspeed that reduced susceptibility to winds. Both contribute to provide a longer range than traditional parachute techniques.

The fuselage was a streamlined body of revolution that used a NASA six-series airfoil profile. The operator would lie prone in the aircraft to minimize frontal area and drag.

The pilot workload had to be minimal with very low training requirements. The operator would line up cross hairs generated by the flight computer, which indicated the appropriate heading and pitch for best glide ratio. The control surfaces were to be actuated by rudder pedals and a hand pitch controller.

The aircraft had a three-surface configuration. The high wing was located midway along the fuselage length. An inverted V-tail had ruddervon control effectors that were controlled by the foot pedals. Ruddervons act as both ailerons and rudders. Because the inverted V-tails acted below the center of gravity (CG), a deflection caused both roll and yaw in the same turn direction. The forward canard surface had variable incidence for pitch control, which was controlled by a hand lever.

The aircraft was folded while aboard the host aircraft to allow clearance when dropped (Figs. 11.6 and 11.7). A one-piece wing pivoted along the top of the fuselage. The inverted V-tails and canards folded out from the fuselage

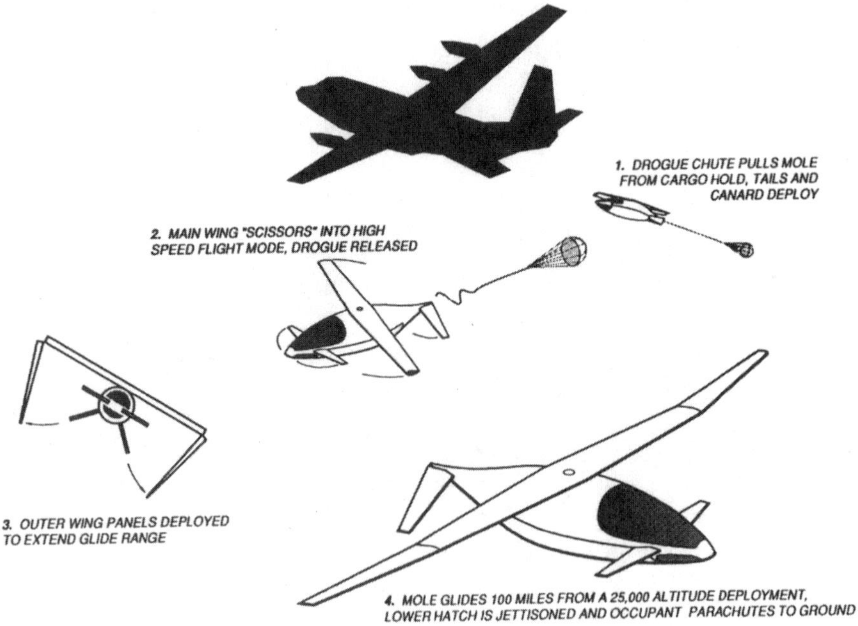

Fig. 11.6 MOLE deployment sequence.

sides. The folded geometry was designed to be stable while the aircraft exited the host aircraft. Once the MOLE was clear of the cargo doors, the tails flipped to the operational positions, and the wing rotated.

NRL built a 1/5-scale radio-controlled (RC) model that was successfully flown. The model was designed to be dropped from an NDM test aircraft. The model was 10 feet long and had a fuselage diameter of 2.5 feet. It had a hollow composite fuselage and composite-covered foam core wings. Springs opened the tails and canard, and these opened automatically. A torsional spring swung the wing, which was initiated by a servo release.

The purposes of the model were to prove that the aircraft could unfold in flight and to test the flight controls. The flight tests were successful. Early flights did not deploy the wing immediately after drop, and the aircraft rolled

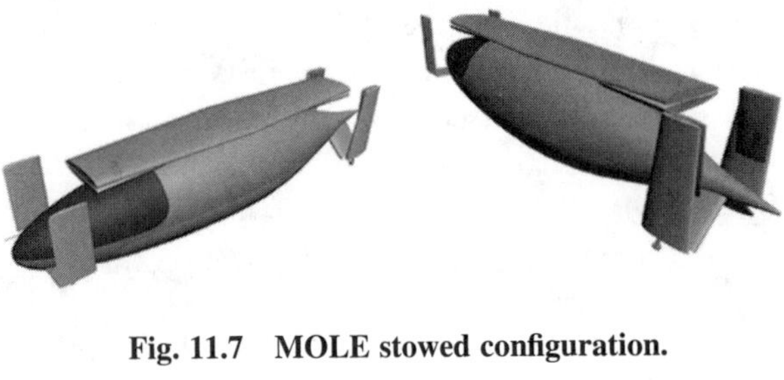

Fig. 11.7 MOLE stowed configuration.

Fig. 11.8 MOLE drop model (top) and MOLE drop model loaded beneath the NDM I (bottom).

at a high rate. Later it was found that rapidly deploying the wing arrested the rolling. A key finding was that the model was able to deploy without the need for a stabilizing drogue parachute.

NRL was concerned that it would be difficult to find people to test a MOLE aircraft, but there was no shortage of volunteers ready to try a full-size version. Some users believed that this design would be safer than the High Altitude Low Opening (HALO) jumps. Sponsorship for the full-size, man-carrying aircraft was not forthcoming. The drop model (Fig. 11.8) was the only flight hardware built. NRL also considered variants with turbine sustainer engines for range extension. A three-view drawing of MOLE can be seen in Fig. 11.9.

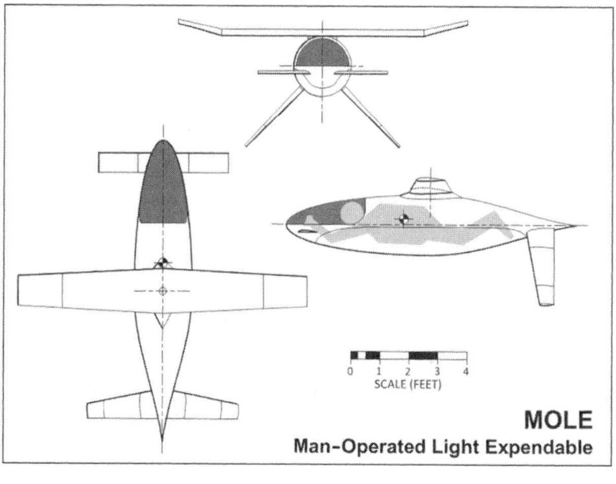

Fig. 11.9 MOLE three-view drawing.

SUBMERSIBLE MANNED AIR-LAUNCHED EXPENDABLE (SMAX), 1993

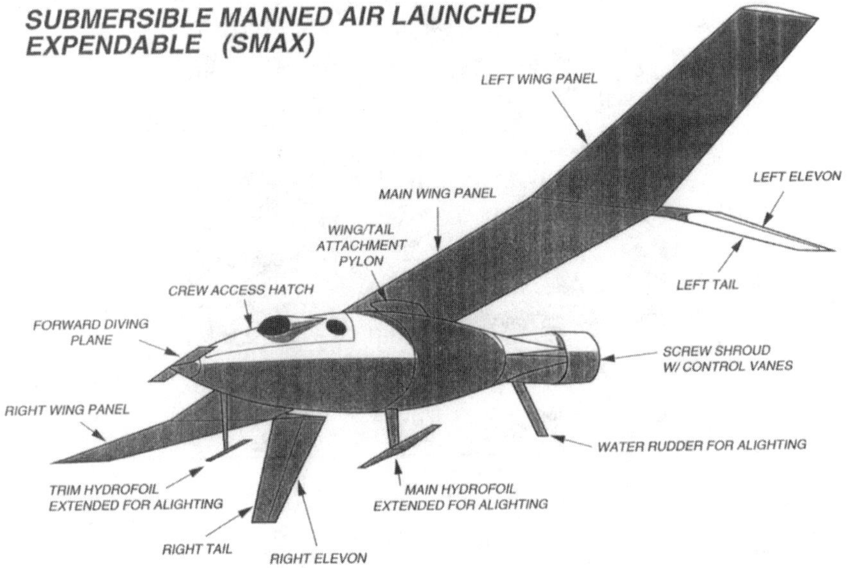

Fig. 11.10 SMAX features.

The Defense Advanced Research Projects Agency (DARPA) asked NRL to explore concepts for an aircraft that could deliver a small submarine for special agents. The vehicle would be deployed out the back of a C-130 or C-17 and carry four people. Through the Submersible Manned Air-Launched Expendable (SMAX) conceptual design study, NRL found that the concept was indeed feasible.

The design was a two-part vehicle consisting of a wing-tail-propulsion module and a submersible fuselage (Fig. 11.10). The wing pivoted along the fuselage and had deployable outer panels. The displaced tail had skewed hinges to allow the tails to rest along the wing prior to launch. A small turbojet engine was mounted above the wings.

The fuselage module could operate as an aircraft fuselage or as a submersible vehicle. The fuselage had hydrofoils to aid water landing without the need for a high-drag step or V-hull. The fuselage was also equipped with a shrouded electric thruster for underwater propulsion and a forward diving plane. After the aircraft landed, the fuselage would flood and separate from the aerodynamic module. The agents would be equipped with breathing apparatus and wetsuits for the submarine operations. It was

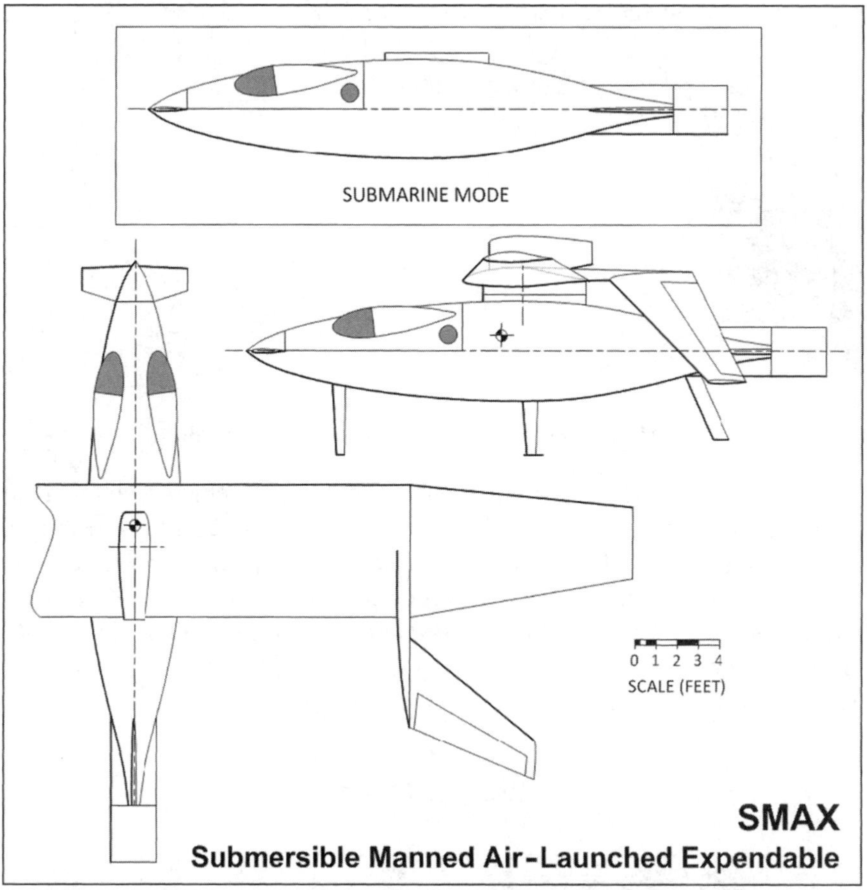

Fig. 11.11 Multiple SMAX views.

necessary to flood the fuselage to reduce the buoyancy for submarine operations without being too heavy for flight. After the operation, it would be possible for the submarine to reattach to the aerodynamic module, purge the water, and take off. Figure 11.11 shows multiple views of SMAX.

NRL built a display model to demonstrate how it could fold up, but NRL did not take the effort further. The program was not funded. Around 2008, DARPA initiated a flying submarine program with nearly identical requirements. NRL is not sure if the SMAX study inspired the program.

PERSONAL EXTRACTION AIRCRAFT (PEA), MID-1990S

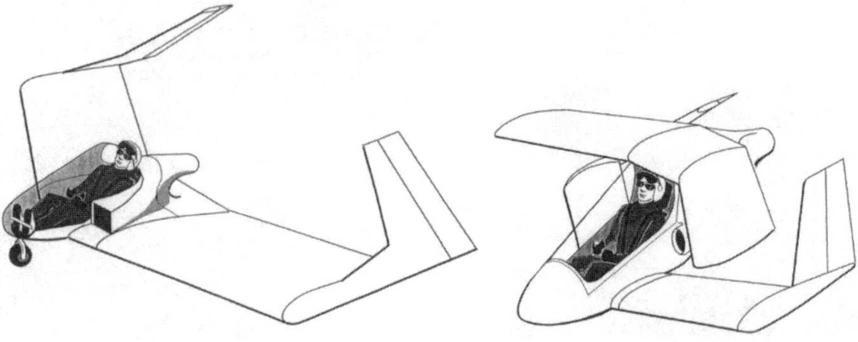

Fig. 11.12 PEA version B (left) and version C (right).

NRL considered concepts for leveraging expendable vehicle technology to extract personnel trapped behind enemy lines. The concept was to drop a crate containing a folded aircraft that could be rapidly assembled. The vehicle could be launched via a rocket or catapult and then fly with a small turbojet engine. Like other NRL manned aircraft concepts, this vehicle would be highly autonomous. The design concepts did not progress beyond the early conceptual design phase (Fig. 11.12).

ION DASCH, 2011–PRESENT

Fig. 11.13 Ion Dasch concept.

Ion Dasch, a fuel-cell-powered general aviation aircraft, exemplifies the significant technology and safety advancements afforded to general aviation by incorporating the latest military unmanned aircraft system (UAS) technologies (Fig. 11.13). Dasch, of Ion Dasch, is a contraction in honor of the late Daniel Schultz, who led the American Society for Testing and Materials (ASTM) electric aircraft rulemaking committee. The program began when Schultz, who was familiar with NRL's expertise in electric flight, asked Foch to advise ASTM on the design implications of the emerging technologies of electric propulsion applied to general aviation aircraft. Unknown to Schultz, Foch was also a private pilot and homebuilt general aviation enthusiast, so Foch was even more qualified for this assignment than expected. As a direct example of how the latest UAS electric propulsion technology could be applied, Foch decided to design an advanced technology manned aircraft. Ion Dasch was designed with support from Jay Gundlach and Francesco Giannini of Aurora Flight Sciences.

NRL was gaining valuable experience with fuel-cell-powered aircraft such as the XFC and Ion Tiger. Studies for larger tactical unmanned aircraft revealed that the propulsion system characteristics were similar to what would be required for small manned aircraft. Foch realized that compressed gaseous hydrogen storage, proton exchange membrane fuel cells, and electric motor technologies were reaching a point at which a practical, hydrogen-powered general aviation aircraft would be feasible.

Beyond civil aviation applications, such an aircraft would have relevant military uses. The Ion Dasch could be used as a military flag officer transport with no dependency on fossil fuels. The Navy is increasingly concerned about its reliance on foreign oil and its impact on the environment. The advanced electric flight control system could also make this aircraft an optionally piloted aircraft or a completely unmanned aircraft. It could carry 400 pounds of payload as an unmanned aircraft, offering capabilities similar to those of a Predator (MQ-1). Ion Dasch could also serve as an advanced fly-by-wire research aircraft.

The Ion Dasch would be a completely "green" aircraft, with no direct use of fossil fuels. Today, field hydrogen sources are often reformed from hydrocarbon fuels. If we look further ahead, it may be possible to use solar energy or wind power to electrolyze water to create the hydrogen. The same electrical power source could also compress the hydrogen for the flight storage tanks. The entire fuel production plant could be part of the hangar facility.

The initial sizing of the aircraft showed some interesting results. The tactical unmanned aircraft fuel cell and hydrogen storage parameters were used directly without the advantages of the square-cubed law where larger systems become more efficient. Even with these conservative numbers, it was

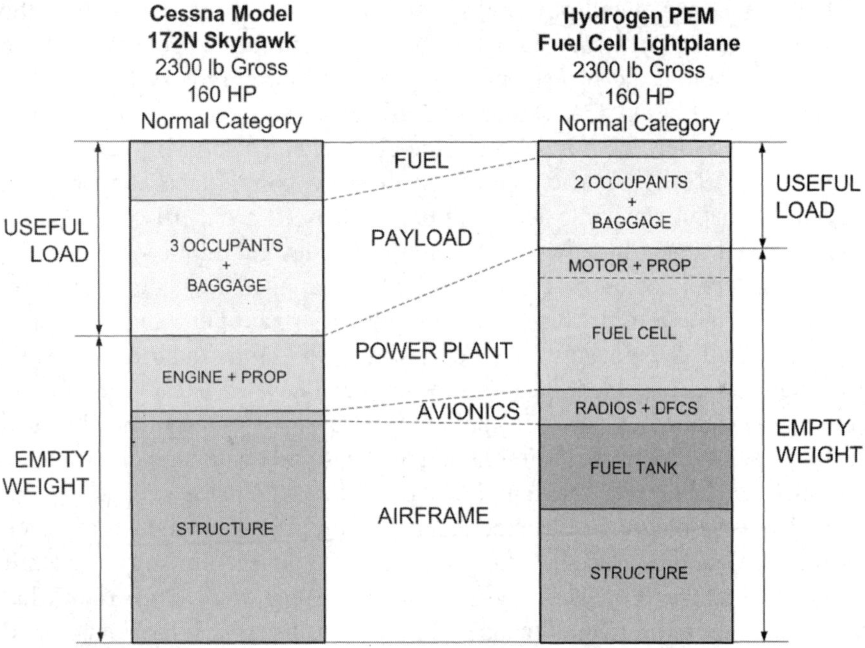

Fig. 11.14 Weight comparison of a Cessna 172 to a two-place, hydrogen-powered general aviation aircraft.

possible to develop a general aviation aircraft that was the size of a Cessna 182 and that weighed less than a Cessna 172. There would be sufficient useful load for two people or a payload. The hydrogen fuel cell and electric propulsors weighed more than a reciprocating engine and propeller. The hydrogen fuel tank consumed a large portion of the empty weight, but the hydrogen itself weighed significantly less than the fuel on a Cessna 172. The new aircraft's composite structure would weigh substantially less than the Cessna 172's metal structure (Fig. 11.14). The digital flight control system, electric actuators, and other avionics on the fuel cell aircraft would weigh more than the Cessna's basic avionics.

The primary design challenge was volumetric. Traditional unpressurized liquid fuels can fit within wing tanks that conform to the airframe shape. But compressed hydrogen is stored in tanks that are large and cumbersome by comparison. It was difficult to find a location in the aircraft to house the large hydrogen tank, which had the shape of a cylinder with hemispherical end caps. The fuel cell stacks and balance of plant also consumed considerable volume.

The primary configuration selection was driven by finding an aerodynamically efficient design that could house the necessary components. It was readily apparent that the only feasible tank location was in the center of the fuselage. The tank was sized for 4 hours of flight with 45 minutes of reserve. To reduce the aircraft's frontal area, the wing spar carry-through structure should not pass above or below the tank. Another consideration was that the cylindrical portion of the tank would be designed to fail before the end caps, so that in the event of a rupture, the tank would burst through the fuselage away from the flight crew and tails. For these reasons, the main spar had to pass ahead of the tank or behind it.

An alternative approach was to use the fuel tank structure as the wing spar carry-through. The hydrogen tank could be built with significant load-carrying capability. The wing bending loads and other loads could be carried around the tank perimeter for a lightweight, multiuse structure. However, the prototype design would have lower development risk if the hydrogen tank was a separate, drop-in unit. The tank could expand and contract with temperature without concern for its flight-load-carrying capability. There would also be fewer concerns over safety and reduced developmental testing requirements.

The original design had a swept-back wing with the carry-through spars located in front of and behind the tank. The area between the two spars housed the fuel cell stacks. The wing was mounted to the lower part of the fuselage for a low wing configuration. The two-person cockpit was located ahead of the tank.

The electric propulsors consisted of the motors and propellers. The propulsors were lightweight relative to a reciprocating gas engine. Traditional reciprocating gas engine propulsion systems serve as both power generator and propulsor. A fuel cell propulsion system segregates the power generation from the propulsor.

NRL developed digital autopilots and electric actuation for most of its modern UASs. While many general aviation aircraft were equipped with autopilots with electric actuators, these were in line with manually driven push rods and cable-pulley systems. The mature general aviation autopilots helped with managing the flight path but were not well suited to high-demand tasks such as takeoff and landing. The Ion Dasch would eliminate the manual control system entirely and commit to all-electric actuation. Such an approach was routine for unmanned aircraft but revolutionary for general aviation. Each control surface would be individually driven by its own actuator, eliminating play in the control system. The control surfaces could be split into multiple smaller control surfaces to provide redundancy. Also, the autopilot could reconfigure the control system for better handling qualities during different tasks or through the phases of flight. The control system could also contend with failures. For example, the flight control system could be automatically reconfigured and retrimmed to deal with a motor failure. This would eliminate the need for pilot twin-engine certification, since the aircraft would manage the event. Such redundancy and fault tolerance would enable improved flight safety.

The flight control system could also be used for multiple levels of autonomy. A pilot-in-the-loop mode would be used for normal piloted flight or for training. The autopilot could also manage all phases of flight from takeoff to landing without pilot interaction. This would theoretically enable the two-place aircraft to carry two passengers who didn't possess pilot licenses! However, Foch would prefer not to request permission from the FAA for nonpilots to fill our skies with Ion Dasches. Since Ion Dasch could also operate without any people aboard, it could serve as an unmanned aircraft for intelligence, surveillance, and reconnaissance and other missions. Ion Dasch could also self-deploy or fly back after delivering passengers. The integrated autopilot and flight control system would not need to undergo a mechanical configuration change to go from piloted to unpiloted modes. The adaptable flight control system would enable the Navy to conduct fly-by-wire flight research without using an expensive tactical aircraft such as an F-18.

Ion Dasch would use an active gust load alleviation system to improve ride quality, extend the structural life, and reduce structural weight. The aircraft would sense gusts using lidar sensors or a multihole air data probe at the nose of the aircraft. The autopilot would command the electrically driven flight control surfaces to respond ahead of the gusts. The core technology for such a system is being developed by ViGYAN Inc.

The Ion Dasch took advantage of the electric propulsion and fly-by-wire design freedom by placing the propulsors with tractor propellers on V-tails (Fig. 11.15). The lightweight propulsors did not shift the CG too far aft, and the loads on the tail structure were not excessive. The propellers were elevated above the ground, which enabled the landing gear length to be shortened. The blown tail surfaces would have increased control authority, especially during low-speed flight. The elevated propellers would cause pitch-down with increased power, but the digital flight control system would automatically retrim. The tail-mounted propellers would reduce cabin noise. The aft-swept tails decreased noise from the propeller blade wake impingement, based on earlier NRL experience. The aft-swept tails also increased the effective tail moment arm for increased stability.

The landing gear had a low stance thanks to the elevated propellers. The main gear protruded from the wings near the wing break (location of the taper change). The nose gear was located below the cabin. The short landing gear was lightweight. There was no need to retract the landing gear, because it was easily faired to reduce drag, thanks to the short length.

The initial Ion Dasch configuration had to be modified. The aft-swept wing had low tip clearance when the aircraft rotated nose up for takeoff or landing. The tips could strike with a slight roll angle. To fix the problem, the landing

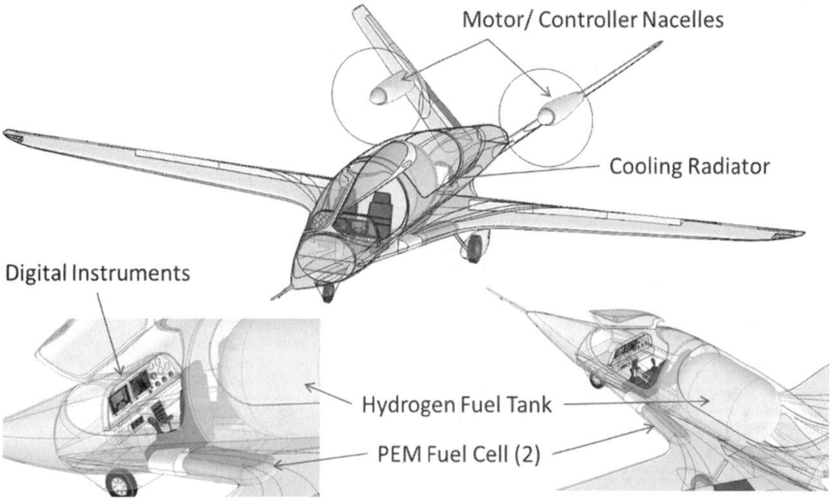

Fig. 11.15 Internal drawing of final Ion Dasch configuration.

gear had to be extended or the wing had to be swept forward. The two authors worked together on a forward-swept configuration. The wing carry-through spar passed behind the hydrogen tank. There was concern that addressing structural divergence would yield a heavier wing. However, a weight and balance analysis revealed that only modest forward sweep was required. The fuel cells had to be positioned just ahead of the hydrogen tank in order to maintain the appropriate CG, but there was no suitable volume available. The solution was to add blunt wing strakes at the wing root. These fairings did not increase the frontal area. The rounded leading edge would prevent formation of destabilizing vortices at high angles of attack. The fuel cell stacks would be readily accessible within the strake. The forward part of the strake included a small fold-down step to facilitate pilot access to the cabin.

The Ion Dasch design solution was possible with existing fuel cell and hydrogen storage technologies, but NRL wanted to push the state of the art. Therefore, the fuel cell stack power density was to be increased by a factor of 2 in the program. Normally a 100% technology improvement would be optimistic for a demonstrator program. However, NRL had already advanced fuel cell power density by a factor of 8 over the previous five years. NRL engineers consulted United Technologies Corp. about this challenge, as that company was conducting research and development on fuel cells for military manned aircraft. The conclusion was that the technology goals were challenging but feasible. Even if the goals were not fully met, existing technologies would enable an Ion Dasch to fly with a single pilot while the

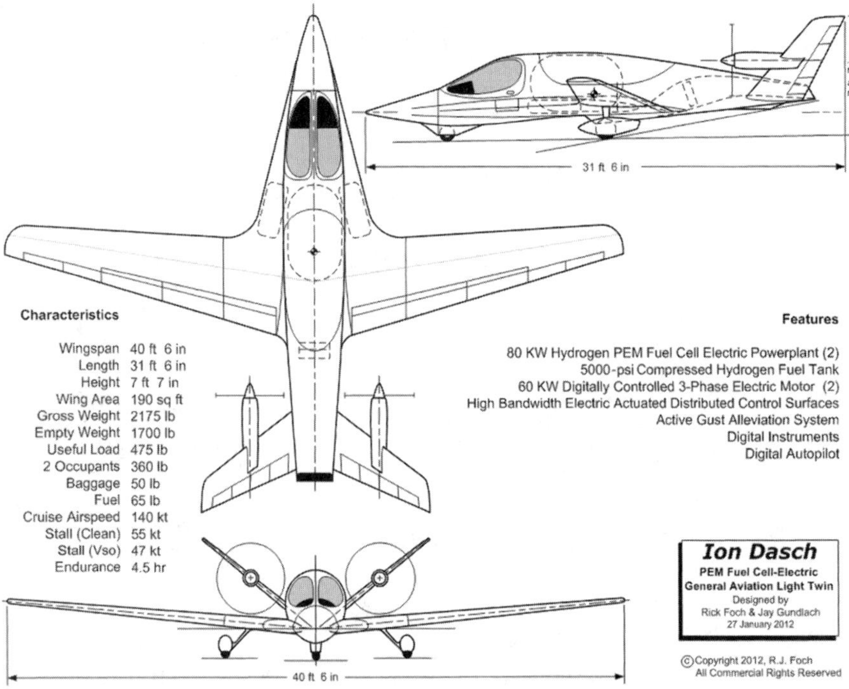

Fig. 11.16 Ion Dasch three-view drawing.

Table 11.1 Ion Dasch Technical Characteristics

Parameter	Value
Wing span	40.5 ft
Wing area	190 ft^2
Length	31.5 ft
Takeoff weight	2175 lb
Empty weight	1700 lb
Payload weight	410 lb
Hydrogen weight	65 lb
Cruise speed	140 KTAS
Stall speed	55 KTAS (clean)
	47 KTAS (VSO)
Endurance	4.5 hr (with two passengers)
Max L/D	12
Launch method	Conventional
Recovery method	Conventional
Propulsion type	PEM fuel cell with electric motors
Power	120 kW

technology advanced. The fuel cell stack and balance of plant would be the primary research focuses.

At the time of this writing, the Ion Dasch program is seeking funding. The program may be funded as a series of technology demonstrations or as a full vehicle-development program. Figure 11.16 shows a three-view drawing of this fuel-cell-powered general aviation aircraft, and Table 11.1 lists its technical characteristics.

Chapter 12

Subsystem Technology

Many significant technologies have been developed at NRL. Those key to high-performance unmanned aircraft systems (UASs), but not necessarily UAS-specific, include neodymium-iron-boron magnets that enable today's powerful and inexpensive electric motors; high-power, fast-switching transistors that were originally developed for miniature microwave amplifiers for electronic warfare (EW) payloads and are used worldwide for today's brushless electric motor controllers; radar; and GPS.

Because of the Navy's long-term research investments in unmanned systems, several UAS-specific subsystem technologies have also been developed at NRL. This chapter highlights some of the most interesting and promising UAS subsystem technologies.

Laser Tethered Decoy System, 1992–1993

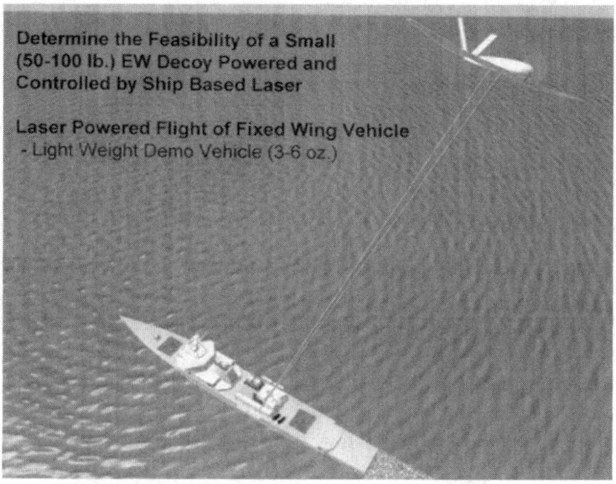

Fig. 12.1 Laser tether concept.

NRL evaluated the feasibility of laser-powered transmission for flying decoys (Fig. 12.1). A laser would transfer power from the ship to the decoy. This would enable very long flight endurances without the vehicle performance limitations imposed by a physical power transmission cable such as used for the Eager Advanced Technology Demonstrator (ATD) program. A laser tethered decoy vehicle (LTDV) was envisioned to be a 50- to 100-pound fixed wing aircraft.

Because of the high cost of developing a large, high-power system, several key aspects of the feasibility of laser power beaming to a UAS were demonstrated initially by a small system. Flown indoors under carefully controlled conditions. Initial investigations used a small, electric motor-powered radio-controlled (RC) blimp. A blimp was used because it was easy to track and did not require much power to propel and navigate. It was tested in a large indoor anechoic chamber. Photovoltaic cells were used for the power conversion. A more refined design could have used a Stirling cycle heat engine or a steam engine for more efficient energy conversion. Next, a few small, RC, fixed wing aircraft were constructed in preparation for repeating the blimp experiments with a more energetic aircraft (Fig. 12.2). However, these fixed wing aircraft were never tested with the laser propulsion.

While the indoor blimp was successfully flown under laser-provided power, it soon became apparent that many technical issues had to be solved

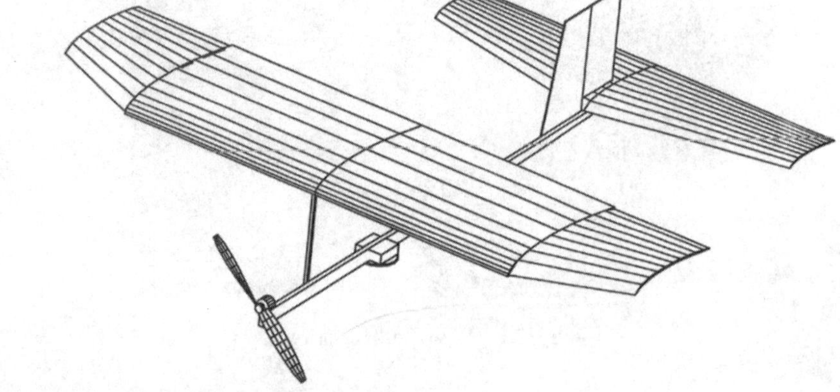

Fig. 12.2 Early fixed wing, laser tethered demonstrator aircraft concept.

and technologies significantly improved before a practical Navy operational system would be possible. The Navy elected not to make those technology investments in 1993. UAS, laser power beaming, and energy conversion technologies have made great strides in the two decades since the LTDV study. Long-endurance EW decoy systems are still applicable to Fleet requirements, and this approach is currently being reinvestigated for future Navy EW decoys and other UAS applications.

ENERGY HARVESTING BY POWER LINE RECHARGING, 2004–2005

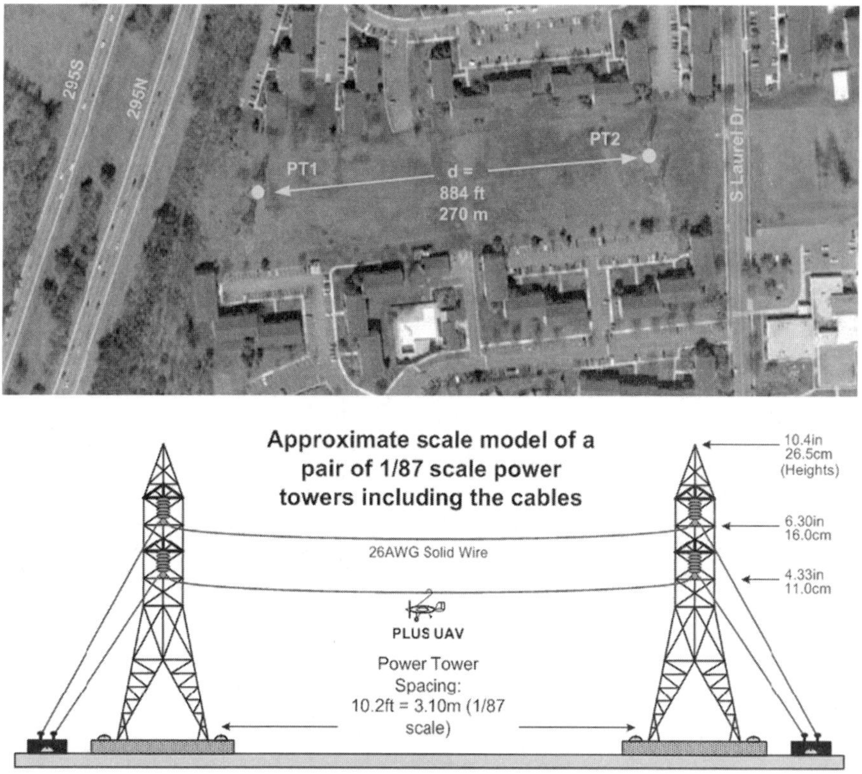

Fig. 12.3 Typical power line and power tower geometry.

The Vehicle Research Section (VRS) explored power line recharging of Micro Air Vehicles (MAVs) and small UASs as part of the SAMPLE MAV program (Chapter 6). The concept was for the unmanned aircraft to land on a power line and inductively recharge its battery (Fig. 12.3). The inductive charging process was inefficient, so the recharging would take a long time. The unmanned aircraft would find, land upon, latch onto, and couple to the power line. The vehicle could serve as an unattended ground sensor while charging. After the battery was replenished, the vehicle would unlatch and fly off. Another option to support MAV operations would be for a small UAS to land on the power line and semipermanently latch on. It would then serve as a docking station for a number of MAVs to latch onto for recharging.

Even if power could be readily extracted, one technological challenge in this concept was to find the power lines themselves. Anyone who flies realizes how difficult power lines are to spot from the air. However, the towers that support the power lines are relatively easy to find visually. So, the navigation scheme would be to find two or more power towers, and then mathematically determine the location of the power lines relative to the towers. The technologies developed for optics-based, non-GPS navigation for MAVs to geolocate radar sites and for the NRL vision-based landing system could be readily adapted. These would enable locating the power towers, determining the location of the power lines, and providing the guidance information needed for the autopilot to perform a landing on a power line.

With the conceptual solution for the geolocation, navigation, and guidance problems in hand, effort focused on the largest uncertainty of how best to inductively couple to the power line and harvest electrical energy. NRL conducted benchtop experiments of the inductive recharging to determine the efficiency at which various magnetic induction "antenna" probe coil configurations could capture enough current to be useful for charging the batteries of a UAS.

A benchtop experiment was designed and built to mimic the electrical and geometric properties of a single power line (Fig. 12.4). This included using a straight section of conducting rod or cable of sufficient length and diameter to simulate the magnetic field around an actual power transmission line. This section of line would also have connectors on either end for safe and convenient hookup.

The short section of transmission line had negligible resistance and hence negligible voltage drop across it. This gave negligible power dissipated by it, thus providing a realistic model of a normal power transmission line. The antenna probes connected to a simple circuit for induced voltage and current measurement, and for impedance matching for maximum power transfer for each probe coil (Fig. 12.5). Data was recorded for the measured quantities of transmission line current, transmission line voltage, antenna probe current, and antenna probe voltage, as well as the calculated values of induced power, power transfer efficiency, and B field magnitude.

This benchtop experimental setup was inherently safe in comparison with using an actual power line, because the transmission line section under test did not have to be at the actual transmission line voltages. Only the line section current had to be of a realistic value to be a good model. The B field

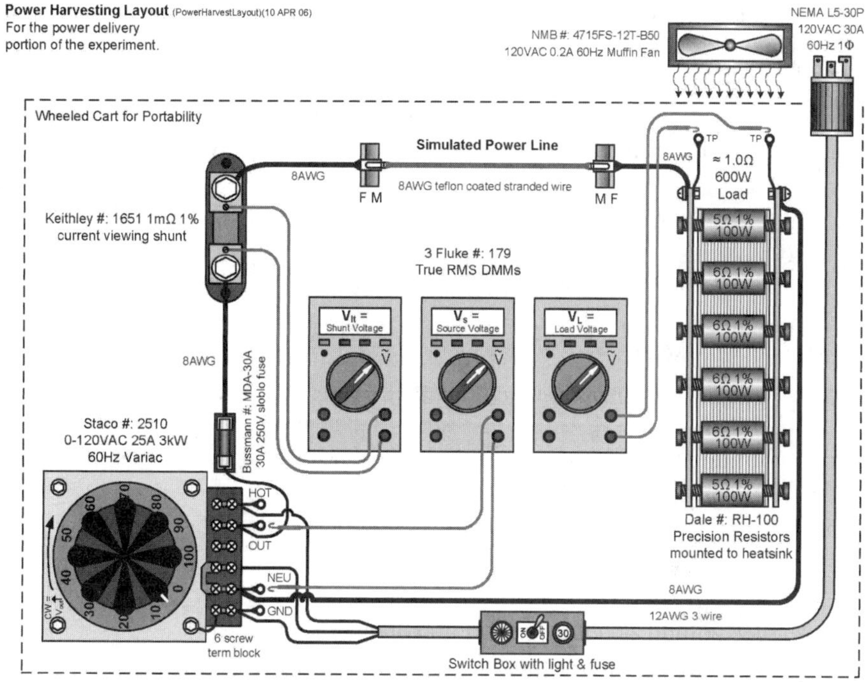

Fig. 12.4 Benchtop experiment schematic for power line recharging.

that the line produced for induction was dependent on the current only and not on the voltage, according to Ampere's law.

Another area of concern in the power line recharging concept was the possible adverse effects of the relatively strong and changing B field, E field,

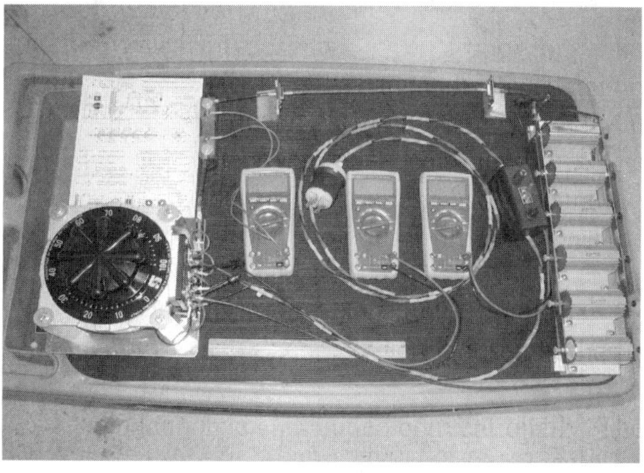

Fig. 12.5 Benchtop experiment hardware for power line recharging.

electromagnetic field, and corona discharges on the UAS's flight control, communication link, navigational, guidance, and onboard computer systems.

Many experiments were conducted, and the requisite engineering data was collected to determine the viability of UAS recharging from power lines. It was shown that the induced current in the antenna probes was low, but given a sufficient period of time, the antenna probes could provide a useful quantity of current for recharging the batteries of a small electric UAS or MAV.

Early in this study, it was discovered that the Air Force Research Laboratory (AFRL) was conducting similar research, so NRL and AFRL shared their design studies.

Although shown to be technically feasible (albeit inefficient), this energy-harvesting approach would probably not be operationally feasible. Under wartime conditions, enemy power infrastructure is usually destroyed soon after hostilities commence, and live power lines are not then available.

Advanced Turbine-Driven Generator, 2004–Present

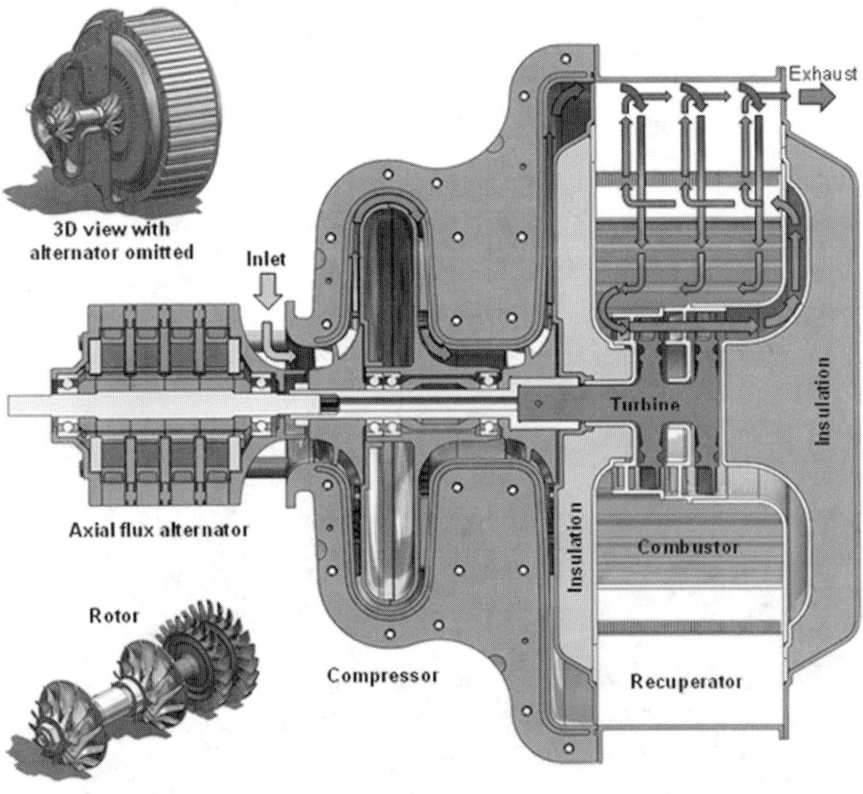

Fig. 12.6 Early conceptual design for the NRL recuperated turbine engine.

NRL developed revolutionary technology to enable high-efficiency small turbine engines and turbine-driven generators (Fig. 12.6). The motivations were to improve operability and performance of small and tactical-class UAS engines and to develop power technologies that could lead to a cleaner environment. The project came about when NRL engineers worked on selecting a heavy fuel engine for the Dragon Warrior system but found that few options were available. Solving this engine technology problem became the doctoral research topic of VRS senior engineer Michael Vick.

Piston engines dominate the small UAS field today. These engines have the appropriate balance between specific power and fuel consumption. Piston engines are well suited to automotive applications where the engine generally operates at low power settings and is throttled frequently. In aircraft, the engine operates steadily at high power, which degrades its life and reliability.

Turbines (turbojets, turbofans, turboprops, and turboshafts) are more reliable than piston engines on manned aircraft. These power plants have long life and high reliability, and are lightweight for their propulsion performance. Relative to piston engines, turbines have negligible vibration because there are no reciprocating masses; all parts operate in pure rotation, and the peak and mean torques are nearly identical. These engines operate on heavy fuel, which is less volatile for Navy applications than aviation gasoline. Turbines also have better performance at higher altitudes than normally aspirated (nonturbocharged) piston engines.

NRL sought to develop turbine technology that would enable turboshafts, turboprops, and turbine-driven generators for small and tactical UASs. The only turbines that were available for UASs and that weighed less than 400 pounds were turbojet, turboprop, and turboshaft engines designed for model aircraft and model helicopters, which typically fly for only 15 to 20 minutes. These engines offered excellent power-to-weight ratios but extremely poor fuel efficiency and short bearing lives on the order of 25 hours. The much larger Northrop Grumman Fire Scout unmanned helicopter uses a turboshaft engine, and the General Atomics Reaper UAV uses a turboprop, but the fuel consumption becomes excessive as the engines scale down.

NRL's program focused on development of a recuperator, development of a generator, and improvements to the turbine engine itself. NRL developed a recuperator to improve fuel consumption. Recuperators use recycled heat from the engine exhaust to preheat the compressed air on its way to the combustion chamber, which reduces the heat required from the combustion. This reduces the fuel flow for a given power. To further improve the efficiency, the turbine rotors and stators of the engine were designed to be made from ceramic materials. This enabled the turbine inlet temperature to be increased to about 400°C beyond the typical value for small engines. The recuperator was ceramic as well, to match the higher exhaust temperature.

This had the added benefits of improving its life due to lower oxidation rates, and reducing its weight due to the ceramic's low density.

Adding a direct drive generator to convert the turbine power to electricity made the system a turbine-driven generator, also known as a turbogenerator. The single-shaft engine had the turbine and compressor on the same shaft. In this configuration, the generator spins at the shaft speed without the need for a heavy gearbox. The generator weight is proportional to torque, and the power input is the product of the torque and the rotational speed. Therefore, a high-speed generator is lighter for a given power.

Making the system a turbogenerator opens up many applications. The electrical power can be distributed to motors that drive propellers or ducted fans on an aircraft. Electric propulsors (motor plus propeller or ducted fan) are lighter than piston engines with propellers, so more configuration options exist. The electrical power can also drive a helicopter rotor or other vertical takeoff and landing vehicles. The system can serve as an auxiliary power unit (APU) on a manned aircraft or ground vehicle.

Other market opportunities for this system outside of aerospace applications include mobile military power generation and residential combined heating and power (CHP). The greater the market, the higher the production quantities, which ultimately reduces the unit price.

The design point changed several times over the life of the project. The project initially focused on a 3-kilowatt peak power, and then it went up to 6 kilowatts for the doctoral research. The current Office of Naval Research (ONR)–sponsored program is for a 12-kilowatt engine that is suitable for the Small Tactical UAS class of vehicles.

The NRL engine technology is suitable for a family of engines that can trade specific power for efficiency across a range of power classes. The degree of recuperation is the key variable. As the recuperator grows, the efficiency increases but the engine weight also increases. The design point depends upon the intended application. For example, ground power units, less restricted by weight concerns, might have larger recuperators than aircraft units.

NRL selected for the research engine a design point that matched the specific fuel consumption for small piston engines. The efficiency record for a recuperated 3-kilowatt turboshaft engine is 19% (0.72 pounds per horsepower-hour), though this system is the size of a refrigerator. To make a flight-weight engine with high power density, it was necessary to adopt new technologies. The engine had to be recuperated but still lightweight.

Several challenges with small turbine engines were dealt with in unusual ways for this program. Compressor and turbine tip clearances are inevitably relatively large at this small size due to bearing stiffness scaling laws and manufacturing tolerance limitations. The compressor and turbine blades have relatively thick trailing edges due to manufacturing and durability considerations at small scale, which creates significant drag. These factors, along with low Reynolds numbers (LRNs), which increase skin friction and other forms of viscous energy dissipation, make it impractical to achieve greater than 80% compressor and turbine efficiencies.

The turbine inlet temperature limit was increased significantly over traditional metal technology by using ceramics throughout the hot section. The higher turbine inlet temperature increases the cycle efficiency. The combustor, turbine, and heat exchanger were all made of ceramics. The turbine exhaust temperature limit (going into the recuperator) with metal alloys is typically around 650°. Using ceramics increased this temperature limit to 1100°C. Metal alloys would oxidize and erode within a few hundred hours at these higher temperatures.

Gas leakage and heat leakage, which are not generally problems for larger engines, become significant at small scales. The scaling laws of volume and surface area drive the phenomena for the small turbines. Gas leakage occurs when air flows along the shaft from the turbine to the compressor stage and then mixes with the cooler air, reducing the compressor efficiency by as much as 10%. The small engine required significant space for thermal insulation to prevent thermal leakage. A quarter of the engine volume was dedicated to porous ceramic rigid insulation and fiber insulation.

The engine's significant use of ceramics was unconventional but by no means new. More than $1 billion have been spent on engine ceramics since the 1970s in the United States, but this technology is not on production engines today. Early problems were brittle fracture and crack growth, but probabilistic life prediction methods and manufacturing technologies to eliminate internal flaws that can initiate cracks were developed to overcome these problems.

However, the U.S. ceramics research appeared to reach a fundamental limit with water vapor erosion. Ceramic materials that can tolerate the turbine environment have silicon in them. The engine burns hydrocarbon fuel, which produces carbon and hydrogen combustion byproducts. The hydrogen and oxygen react to form water vapor in the exhaust. The water vapor reacts with silicon at temperatures above approximately 1000°C. The reaction is also a function of pressure. The silicon and water vapor forms silicon hydroxide gas. The blades are slowly eaten away by this reaction. The problems were exemplified by a key test on a Rolls Royce RR501K engine modified with ceramic components, which experienced heavy erosion after only 1000 hours. Significant research effort focused on environmental barrier coatings (EBCs), but these would crack and expose the underlying silicon ceramic components to the water vapor. As a result of this problem, funding for ceramics research nearly dried up after the year 2000.

NRL's solution to this water vapor erosion problem was to avoid it altogether, which was a key contribution of this VRS research. The recuperator recycles heat from the exhaust and requires less fuel to be burned for a given power output. This reduces the amount of water vapor concentration in the combustion products. Also, the optimum pressure ratio for best efficiency is lower for recuperated engines. These two factors resulted in water vapor erosion conditions that were nearly two orders of magnitude less aggressive than for the RR501K test. Thus, NRL showed that ceramic turbines were practical in low-pressure recuperated turbine engines. At several recent turbine and ceramic materials conferences, many well-known researchers expressed surprise when these results were presented.

Another key contribution was an improved method of joining the ceramic turbine to the metal shaft. Ceramics have a much lower coefficient of thermal expansion (CTE) than metals. Bonding a ceramic turbine to a metal shaft has traditionally been a major challenge. As the turbine temperature rises, the joint may become loose, slip, or break. One technique used successfully in automotive turbochargers is to use brazing with a compliant alloy between the components, but this does not accommodate very high temperatures, and it requires special braze alloys and equipment, as well as strain relief features in the shaft that are tricky and time consuming to design. NRL invented an alternative approach that used a Hirth coupling. This arrangement allows radial sliding but keeps the turbine centered. The V-shaped grooves transmit torque but allow the turbine to grow or shrink without breaking. This simple approach had been used for metal turbines, but not for ceramics.

The ceramic recuperator was another major contribution to the state of the art. Traditional heat exchangers have surface features that create turbulent boundary layers to promote heat transfer. However, this approach does not work for the LRNs associated with small recuperators. Rather, the heat exchanger had to work with laminar flow. The passage diameter was small, and the stream-wise length was short. The recuperator walls were smooth, flat, and simple. The challenges were to design a low-loss manifold system that distributed the air and exhaust inlets into thousands of short-length passages that interweave with one another, and to collect them together again at the outlets. The computational fluid dynamics (CFD) tool was used to create a design that reduced losses and improved thermal strain tolerance. The early research focused on how to produce this design. Test results indicated that the heat exchanger efficiency had a 40% improvement relative to the state of the art.

The recuperator's ceramic base material was mullite, which has moderate thermal expansion and is affordable, has low density, and is thermally stable up to 1500°C. The NRL Materials Science and Technology Division supported the research by finding additives to make the material tougher and crack resistant. Mullite is not usually considered to be a good heat exchanger material because it has low thermal conductivity. However, mullite's value turned out to be very close to the optimum for this application. Perhaps surprisingly, high thermal conductivity does not always yield the best heat exchanger design.

The NRL turbine engine was designed to operate with a cantilevered shaft. The axial thrusts of the compressor and turbine balanced each other to reduce the loads on the bearings. Bearing life is proportional to the cube of the axial load, so this approach was very significant. Bearings in small turbojets often last only 25 hours because the engines have large axial thrust loads. Most model aircraft turbojets and turboprop engines also place the bearings in the harsh environment of the combustion chamber. The NRL turbine placed the bearings outside the hot section upstream of the compressor. The shaft dynamics of the cantilevered shaft were found to be benign.

The research required development of customized engine cycle analysis tools. Legacy tools did not have the capability to model all the unusual features. The analysis used a one-dimensional design point model with real gas properties, including the effects of fuel combustion byproducts on the specific heats and viscosity of the burned gases. The fuel-to-air ratio could be modified as well. The detailed design and analysis of the compressor and turbine used industry-standard 3D viscous compressible CFD tools with LRN turbulence and transition models.

Several components were built for the project. Two iterations of the turbine were built as part of a manufacturability study but were not tested. The compressor rotors were designed and built using best practices for turbochargers. Recuperator heat exchanger segments were built and tested (Fig. 12.7).

Based on the success of the early research, NRL was awarded a four-year program from ONR to build a flight-weight, 16-horsepower (12-kilowatt) engine for UAS applications. This engine will have specific power and fuel efficiency characteristics of mature and emerging piston engines, but with potentially much higher reliability, longer life, and lower life-cycle cost. This program builds upon prior research, including ceramic recuperators and turbines, shaft couplings, and noncontact thrust bearings. The program aims to conduct laboratory tests of the complete engine. Future engines using this technology could achieve four times the fuel efficiency of today's miniature turboprops and five times the life and reliability of current piston engines, at about the same production cost per kilowatt.

The engine characteristics will be revolutionary for UAS propulsion. The engine will weigh 20 pounds without a gearbox. A 4-pound gearbox will reduce the rotational speed from 120,000 rpm at the turbine shaft to 5000 rpm

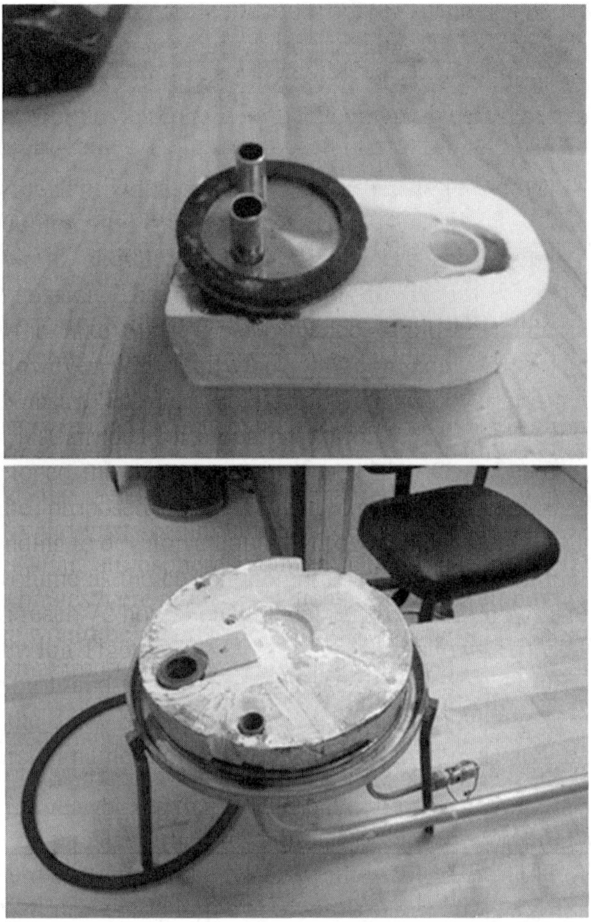

Fig. 12.7 A ceramic heat exchanger sector and how it mounts to the test rig.

at the prop shaft. A high-speed, 12-kilowatt generator integral with the turbine shaft will allow full conversion of the shaft power to electricity. This Commercial Off-The-Shelf (COTS) generator has a voltage between 200 and 400 volts DC. The brake-specific fuel consumption goal is 0.6 pounds per horsepower-hour across a power range of 4 to 16 horsepower. The engine will be capable of balancing power between the generator and a variable-pitch propeller, which will enable parallel hybrid operation. The engine is designed to operate for 1000 hours mean time before overhauls. The reliability target is a 1% chance of critical failure in 1000 hours of operation. This engine will operate on JP5, JP8, and Jet-A/Jet-A1 fuels.

TURBINE ENGINES: FURTHER READING

Vick, M.J., "High Efficiency Recuperated Ceramic Gas Turbine Engines for Small Unmanned Air Vehicle Propulsion," doctoral thesis, Imperial College, London, 2013.

VISION-BASED RECOVERY SYSTEM, 2006–2008

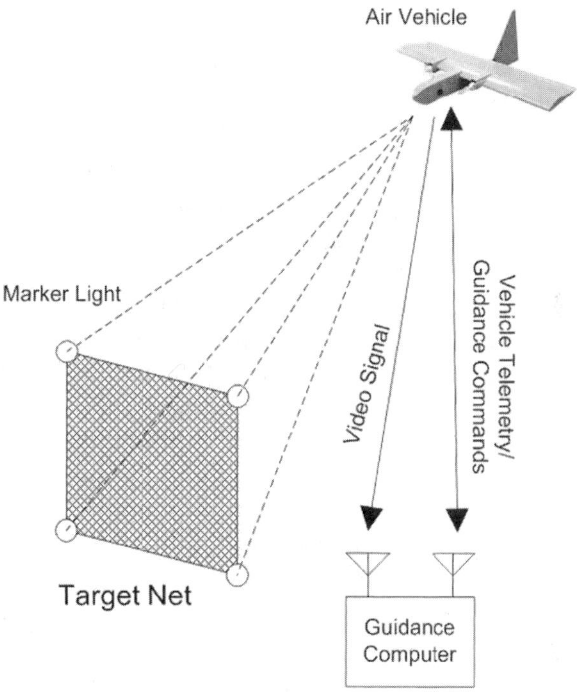

Fig. 12.8 Vision-based recovery system components.

Current military operations have shown the tactical benefits that small UASs can provide. These systems are typically used for video reconnaissance and are therefore equipped with one or more video imagers and transmitters (Fig. 12.8). They are also autonomous, with flight control systems designed to fly preprogrammed GPS waypoints. Although capable of autonomous flight, these small platforms are generally not capable of performing precision automatic recovery.

Present methods to enable the automatic recovery of unmanned aircraft require that additional equipment be installed on the aircraft. This equipment is typically costly, is large, or requires substantial power, limiting these current methods to large vehicles. NRL's Vision-Based Recovery System was developed to facilitate automatic recovery of small UAS platforms aboard ships, using the existing video camera and transmitter typically found on small unmanned aircraft as the source for the additional guidance information required. Information from the vehicle autopilot is coupled with machine vision to generate guidance commands for the vehicle to track.

NRL's recovery system was designed to utilize the existing sensor equipment already installed on the air vehicle. A number of assumptions were made in the design of the system. First, the vehicle could be flown using the internal waypoint guidance system to a position where the camera could view the target. Second, the vehicle was equipped with a body-fixed analog camera, analog video transmitter, autopilot, and command-and-control bidirectional digital data link. Third, the vehicle's autopilot and navigation system were estimating the inertial vehicle attitude and velocity. Finally, it was assumed that the autopilot allowed for the injection of external roll, pitch, and airspeed commands. All the machine vision and guidance computational equipment, along with the UAS ground station, were located near the recovery target, which nominally would be aboard the naval vessel.

The machine vision portion of the recovery system was responsible for the acquisition and tracking of the target. The target was defined by four high-intensity lights. The acquisition portion used the known geometry of these marker lights to automatically find the target. Although a planar geometry can be readily defined by three points, natural reflections can also cause three points of light that will be interpretable as a geometric plane, while seldom will there be four points of light that will lie in the same resolved plane. This was the rationale for using four lights, since they represent the planar geometry of an angled flat net without ambiguity for determining the location and orientation. Once the target was located, the tracker algorithm gated the target and used a fixed gain observer estimate for the velocity of the target in pixels per frame.

Experimental tests indicated that simple charge coupled device (CCD) cameras had a narrow dynamic range when compared to the outdoor lighting environment. It was found that placing a dark filter over the lens could shift the dynamic range of the camera to the extreme end of the exposure range of the camera. In effect, this filter shifted the average intensity of background out of the dynamic range of the camera, allowing only the brightest of objects to be detected. High-intensity lights used as the markers could then be easily detected by the camera. In the NRL design, the filter was fixed in the optical path of the camera, but a servo actuator could be used to rotate the filter into place when needed.

The output of the camera was transmitted to the ground computer by an analog video link. Once at the ground, the video signal was received and digitized by a frame grabber card within the guidance computer. The image was then contrast enhanced by gamma-correction and thresholded to produce a simple binary image. Almost all the background clutter had already been removed via the intensity filtering. After the thresholding was performed, the centroids of all the white blobs were found and listed.

A simulation environment was created to evaluate the robustness and performance of all portions of the vision guidance system. It was important to design the simulator to allow the cross-compilation of all the flight software, machine vision algorithms, and vehicle model into a single executable file. This was desired to guarantee time synchronization between the tasks and to allow for logging of any state in the system. Cross-compiling the flight code would allow any dynamics associated with the navigation system and controller to be included in the overall performance.

Simulation results indicated that the system should perform as envisioned and should be robust to dynamic position errors caused by ship movement and wind. This robustness to wind is an especially important feature for a system that will be operated in a maritime environment. Simulation indicated that the performance of the guidance system in the presence of wind and other sensor errors was sufficient to enable recovery of these vehicles with accuracy within one wingspan of error. So, for a Dragon Eye–sized vehicle with a wingspan of 4 feet, a shipboard recovery net would at most need to have a recovery net width of 10 feet and height of 6 feet for safe operation.

To verify the validity of the simulation results and to obtain hands-on experience, flight tests were performed (Fig. 12.9). A small model airplane was equipped with a nose-mounted CCD camera and an autopilot with GPS waypoint navigation. To represent the lighted recovery net, an "X" was constructed from two wooden beams with a selectable array of multicolored LEDs attached to the ends of each beam. The use of selectable LED colors enabled measurements to be taken to determine which color provided the best response to given background lighting for the selected CCD camera.

Fig. 12.9 Test aircraft beginning its wave-off maneuver (left), and image from aircraft camera during an automated approach (right).

Since there was no actual recovery net, the aircraft was programmed to perform a wave-off manuever just prior to what would have been net impact.

The system performed well and indeed verified the system's ability to enable small UAS platforms to perform recovery into a small net, with the needed robustness to wind.

This vision-based recovery approach does not require the addition of large or expensive equipment beyond what is already present on the baseline air vehicle. So, small UASs can be readily recovered aboard ship with the addition of a small shipboard system. Development of an operational system awaits a Navy desire to do so.

AUTONOMOUS LOCATOR OF UPDRAFTS (ALOFT), 2007–2009

Fig. 12.10 RnR Products SB-XC aircraft used for the ALOFT project (left) and internal hardware (right).

The purpose of the ALOFT program was to develop autonomous soaring algorithms that could extend the endurance of unmanned aircraft by harvesting atmospheric energy (Fig. 12.10). Sailplanes operate by gaining altitude through thermal updrafts (thermals), slope lift from terrain, or dynamic soaring through velocity gradients. ALOFT used thermals to extend the sailplane endurance and range in a road-following mission. This effort was the doctoral research topic of VRS senior engineer Daniel Edwards.

Thermals are columns of rising hot air that are caused by uneven heating of the earth's surface. Their size, structure, and intensity are a function of temperature, time of day, terrain, humidity, and winds. Predicting thermals is challenging, but many manned and RC sailplane pilots are skilled at locating thermals and climbing within them. To gain altitude in a thermal, the pilot must circle within the thermal, ideally around its center. ALOFT used sensors and software to autonomously replicate this task.

A typical unmanned aircraft executes a mission plan without consideration for atmospheric lift. The aircraft flies in a straight line or turning flight without reacting to atmospheric lift or sink. When the aircraft flies through a thermal, the resulting higher-than-commanded rate of climb is countered by the autopilot by reducing power. Similarly, flying through a sink region between thermals requires additional power. The atmospheric energy tends to average out over a flight when the autopilot rejects the atmospheric perturbations in this way. A typical mission profile includes ingress from the launch location to the target area, loiter over the target, and egress to the recovery location. No attempt is made to use the thermals to enhance endurance.

Using thermals for sustained flight required a fundamental change in flight behavior. The unmanned aircraft would simultaneously seek lift from thermals while attempting to satisfy the mission objectives. Rather than flying in a straight line for transit, the vehicle would find thermals along the way. The aircraft would circle within the thermal to gain altitude, and then fly along the mission route until reaching a new thermal. In the mission area, the aircraft would find thermals to stay aloft while achieving the mission objectives. The penalties for atmospheric energy harvesting are longer time to target, and any data collection penalties incurred while finding thermals.

The ALOFT program used a commercially available RnR Products SB-XC cross-country model sailplane to research autonomous soaring. This high-performance model sailplane was constructed of composite materials and had advanced LRN airfoils. It used a Cloud Cap Piccolo II autopilot for autonomous flight. The flight control system allowed an RC pilot to control landing. An 87-watt-hour lithium-ion battery provided power for the avionics and flight controls for up to 8 hours. The aircraft was launched using a model sailplane electric winch. In still air with no thermals or sink, the aircraft could fly for approximately 3 minutes from a winch launch. The system used a ground station to allow the operators to monitor and re-task the aircraft. The communication system consisted of a 900 MHz spread spectrum radio, which provided a two-way link between the aircraft and ground control station.

NRL developed a simple thermal model that could be used by the onboard processors. The model made several simplifying assumptions to decrease computational burden. The model assumed that the thermal is radially symmetric about a center point, the thermal produces no horizontal wind, the thermal drifts with the ambient wind, and the updraft strength has a modified Gaussian distribution. The only parameters for the thermal model were horizontal center location, peak updraft strength, and characteristic radius.

The thermal parameters were identified based on a power model. The energy and power applied to the aircraft system must be conserved. One element was the power state of the aircraft itself, which was a function of the drag, rate of climb, and velocity. The drag polar of the aircraft combined with a dynamics model helped determine the aircraft contributions. The other contribution was from the atmosphere. The total power state was measured with an onboard inertial measurement unit for accelerations and GPS for velocity and position. These sensors were already built into the Cloud Cap autopilot. Any power that was not accounted for by the aerodynamic performance of the aircraft comes from the atmosphere. The vertical wind speed was estimated from the atmospheric power.

The thermal model parameter identification used a recursive unscented Kalman filter (UKF) method. This method used recent data to rapidly update the previous set of parameters. This approach had a lower computational burden than that of adaptive neural network batch methods that accumulate sensor time history and iteratively solve for the model parameters. The objective of the thermal model was to extract atmospheric energy rather than model the thermal with high accuracy, so the faster-running recursive methods were appropriate. The batch method had a lower error rate than the UKF method, but the UKF had sufficient accuracy for the task. All the sensor data came directly from the autopilot, so the thermal model was implemented in software only.

The Cloud Cap autopilot already had the ability to circle around a waypoint. The autonomous soaring algorithms simply adjusted the center point of the orbit. In this way, the autopilot served as the inner loop, and the soaring algorithms formed part of the outer loop. The autopilot inner loop used high rate sensor data and control commands, which ran at 50 Hz. The soaring estimation ran at 2 Hz.

The mission management system had to balance the competing objectives of soaring effectiveness and mission satisfaction. The system used the MacCready speed ring theory to determine the minimum thermal strength to support the cross-country goals. If the aircraft was at high altitude, the need to catch a thermal was diminished. High-altitude thermals had to be strong to justify latching. If the altitude was low, catching thermals to sustain flight was more important. The mission management system also had to account for airspace boundaries.

Table 12.1 RnR Products SB-XC Technical Characteristics

Parameter	Value
Wing span	14.2 ft
Length	6.5 ft
Takeoff weight	12.5 lb
Payload weight	4 lb
Loiter speed	25–30 KTAS
Endurance	5.3 hr demonstrated
Max L/D	24
Airfoil	SD-2048
Launch method	Winch launch
Recovery method	Skid landing
Propulsion type	Sailplane

The ALOFT glider completed over 170 flights with more than 70 hours of flight time. The aircraft typically operated at altitudes of 500 to 1500 meters above ground level. The first autonomous soaring demonstration occurred on September 30, 2007. ALOFT unofficially broke the world record for goal-and-return cross-country soaring for the 5-kilogram-class sailplane by flying 48.6 kilometers each way. The longest-endurance flight was 5.3 hours. This endurance was significantly higher than the 3-minute unassisted glide time from the winch launch!

The autonomous soaring technology has important applications to future unmanned aircraft. While autonomous soaring with an unpowered aircraft demonstrates the performance potential, the system could be applied to powered aircraft as well. Efficient battery electric–, hybrid electric–, and fuel cell–powered aircraft tend to have high lift-to-drag ratios (L/Ds) and light wing loadings. These attributes are conducive to soaring. Application of autonomous soaring algorithms to these vehicles could greatly extend the endurance by conserving the power source. The technical characteristics for the RnR Products SB-XC are listed in Table 12.1.

ALOFT: FURTHER READING

Edwards, D.J., and Hazard, M., "Autonomous Soaring for Unmanned Aerial Vehicles," *2011 NRL Review*, Naval Research Laboratory, Washington, DC, 2011, pp. 82–88.

Edwards, D.J., and Silverberg, L., "Autonomous Soaring: The Montague Cross Country Challenge," *AIAA Journal of Aircraft,* Vol. 47, No. 5, 2010, pp. 1763–1769.

Hazard, M., "Unscented Kalman Filtering for Real-Time Atmospheric Thermal Tracking," M.S. thesis, Mechanical and Aerospace Engineering

Dept., North Carolina State University, Raleigh, NC, 2010, http://www.lib.ncsu.edu/resolver/1840.16/6353.

LOW COST MICRO AUTOPILOT (LCMA), 2008–2012

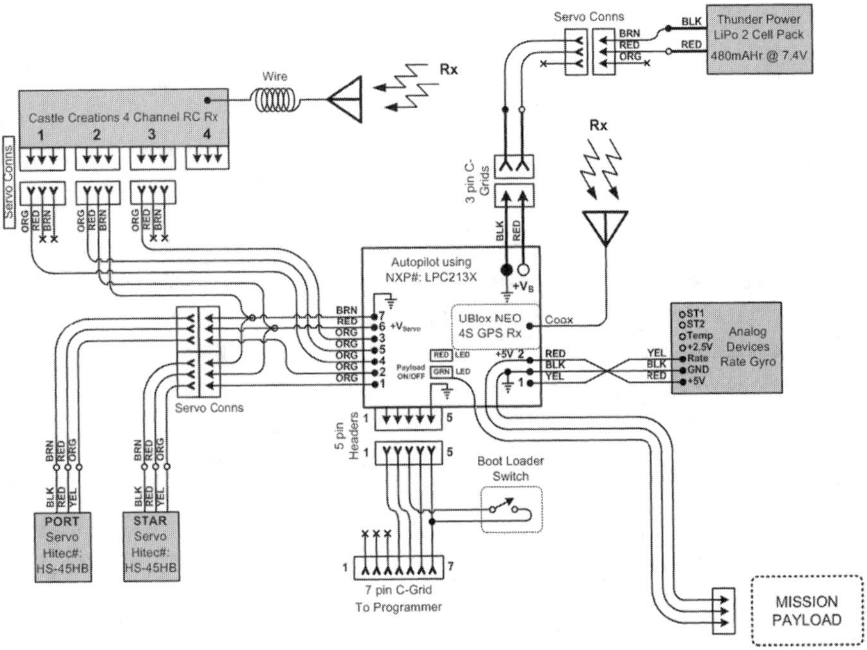

Fig. 12.11 Schematic of the first LCMA for CICADA.

During the development of the CICADA MAV, it was necessary to develop a guidance, navigation, and control system that was extremely small and lightweight and that could be produced in large quantities for less than \$100 US per unit. This seemed incredible to the pioneering members of the VRS who remembered that three decades earlier, one could only dream of autonomous flight with a UAS as small as 50 pounds and a \$10,000 autopilot. Now, even an affordable, 200-gram, disposable MAV could be made fully autonomous.

An automatic flight control system (ACFS) was developed that provided two-axis control (pitch and roll/yaw), a radio command uplink for control override or in-flight reprogramming, GPS navigation, and the power and signal architecture to support various miniature mission payloads. In addition, a four-channel RC receiver was incorporated to enable the capability for in-flight GPS waypoint reprogramming, payload operator control, or aircraft control override. The heart of the Low Cost Micro

Autopilot (LCMA) was a microcontroller coupled to a uBlox NEO-6 GPS receiver and a miniature rate gyro. All required software code was developed within the VRS. The complete prototype system, not including the mission payload, weighed less than 100 grams (Fig. 12.11).

Three aspects of the initial version of the LCMA pleasantly surprised everyone. First, it could be assembled solely from commercially available, off-the-shelf, modular, plug-together components and required no custom fabrication. Second, it worked reliably from the start. Third, even built in small quantities, it cost only slightly over $100 per unit.

Industry was contacted for quotes to estimate the unit cost to produce large quantities (10,000 to 100,000 units per year) of a fully integrated LCMA unit that could be imbedded into a CICADA airframe during assembly. Estimated production cost was quoted to be less than $50 per unit. One vendor quoted a cost of $29.95 per unit and was apologetic that it could not produce the unit for less!

As this is being written, CICADA development for various applications continues, and the LCMA continues to evolve and be refined. Volume production awaits only the requirement to develop a production MAV program.

LCMA: Further Reading

Kahn, A.D., and Edwards, D.J., "Navigation, Guidance and Control for the CICADA Expendable Micro Air Vehicle," *AIAA Guidance, Navigation and Control Conference*, AIAA Paper 2012-4536, 2012.

Chapter 13

Conclusion

Few organizations that develop unmanned aircraft have fostered and retained an integrated team for as long as NRL's Vehicle Research Section (VRS). This team of professionals has seamlessly combined skills covering all requisite technologies, including low Reynolds number (LRN) aerodynamics; sophisticated composite structures; fuel cell, battery electric, and turbine propulsion; digital avionics; flight-proven software codes for autonomous guidance, navigation, and control; six-plus-degree-of-freedom digital flight simulations with in-flight morphing from a ballistic body into an aircraft; multidisciplinary design optimization; airframe and all subsystem fabrication, assembly, and checkout; affordable expendability; and, of course, flight research. Many lessons were learned and several guiding principles emerged during the four-decade-long development of the unmanned aircraft presented in this volume.

Lessons Learned and Guiding Principles

Expect Research Airplanes to Crash

When conducting cutting-edge research involving aircraft, especially in a technologically immature area such as the development of early-generation unmanned aircraft systems (UASs), plan the resources and time to continue the project despite crashes and the devastating loss of hardware. It is also crucial that sponsors be fully aware that prior to successfully developing any advanced-technology unmanned aircraft, there will be crashes. Most engineers and technicians find it emotionally disheartening to slave over a pet project for months or years only to see it reduced to tiny parts in a few seconds. Unfortunately, this is the price of the game. Sometimes the innate fear of crashing drives the developers to become too cautious and slow in the project's development. However, it is far worse to have a project terminated because flight was never attempted from fear of crashing, than because of a crash when attempting a test flight. No member of NRL's VRS was ever reprimanded or denied a promotion due to a crash caused by an unforeseen problem or even a human mistake, so long as the mistake was not repeated.

On the positive side, more valuable engineering lessons are learned from crashing than from routine test flights, even though less data is collected. Upon seeing the flight test team return with the LODED aircraft remains stuffed in a garbage bag following the project's seventh crash during the seventh flight attempt, one technician remarked, "At least it broke the monotony of success." The team nearly died of laughter, since success had yet to be tasted. Instantly, everyone realized that new, radical-design airplanes sometimes just crash until all the problems are resolved. Incidentally, the eighth flight of the LODED was successful.

Unmanned aviation is still young and rapidly evolving. When manned aviation was young, crashes were common and expected, too. Many years ago, one of the coauthors (Foch) had the good fortune to befriend a neighbor who was a World War I Sopwith Camel fighter pilot. When asked what it was like to fly in the pioneering age of military aviation, he said, without a hint of hesitation, "We crashed a lot!"

Persevere

If the engineers and technicians lose hope and give up believing in a project, the ability to achieve success is lost. Many complex projects often fail to reach a successful conclusion because the project is abandoned moments before success. So, the trick is not to give up too early. One day, one of the engineers showed everyone a motivation/demotivation poster that stated, "Winners never quit; quitters never win; someone who never wins and never quits is an idiot!" Many people find it easier to walk away from a series of failures than carry enough faith to persevere to reach success. They fear ridicule or looking like the aforementioned idiot.

Unfortunately, financial sponsors must operate in a business environment and have less patience than engineers. They often have to withdraw their support before the project is perfected. One important thing that engineers can do about this is to lay out the project in a series of small steps that regularly demonstrate small, mutually supportive technical successes, building up to the final project demonstration.

Positive attitude is critically important to achieving success. More than 2000 years ago, the great Roman poet Virgil was asked why he thought that Rome was able to conquer the world. Virgil penned, "They can, because they think they can."

Manage Time

Time is the most precious commodity for any project. Developing an aircraft is a relatively expensive undertaking that places a strong emphasis on obtaining the financial resources needed. However, the authors have noted that a greater number of advanced-aircraft projects fail due to severe schedule slips than due to cost overruns. Sponsors are bound to work within a schedule-based fiscal system and can more easily find additional funding for a promising project evolving on schedule with unforeseen cost escalation, than allocate more time, even at the original cost. Incidentally, when more time is added to a schedule without additional funding, the product suffers since additional costs that accrue for storage and labor must be taken away from the material resources.

So, time management for any project, especially advanced-technology research efforts, is crucial. Engineers and managers typically underestimate the required development time by a factor of two! Any project must be accomplished within some reasonable fiscal schedule, and sponsors can seldom provide as much total time as desirable. Time must be used efficiently because only a finite amount is ever available.

Flight research is a labor-intensive process; this must be anticipated and planned for in the development of a realistic schedule. When visiting the Kaman Aerospace flight test facility, one of the authors noticed a sign prominently displayed on the wall of the hangar. It said, "Remember, 12 hours is only half a day!"

Prepare Fully, and Don't Count on Luck

Whenever a complex advanced system is developed and begins testing, many unforeseen problems arise. Even when the subsystem functions are verified and deemed ready for flight, there always seem to be small bugs and strange difficulties that occasionally lead to catastrophic results. As any development team matures through experience, the success rate improves and the number of failures becomes less frequent. Unfortunately, it is also quite easy to become complacent and begin making simple human mistakes. The protection against this is to carefully develop detailed test plans with extensive checklists that must be followed religiously. The authors do not believe in blind luck. They have learned that good luck happens when preparation meets opportunity.

Choose Good Project Names

Coming up with a name for a project or an airplane design often seems to be nearly as distasteful to engineers as cost determination. While of no technical value whatsoever, choosing an appropriate and memorable name often has a large impact on marketing the proposal, obtaining support from the sponsor, and being embraced as an important system by the users. A good name can also foster cohesion and pride within the development team. As shown in this book, military systems often use names that are clever acronyms describing their functions, but they are sometimes just neat words that the sponsor or user likes. For example, FLYRT stood for Flying Radar Target, and LAURA stood for Low Altitude/Airspeed Unmanned Research Aircraft. The name LAURA was selected from a list of possibilities by the Office of Naval Research (ONR) sponsor because in addition to being a good acronym to describe the project, it was the name of Foch's first child. Eager was named as such because the VRS was eager to do the project, and Dragon Warrior was picked by the U.S. Marine Corps generals because they wanted their Marines equipped with a system having a warlike name.

Enjoy Flying

Nothing beats actually flying. There are many extremely advanced and proven tools for designing and fabricating aircraft. It is relatively easy to jump into the UAS field, especially for aero modelers. However, the indisputable best learning tool and often a rather shocking reality check is actually flying the design. Book learning, wind tunnel testing, and computer simulation provide a high confidence level, but one is not actually an airplane designer until three things happen. First, a design must fly entire flights and demonstrate all aspects of the intended application without crashing. Second, it must demonstrate performance close to that predicted. Third, it must be capable of being reproduced by others, which also includes the required documentation. When one accomplishes those things, he or she is an airplane designer. Despite everything, nothing beats the reality check of actually flying the thing.

Document

Documentation is costly to produce but invaluable to have. Just the mention elicits groans from most scientists and engineers, who would just prefer to do the experiment, solve the technical challenges, and demonstrate the successful operation of their creations. NRL develops Mil-Spec Level 2 drawing packages for all its major projects. These drawings are equivalent to "best commercial practice" drawings and do not detail every fastener or connector, but they do provide sufficient details for others to reproduce the system. Likewise, software is written in a standard Navy format to Naval Air Systems Command (NAVAIR) requirements. While such a level of documentation might seem excessive, especially for a single airframe for an experiment, without this level of documentation, there will be no chance that the design will ever become a production line item. Many industry vendors have demonstrated good UAS systems to the military but had to be turned down for future development using military funds because of poor documentation. While it is possible to backfill missing documentation, doing so is much more expensive than documenting in real time and usually too expensive for most sponsors.

When UAS development projects are being planned and performed, 10% to 15% of the total estimated project cost should be allocated for documentation. Typical system development documentation includes schedules, drawings, photographs, videos, data lists, wiring diagrams, software code, periodic progress reports, a complete drawing package, and a final report.

When the developers and their hardware are long gone, documentation preserves their efforts for posterity and adds to mankind's knowledge base. Indeed, without documentation, this book would not have been possible.

Appendix A

Engineering Background

A few engineering principles may help the reader understand the design challenges of the various NRL aircraft designs. A complete coverage of all unmanned aircraft design drivers is not possible in this book. For more in-depth treatment of unmanned aircraft design, the reader is encouraged to consult Gundlach's unmanned aircraft design book (Gundlach, 2012).

The aerodynamic flight regime can be largely characterized by the Reynolds number and Mach number. The Reynolds number is a measure of the relative influence of inertial forces to the viscous forces. Reynolds number, *Re*, is defined by

$$Re = \frac{\rho \bullet V \bullet l}{\mu}$$

where ρ is the air density, V is the flight velocity, l is a characteristic length, and μ is the kinematic viscosity of the fluid.

Low-speed aircraft with small reference length operate at low Reynolds numbers. Air density decreases with altitude, so high-altitude aircraft tend to operate at a lower Reynolds number than an equivalent low-altitude aircraft. Viscosity is mostly a function of temperature. The definition of low Reynolds number (LRN) is subject to debate, but typically this regime occurs when the chord Reynolds number is below 500,000. Most NRL unmanned aircraft operate in the LRN regime.

Mach number, M, is the ratio of the flight speed, V, to the speed of sound, a:

$$M = \frac{V}{a}$$

Aircraft that operate below Mach 1 are considered subsonic, and those operating above Mach 1 are supersonic. The transonic regime is typically given as Mach 0.8 to 1.2. The compressibility effects are important in the transonic and supersonic regimes. Most NRL unmanned aircraft operate in the subsonic regime. The Mars flyers, TMET, and AAED-FF operated in the LRN transonic regime, which posed aerodynamics challenges that required groundbreaking research to solve.

The most important aerodynamic parameter for range-optimized aircraft is the lift-to-drag ratio (L/D). The lift coefficient for LRN range-optimized aircraft tends to be less than 1. Flying wing designs tend to have a low achievable lift coefficient due to the combined requirements of lift and longitudinal trim placed on the wing. Flying wing and conventional configurations tend to have equivalent range performance.

For endurance-optimized aircraft, the aerodynamic parameter of interest is the endurance parameter ($C_L^{3/2}/C_D$). This parameter favors lift coefficients greater than 1. Here the conventional configuration has the advantage due to its greater lifting capability. A light wing loading (W/S) is also important for endurance aircraft.

Many of the designs covered in this book have unconventional configurations or unique design features. The unusual designs were driven by many factors such as packaging, in-flight deployment, or payload integration. Conventional configurations can be designed for good stability and control characteristics with little effort and low risk. This is due to a vast body of knowledge gained from thousands of conventional aircraft. Unconventional designs do not have this luxury, and the system development efforts often experience challenges with the handling qualities. This book describes many such problems and solutions.

The reader will observe that the lateral-directional handling qualities are particularly challenging for many unusual aircraft configurations. It is worthwhile to discuss a few lateral-directional terms here:

- *Geometric dihedral*: Dihedral is the angle that the wing makes relative to the horizontal plane when viewed from the front. A positive dihedral angle has wing tips elevated above the wing root. A negative dihedral angle is known as "anhedral."
- *Dihedral effect*: Dihedral effect produces roll stability. Roll stability occurs when the aircraft returns to wings level after a lateral disturbance. Roll stability occurs when the roll generates a yaw (sideslip) angle and yaw rate that is accompanied by a restoring rolling moment. Dihedral effect can be generated by wing geometric dihedral, a high wing mounting, wing sweep, low wing aspect ratio, and the vertical location of the center of gravity (CG). The roll stability term is $C_{l\beta}$, and a negative value provides positive roll stability.
- *Directional stability*: Directional stability occurs when the aircraft produces a restoring yaw moment when sideslip is introduced. A large vertical tail typically provides directional stability. However, if the vertical tail is too large, an excess of directional stability can overwhelm the dihedral effect and induce spiral instability, causing the airplane to enter a spiral dive. Designers must create a balance between directional stability

and roll stability. An overabundance of roll stability tends to create an oscillatory coupled yaw/roll motion termed "Dutch roll." An overabundance of directional stability tends to create spiral instability resulting in a spiral dive. The directional stability derivative term is $C_{n\beta}$, and a positive value provides positive directional stability. In general, $C_{n\beta}$ and $C_{l\beta}$ values should be the same order of magnitude for acceptable lateral/directional stability characteristics.

- *Ruddervator*: A ruddervator combines the functions of a rudder and an elevator. These devices are typically used on V-tails or other angled stabilizers such as displaced tails.
- *Elevon*: An elevon is a control surface that provides the functions of an elevator and an aileron. Elevons are often used on flying wings.

METHODS FOR QUICK INITIAL ESTIMATES OF PERFORMANCE AND SIZE

Foch utilized several quick and informal methods for initially estimating performance and size (and therefore cost) of an unmanned aircraft without the need to perform in-depth analysis or extensive computations. For example, all the calculations in the world cannot solve a problem when in the first order, a solution is beyond the laws of physics. These methods are not taught in school or well embraced by the mainstream aerospace community. Nevertheless, they are based on actual observation, and their validity has been repeatedly demonstrated.

WEIGHT ESTIMATION

All aerodynamic and sizing calculations for an airplane begin with an estimation of its gross weight. Since the fundamental function of the unmanned aircraft system (UAS) is to carry a mission payload, the entire airplane design must begin with establishing the payload weight. Once this is known, multiply this value by 5 to obtain the estimated total weight for short-endurance or short-range, fixed wing designs. For long-endurance or long-range airplanes, multiply the payload weight by 10 for the total estimated airplane weight. For helicopters, multiply the payload weight by 8 for fuel-powered designs and by 10 for battery electric designs to obtain the total weight estimate. If designing a battery electric research aircraft with minimal payload weight, assume the total aircraft estimated weight to be 2.5 times the weight of the propulsion system (batteries, motor, controller, wiring, and connectors).

INITIAL CONDITIONS FOR AERODYNAMIC CALCULATIONS

Use a density altitude of 5000 feet instead of standard sea level conditions. On a hot day, even at sea level you can easily have a density altitude of 4000 to 5000 feet. Using 5000 feet is a more realistic condition to calculate the airplane's performance. In addition, the math becomes much more simplified since the density (ρ) is 0.002 slugs per cubic foot, so 0.5 ρ is 0.001 or 1/1000. This allows a quick, in-your-head calculation of dynamic pressure by simply squaring the speed in feet per second and dividing by 1000. For example, the dynamic pressure at 60 feet per second is 3.6 pounds per square foot ($60 \times 60 = 3600$, and $3600 \div 1000 = 3.6$).

L/D ESTIMATION

If there exists another airplane of similar configuration and number of drag-producing components, assume your design will have the same L/D value.

Or, if you have a new configuration that is aerodynamically clean with a fairly high aspect ratio wing and a body in a reasonable proportion to the wing, select a candidate airfoil and divide the section L/D at the cruise angle of attack by 4 to get the best airplane L/D that an experienced designer could ever achieve. This L/D value should then be the ultimate goal, and the realistic value will be about 80% of this. The reason this method works is that if you begin with an infinite wingspan that is providing lift, the drag of a high aspect ratio wing is about twice as high due to induced drag caused by the finite wingspan. Then by definition, for an airplane flying at its best L/D, the total drag is twice the induced drag. It is virtually impossible to develop an airplane that has an L/D value higher than the airfoil section drag divided by 4, unless one could build a flying wing having a high aspect ratio, no sweep, no bulges, no antennas, and no trim drag, which is very improbable. Also, if the design must carry external antennas and have a fixed landing gear, exposed engine cylinders, and gaps and surface discontinuities, all of which are found on most unmanned airplanes, divide the section L/D by 6 to obtain a realistic value of the aircraft's best L/D.

DRAG ESTIMATION

To calculate the power required for level flight, the total drag must be estimated as accurately as possible. Classically, drag build-up is a process that involves estimating skin area and skin drag coefficient, wing sizing, and airfoil selection; adding up all the parasite drag components (fuselage, tails, landing gear, etc.); estimating the regions of separated flow (for base drag); and then summing all these values. However, there are a few quick methods that are sufficiently accurate for the initial estimate.

Since lift must equal weight, take your estimated weight and divide by the L/D value. At the best range condition, use the maximum L/D.

Or, the standard airplane drag equation for a parabolic drag polar is

$$C_D = C_{D0} + \frac{C_L^2}{\pi \cdot AR \cdot e}$$

Where C_D is the vehicle drag coefficient, C_{DO} is the zero-lift drag coefficient, AR is the aspect ratio, and e is the Oswald efficiency factor.

From the standard airplane drag equation, use a value of 0.68 (for an aerodynamically dirty airplane) to 0.72 (for an aerodynamically clean airplane) for e and calculate the induced drag coefficient. Then select a value for C_{D0} equal to the induced drag coefficient (C_{Di}). Textbooks teach one to use values of 0.90 to 1.00 for e based on a perfect wing design relative to the ideal lift distribution. However, if you use an e value of 0.68 to 0.72, the resulting estimate will be closer to that of a real wing, and the resulting slightly higher value of minimum airplane drag coefficient (C_{D0}) will be more realistic as well. The parabolic nature of the drag polar accounts for not just induced drag but also profile drag behavior and trim drag. This technique was taught to Foch by the late Dr. Bernard Carson of the U.S. Naval Academy many years ago. Dr. Carson discovered this realistic value of e by back-calculating it based on characteristics for aircraft published in *Jane's All the World's Aircraft*. Using these values of e actually slightly overpredicts the induced drag, but this overprediction in summation with the higher than theoretical, but still slightly underpredicted, C_{D0} provides a realistic estimation of total airplane drag coefficient (C_D) with one simple algebraic equation. Finally, using the calculated C_D value, re-solve the airplane drag equation for C_{D0} using 0.90 for e to obtain a more accurate estimate for C_{D0}.

REQUIRED POWER

In level flight, aerodynamic power, in watts, is drag (pounds) multiplied by velocity (feet per second) multiplied by 1.358. Divide this by 0.75 to allow for actual propeller efficiency to determine the power the motor must produce. For electric flight, divide the required motor power by 0.85 for motor efficiency and again by 0.85 for controller efficiency to determine the power that the battery or fuel cell must deliver for level flight.

For an aircraft to operate in realistic environmental conditions and to be able to maneuver, it must have reserve power beyond that required for level flight. Reserve power should be that required for climbing at a rate of 10 feet per second (600 feet per minute). Climb rates much lower than 10 feet per second are not recommended for ground takeoff systems. An air-launched or assisted launch system should not have a rate of climb lower than 5 feet per second in order to enable operations in less than perfect weather conditions. When computing the climb power, remember to include propeller, motor, and controller efficiencies.

To the required climb power, add the avionics power (watts) plus the payload power (watts) to obtain the value for reserve power. The total power required from the energy source is then the sum of the power to fly level and the reserve power. This should be considered the continuous power rating, since under adverse weather conditions it is sometimes required to operate at full throttle for the entire flight.

If, for some reason, the payload power, avionics power, and required climb rate are unknown, experience has shown that for aerodynamically efficient long-range or long-endurance designs, a reasonable estimation of total power required is twice that required to fly level.

Dihedral Effect

As an airplane design evolves, it becomes necessary to determine the dihedral effect, which is the stability derivative coefficient of roll with respect to yaw ($C_{l\beta}$). Textbooks correctly point out that the factors for determining dihedral effect include geometric dihedral of the wing, wing sweep angle, vertical location of the wing on the fuselage, and height of the vertical tail. However, for small UASs or any other aircraft having a low aspect ratio wing, such as Micro Air Vehicles (MAVs), wing aspect ratio also plays a major role in producing dihedral effect. Often for MAVs, aspect ratio provides the strongest dihedral effect of all factors. To calculate the value of this effect, rotate the wing planform in yaw by 10 degrees and compute the difference in wing area on each side of the new centerline. Ratio the difference in area between the two sides and calculate the lift contribution of each individual side such that their sum equals the weight (lift) of the airplane. Next, calculate the distance from the yawed centerline to the spanwise location of the center of lift for each panel. Multiply each panel's lift contribution by the spanwise location of its center of lift to obtain its roll moment. Sum the roll moment created by both wings and divide this value by 10 to obtain the roll moment per degree of yaw. Finally, nondimensionalize the moment to become the moment coefficient to obtain the aspect ratio contribution to $C_{l\beta}$. As with all stability derivatives, remember to keep track of signs per your coordinate system, with dihedral effect becoming a larger negative value as dihedral effect increases.

Wing Sweep

Classically, wing sweep is taught to be desirable for reducing transonic wave drag and detrimental to performance at low subsonic airspeeds. Sweep tends to reduce the maximum lift coefficient and lowers the lift curve slope, but sweep is sometimes necessary for obtaining adequate pitch and lateral-directional stability and control characteristics, such as for a flying wing configuration. During the XFC program, it was discovered that when an airfoil with a thickness of at least 10% and operating at a chord Reynolds number of 200,000 or less, sweep-induced spanwise flow in the boundary layer facilitates transition from laminar to turbulent flow. This reduces the size of the laminar separation bubble and sometimes completely eliminates it. In both cases the wing profile drag is reduced. For wing quarter chord sweep angles between 15 and 30 degrees, the reduction in profile drag compensates for the reduction in lift due to sweep such that the unswept versus swept wing L/D is the same value! This means that if the designer finds it desirable to add a moderate sweep to the wing for stability, control, packaging, or reducing the radar cross section, there will be essentially no reduction in aerodynamic performance at cruise. This effect begs further research. For the XFC, which has an airfoil thickness of 12%, the optimum sweep angle was found to be 25 degrees, which was close indeed to the design sweep angle of 25.29 degrees.

Airframe Folding

One thing learned with respect to folding or especially unfolding an airframe in flight is that the reliability of deployment is inversely proportional to the number of hinges, latches, and sequences. So rather than designing the most aerodynamically efficient airframe and using clever mechanical engineering to make it unfold from a compact package, it is better to design a configuration that is easy to unfold with the minimum number of gadgets and then maximize its aerodynamic efficiency. The final airplane performance of the folding-optimized aircraft is usually comparable to the performance of the aircraft that was designed with folding as a later consideration, since a large number of hinges and latches both increases the airplane weight and adds drag, rendering the aerodynamically driven design much less efficient than expected. Although this approach might lead to an unusual configuration, such as in the case of the Shearwing and XFC designs, the Vehicle Research Section (VRS) has determined through extensive wind tunnel testing and flight experiments that there are no aircraft configurations that provide undisputable aerodynamic advantages for small aircraft. In addition, there is little or no limitation on successful aircraft configurations, especially with digital stability augmentation systems.

When one is designing the packaging/unfolding mechanisms, pivots and hinges are much more reliable and easier to implement than telescoping. Telescoping structural components tend to jam under high vibration and high load environments and often have problems with extreme temperature ranges.

Rigid airframes that fold into containers become volume limited before they become weight limited. Inflatable aircraft that pack into containers usually become weight limited before they become volume limited.

The type of deployable aero surface (wings, tails, rotors, and propellers) is usually driven by packaging constraints. For a given lift, rigid wings give the highest L/D. Rotary wings give the next highest L/D. Inflatable/flexible wings have the lowest L/D. Folding constraints are usually driven by the shape of the available packaged volume. Rigid wings require a long, moderately thin volume. Nontelescoping rotary wings require a long, very thin volume. Inflatable/flexible wings are relatively volume and shape independent. Unlike for conventional aircraft, the transient loads and dynamic loads, such as shock, often determine the structural limit loads for in-flight-deployable designs.

System Cost

Probably the least attractive necessary aspect of airplane design is estimating development and eventual production cost. Based on experience, for the cost of a production UAS to be considered reasonable, the cost of its airframe and requisite subsystems must be within the same order of magnitude as the cost of its payload. So if the payload cost is $100K, the aircraft system cost target must be between $100K and $1.0M. If the aircraft cost is higher than this, it will be considered too expensive to risk carrying the payload. Conversely, if the aircraft costs less than the payload, the payload will be considered too valuable to risk in such an inexpensive (and perceived less reliable) aircraft. For estimating development cost, refer to Gundlach (2012).

Reference

Gundlach, J., *Designing Unmanned Aircraft Systems: A Comprehensive Approach*, AIAA Education Series, AIAA, Reston, VA, 2012.

Appendix B

AIRFOIL TECHNOLOGY

RIBLESS WING CONSTRUCTION TECHNIQUE, 1977–PRESENT

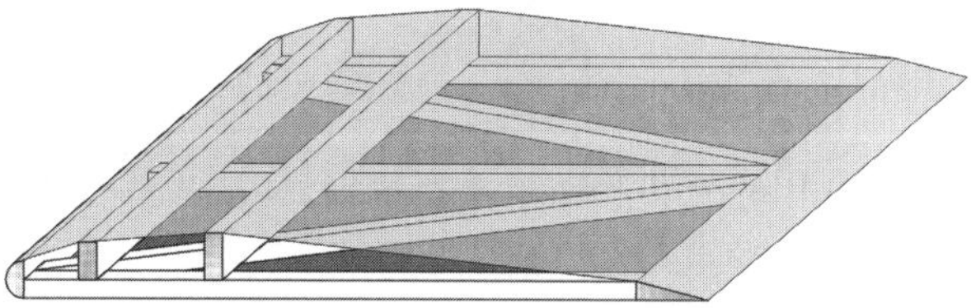

Fig. B.1 Ribless wing section.

The Shearwing research vehicle (Chapter 4), the subscale Pendulus (Chapter 3), and the subscale XFC (Chapter 10) used a ribless wing construction technique invented by Foch in 1977 when he was in college. With this technique, the wing is constructed exclusively of wooden strips. A balsa wood frame consisting of a leading edge, a trailing edge, and chord-wise connecting pieces with diagonal bracing is assembled to form a flat warren truss surface (Fig. B.1). A large balsa or spruce spar is placed on top of this frame at the point of maximum thickness. One or two secondary spars with reduced heights are placed ahead of the main spar. The covering stretches from the leading edge, over the spars to the trailing edge, and forms a simple airfoil that can approximate the shape of any desired airfoil. Many model aircraft that operate at low Reynolds numbers (LRNs), such as free-flight gliders and rubber-powered models, have turbulator spars between airfoil ribs, and the covering forms a nearly straight line between the spars. To close out a wing panel, a thin plate or block is glued to each outboard end and shaped to the airfoil contour to form a wing tip. At the center dihedral joint, after the dihedral braces are installed, polygon- and triangle-shaped balsa members are glued into position to fill the spaces between the spars, leading edge, and trailing edge.

The ribless method provides an external geometry similar to that of ribbed wing construction, but without the need to fabricate and assemble a number of precise curved wing ribs, and so it is a faster and easier method. The drag

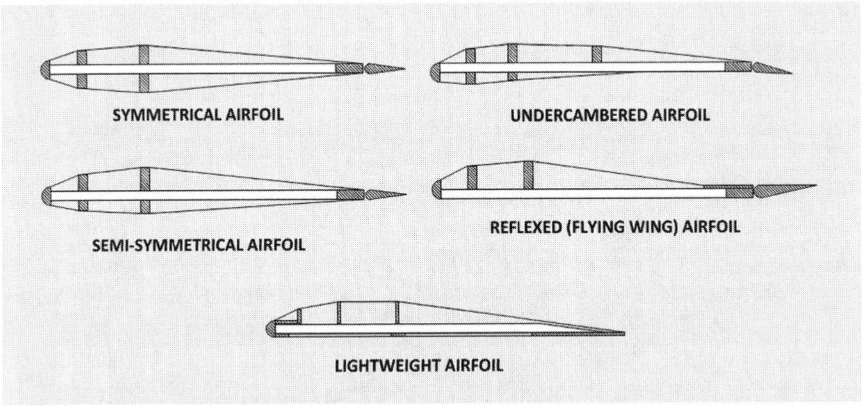

Fig. B.2 Examples of airfoil sections using the ribless construction technique.

penalty is low, because the faceted surfaces cause transition from laminar to turbulent flow before reaching the adverse pressure gradient. This prevents laminar separation bubbles. Figure B.2 shows examples of airfoil sections using the ribless construction technique.

AFTERWORD

JAY GUNDLACH, PRESIDENT, GUNDLACH AEROSPACE LLC, DECEMBER 2013

It is a true honor and pleasure to write this book with Rick Foch. Rick has been many things to me over the years: a supervisor, mentor, collaborator, and dear friend.

I met Rick when I was a senior in high school. Chris Bovais, a Naval Research Laboratory (NRL) engineer and model aviation enthusiast, was helping me as a test pilot for my student design project. Chris was going to fly my design for the world's largest radio-controlled (RC) glider, spanning 50 feet. During our collaboration, Chris introduced me to Rick at NRL's Vehicle Research Section (VRS). The meeting lasted several hours, and it set the path of my career.

I came to work at NRL in December 1994 as an intern working for Kaman Corp., then an engineering support contractor for NRL. I continued to work at the Laboratory during most summers and winter breaks through 1998. During those eventful years, I helped develop more than 17 unmanned aircraft, including air-launched systems, towed vehicles, rotorcraft, Micro Air Vehicles (MAVs), and Mars flyers. Even as an intern, I was able to build prototypes, design aircraft and aircraft components, and support flight tests. In this innovative environment, the only limitations to my involvement were my skill and personal effort. Experienced engineers and technicians generously offered guidance and support.

Rick helped me tremendously in my professional development as an aircraft designer. His support went well beyond what could be expected. When I was an intern, he patiently answered my innumerable questions about aircraft design and aviation history. Rick invested a considerable amount of his valuable time in my professional growth. I was very fortunate that the most senior leader of the VRS would spend hundreds of hours teaching the most junior intern. Rick served on my Ph.D. advisory committee, and meeting his high standards was a welcome challenge. Rick has encouraged me to think big and take on major design challenges. I hope this level of mentorship can be enjoyed by other aspiring engineers. Such knowledge transfer from senior engineers to new engineers is critical for the future of the U.S. aerospace industry.

I have enjoyed our professional interactions over the years. Rick's designs have influenced my own. A key example: Rick's modular construction of SENDER and Nexus influenced my work on the Insitu Integrator, which won the Navy and Marine Corps Small Tactical Unmanned Aircraft System (STUAS) program of record competition. We recently collaborated on the Titan Flyer and Ion Dasch designs.

Through this book, I hope to share with a broader audience the knowledge I gained from my interactions with Rick and other NRL staff. Since many of the senior NRL engineers are in the final years of their remarkable careers, it is timely to document the incredible accomplishments of the NRL VRS. Many valuable lessons about these aircraft programs reside within the minds of the engineers and technicians. This experience can now be enjoyed by others.

RICHARD J. "RICK" FOCH, SENIOR SCIENTIST FOR EXPENDABLE VEHICLES, NAVAL RESEARCH LABORATORY, WASHINGTON, DC, MAY 2013

My love of aviation began in one defining moment. When I was a young child, my father took me to Bartow Air Base, a U.S. Air Force Air Training Command facility, for an Armed Services Day air show. He was responsible for the base avionics shop and set up their jet engine test facility. As we walked across the apron from our car to the displays, my senses were filled with the sights, sounds, and smells of North American T-28s, Beechcraft T-34s, and Cessna T-37s. We stepped out of the bright Florida sun for the welcome shade of a hangar and walked right up to a pristine T-37 jet trainer, resplendent in its natural aluminum finish with fluorescent orange markings and red, white, and dark blue USAF insignia. My father asked me if I would like to sit in it, which seemed to me like asking a starving man if he'd like something to eat. In a few seconds, I was in the cockpit, immersed in a marvelous new, foreign, yet strangely exciting world. At that instant, I fell in love with anything and everything involving airplanes.

Shortly thereafter, my father took a job with McDonnell Aircraft to support the new, exciting Mercury and Gemini spacecraft programs, so we moved to Titusville, Florida, near Cape Canaveral. Titusville was, and remains, an unremarkable, quiet little town. But to me, the one endearing feature was its small airport. My grandparents as well as my parents bought houses next to the airfield, much to my delight. Our house was just off the end of one of the runways, which may not have been the safest place to live, but it certainly was exciting for a little boy who loved airplanes. Many of my Saturdays began as silver Stearman biplane crop dusters roared over my house to land and reload their hoppers for another sortie to eradicate the local mosquito population. I would pop out of bed, dress fast, and quickly run outside to watch all the action. The Stearmans, the shiny new Cessnas, as well as the Cubs, Champs, and other light aircraft were interesting and fun to study. However, a few tiny homebuilt aircraft impressed me the most, especially the 16-foot-wingspan Knight Twister biplane and the 20-foot-wingspan Wittman Tailwind. These were real airplanes built by ordinary people, not by huge airplane factories. I really wanted Dad to build a homebuilt airplane, and I was not deterred by the fact that he wasn't a licensed pilot. He was, and still is, a superb craftsman and electronics technician, so I knew he had the needed skills. I also continuously asked questions pertaining to why each airplane had particular design features and the functions for each part. It seemed to me that most airplanes were more complicated than they needed to be. Of course, I didn't appreciate things like airfoil shape, wing dihedral, center of gravity location, the sizing of control surfaces, and the many dozen other key factors of airplane design.

Dad, an accomplished model airplane builder, suggested that if I really wanted to learn all those things, I should design an airplane and build a small-scale flying model. So, we visited the local hobby shop to buy some balsa wood, tissue paper, airplane dope, and glue, all for less than one dollar. My airplane's wheels, propeller, and rubber band power plant would be scavenged from one of my ready-built balsa rubber band models. I designed a crudely simple high wing monoplane devoid of wing struts or dihedral with tiny tail surfaces. In my eye, it was a thing of beauty. In a few weekends, with considerable help from Dad, my scale model was complete with a pretty green and white color scheme, ready for flight testing. It flew very poorly under power but glided a little better. Nevertheless, it flew. I had designed an airplane! This project led to hundreds of model airplanes, first free-flight and then control line aerobatic and scale models. By the time I was in high school, I was building and flying radio-controlled model airplanes including unusual configurations of my own design. Two of these designs enabled me to place first in the engineering category at the 1974 and 1975 International Science and Engineering Fairs. Upon graduation from high school, I had earned a full-tuition scholarship to the Florida Institute of Technology (FIT) for an undergraduate degree in mechanical engineering. I also learned that model aircraft could play a vital role in research.

While attending college, I scrimped and saved enough to earn a private pilot's license and began enjoying aviation as a pilot. I also started to dream about designing small unmanned airplanes to perform many civilian and military missions. Most people, including my professors, thought this was a little crazy, since model airplanes and the few first-generation unmanned aircraft (then called "remotely piloted vehicles") were far too unreliable or even dangerous for widespread use at that time.

In July 1976, I opened the morning newspaper to see a most incredible color photo. The NASA Viking had just landed on Mars, and there in front of me was a view of another world! After spending a very long time studying the picture while ignoring my breakfast, I read the newspaper article, which stated that during atmospheric entry, the Viking was slowed by parachute. It occurred to me that if there was enough atmosphere on Mars for a parachute, it should be possible to fly an airplane there! I wrote to the NASA Langley Research Center, which managed the Viking program, and requested any information they could provide on the Martian atmosphere. Much to my surprise and delight, I received a prompt reply from a Viking engineer who enclosed three pages of accelerometer data taken during atmospheric entry and descent, with instructions and coefficients to enable me to calculate the atmospheric density versus altitude. My rather dismaying calculations indicated that the density a kilometer or two above the Martian surface was similar to that of Earth's atmosphere at over 100,000 feet. One would need a very flimsy airplane for Mars flight. I doubted that such a machine could survive the journey and deployment. So, I wrote back to the Viking engineer to thank him for the data and mentioned my disappointment with regard to a Mars airplane. He wrote back and said I needed to consider in my airplane calculations that Mars has only 37% of Earth's gravity. Instantly, an impossible design problem seemed possible, albeit challenging.

The next year of my spare time was spent designing Mars airplanes between classes, work, and flying. Critical factors included aerodynamics, navigation, communications, unfolding during atmospheric entry, and non-air-breathing propulsion. I eventually decided to focus on aerodynamics and deployment. Because of the intrinsic low Reynolds numbers, the aerodynamic conditions would be similar to those of my free-flight model airplanes. Designing gliders that would fly for several minutes set aside the propulsion problem. NASA and its resources could solve the communications and navigation problems.

During my junior year, the opportunity came to propose the topic of my senior engineering project to my faculty advisor and staff of the mechanical engineering department. I submitted a paper proposing the design of a Mars airplane. The effort would include performance verification using a wind tunnel model and a subscale radio-controlled demonstrator to show deployment. The proposal was rejected for being too ambitious. My professors politely told me that it was not only too ambitious for an undergraduate to design a Mars airplane, it was too ambitious for a graduate student and probably even too ambitious for NASA! Nevertheless, I still needed a senior project, so I proposed wind tunnel testing and radio control model testing a stall-resistant lightplane of my own design. This less-imaginative project was accepted by my professors.

One memorable spring day, as I was starting my senior project, my faculty advisor introduced me to a visiting former FIT undergraduate student, Dr. Frank Klemm. He was employed as an aerospace engineer by the Naval Research Laboratory in Washington, DC. Dr. Klemm was job recruiting to form a new group at NRL called the Vehicle Research Section. Specifically, he was looking for a new engineer who was also a pilot and model airplane enthusiast. Dr. Klemm wanted to develop small unmanned and autonomous airplanes to perform crucial Navy missions. When he offered me a full-time engineering job upon graduation, essentially doing what I dreamed of, it was like when Dad asked me if I'd like to sit in that T-37.

Debbie, my college girlfriend, earned her degree in space science physics and also became a private pilot. She was offered and accepted a job as a physicist at NRL, so with our combined incomes, living in the expensive Washington, DC, area became affordable. One week after graduation, we were married. After a short honeymoon, we packed all our belongings into our car, drove to Washington, DC, rented our first apartment, and began our careers. Little did I know that I was embarking on a lifelong adventure raising a family with a wonderful woman who still loves me as much as I love her… all while I earned a living forming and supporting a second, professional family of brilliant, talented NRL engineers, technicians, and scientists.

Jay Gundlach, with whom I have had the pleasure to co-author this book, began his career as part of the NRL family. Today, he remains a close colleague and will always be a lifelong friend.

The well-integrated NRL team accomplished many technical dreams of new advanced-technology airplanes. The work was seldom easy, but the results were very satisfying. And we all love aviation.

Subject Index

For a listing of just aircraft, see the Aircraft Index.

Note: Page numbers in italics refer to figures, those in bold to tables.

A

Abzug, Mal, 27
Active Electronic Decoy (AED)
  designs, 9–10
  DSI AED, 15–16
  E-Systems AED, 19–20
  Interim Active Electronic Decoy (IAED), 30–33
  Sailwing AED, 12–14
  Spinner AED, 10–12
  Teledyne Brown AED, 17–18
Advanced turbine-driven generator
  applications, 434
  cantilevered shaft, 437
  ceramics, 435, 436, 438
  compressor, 434–435
  early conceptual design, 432
  heat exchanger sector, 438
  piston engines, 433
  propulsion, 437
  recuperator, 433
  turbine tip clearances, 435
Aerodynamic calculations, initial conditions for, 458
Aerodynamic design, 7
Air Deployed Long Endurance Reconnaissance (ADLER)
  deployment sequence, *96*
  propulsion system, 97
  technical characteristics, 96, **98**
  three-view drawing, *97*
Air Launched Integrated Countermeasure, Expendable (ALICE)
  deployment sequence, 92
  propulsion system, 92–93
  technical characteristics, **94**
  three-view drawing, *93*
  wing, 92
Air-launched fixed wing aircraft
  AAED-FF (*See* Airborne Active Expendable Decoy–Forward Fired (AAED-FF))
  ADLER (*See* Air Deployed Long Endurance Reconnaissance (ADLER))
  ALICE (*See* Air Launched Integrated Countermeasure Expendable (ALICE))
  DAVE (*See* Disposable Air Vehicle Experiment (DAVE))
  Duster, 94–96
  EMMA (*See* Expendable Multi-mission Marine Aircraft (EMMA))
  Extender, 79–85
  FINDER (*See* Flight-Insertable Detection Expendable for Reconnaissance (FINDER))
  Insertion of Cargo Expendable (ICE), 98–100
  NRL Raptor, 71–75
  Pendulus, 66–71
Airborne Active Expendable Decoy-Forward Fired (AAED-FF)
  air-launching, 67
  canard control surfaces, 64
  demonstrator systems, 65
  deployment tests, 70
  magnetic anomaly detector, 67
  payload, 63
  technical characteristics, **66**
  tethered system, 64
  three-view drawing, *65*
  unconventional fluidic system, 64
  weight shifting/CG shifting, 68
Airborne Biological/Chemical Detection (ABCD), 178, 179
Airfoil technology, 463–464

Airframe folding, 461–462
Airplane for Mars Exploration (AME) (*See* Kitty Hawk 2)
Arcturus T-15, T-16, and T-20, 293–294
Arsea
  airframe, 262
  radar calibration testing, 262
  reflector fairing, 262
  technical characteristics, **263**
  three-view drawing, *263*
Autonomous Locator of Updrafts (ALOFT)
  onboard processors, 443–444
  power model, 444
  SB-XC cross-country model sailplane, 443
  thermals, 443
Autonomous operation, 5
Autonomous soaring technology, 442, 445
AVIATR, 368–378

B
BAI Tern, 295
Baseline Aircraft Radio Controlled (BARC)
  airframe, 268
  flight test, 268
  technical characteristics, **269**
  three-view drawing, *269*
  wings and tails, 267–268
Battery technology, 7, 382
Biplane Insectoid Travel Engine (BITE)-Wing
  front and rear pair wings, 247
  ground locomotion, 248, 249
  Mk 1, *247*
  Mk 3, *248*
  Mk 4, *249*
Bird Wing AED (*See* Teledyne Brown AED)
Blimp, 428

C
CG Shift Bird aircraft
  control system, 152
  electric-powered mission concept, *152*
  flight test program, 153
  simulator, 153
  technical characteristics, **154**
  three-view drawing, *154*
  wing design, 152
CG Shift Demonstrators
  longitudinal (pitch) control, 143–144
  payload, 143
  wing warping and ailerons, 144
CG Shift DT, 150–151
Charge coupled device (CCD) camera, 440
Close-In Covert Autonomous Disposable Aircraft (CICADA)
  CICADA I, 252–254
  CICADA II, 255–256
  CICADA III, 257–258
Cloud Cap Piccolo II autopilot, 337
Commander's Overhead Vehicle for Elevated Reconnaissance (COVER)
  camera pointing, 310–311
  control system, 310
  flight tests, 311
  flying duct system, 311
  propulsion system, 309
  technical characteristics, **313**
  three-view drawing, *312*
Configuration design, 6
Counter-UAS UAS, 215–216
Crashes, research airplanes, 449–450
Crawler Delivery Aircraft, *177*
Cube of Sky, 241–242

D
Dakota
  airframe, 281
  balsa wood and fiberglass floats, 282
  launch, 283
  payload, 280, 281
  technical characteristics, **284**
  three-view drawing, *284*
  V-tail modification, *283*
Dakota 2
  hangar storage, 287
  in folded configuration, *286*
  internal fuselage, *286*
  payload, 285
  technical characteristics, **288**
  three-view drawing, *287*
  wood and fabric structure, 285
Defense Advanced Research Projects Agency (DARPA), 227–228
Defense Threat Reduction Agency (DTRA), 85
Delphinopter, 250
Developmental Sciences International (DSI) AED
  launch sequence, *15*
  tail design, 15
  technical characteristics, **16**
  three-view drawing, *16*
Dihedral effect, 456, 460
Directional stability, 456–457
Disposable Air Vehicle Experiment (DAVE)
  conceptual design, 100–101
  technical characteristics, 101–102
  three-view drawing, *101*
Documentation, 452–453

Drag estimation, 458–459
Dragon Eye
  AeroVironment-produced, 202
  aircraft configuration, 201
  airframe modules, 203
  flight tests, 204–205
  internal layout, *203*
  launch methods, 203–204
  MCWL, 201
  plank flying wing, 202
  propulsion system, 204
  technical characteristics, **206**
  three-view drawing, *206*
Dragon Warrior/Vantage
  internal arrangement, *333*
  iron bird testing, 334
  propulsion system, 332
  rotor guard, 331
  tail rotor, 332, *333*
  technical characteristics, **336**
  three-view drawing, *335*
Dragonfly, 239–240
Drone Aircraft Radio Controlled (DARC)
  fuselage, 264
  O.S. 61 2-stroke model engine, 265
  payload, 264
  tail surfaces, 266
  technical characteristics, **267**
  three-view drawing, *266*
Duster
  design features, *94*
  tails, 95
  technical characteristics, **96**
  wing, 95

E

E-Systems AED
  flexible airfoil concept model, *19*
  launch sequence, *19*
  three-view drawing, *20*
Eager
  capstan reel system, 325
  flight tests, 325
  payload, 323–324
  quad rotor concept, 326
  technical characteristics, **327**
  tether system, 324, 326
  three-view drawing, *327*
Electric Assisted Take-Off (EATO), 391–393
Electric motors, 6–7
Elevon, 457
Endurance parameter, 456
Energy harvesting, by power line recharging, 429–432
Existing sensor equipment, 440
Expendable Multi-mission Marine Aircraft (EMMA)
  high-speed and low-speed configuration, *76*
  performance capabilities, 77
  technical characteristics, **78**
  three-view drawing, *77*
Expendable Radar Active Countermeasure Electric Rotorcraft (ERACER), 340–341
Experimental Fuel Cell (XFC)
  compressed hydrogen storage tank development, 390
  configuration and internal layout, *386*
  custom pressure regulator, *391*
  deployment, *386*
  EATO, 391–393
  flight tests, 388
  foam and fiberglass, 388
  integrated fuel cell system, *390*
  Navy launches UAS, 395–397
  single-stage pressure regulator, 389
  skin radiator, 389
  technical characteristics, **394**
  three-view drawing, *393*
  vertical edge/knife edge, 388
  wind tunnel testing, 387
Extender
  airframe, 83
  and P-3, 79
  balloon deployment tests, 83
  deployment concept, *81*
  folded in life raft container, 80
  in WPAFB vertical wind tunnel, *83*
  NAVAIR, 79
  RC demonstrator, *82*
  technical characteristics, **84**
  three-view drawing, *84*

F

Fairchild Sailwing AED (*See* Sailwing AED)
Fairchild Spinner AED (*See* Spinner AED)
Fiber-optic laser rate gyro, 45
First Skew Wing CG Shift, 145–146
Flexible airfoil concept model, 19
Flight-Insertable Detection Expendable for Reconnaissance (FINDER)
  air launch from Predator, *89*
  air launch sequence, 88
  airframe design, 86–87
  CP 2 ACTD requirements, 86

- electric aircraft, 85
- internal arrangement, *87*
- mission profile, 85
- on Predator wing pylon, *88*
- payload, 87
- technical characteristics, **90**
- three-view drawing, *90*
- wind tunnel testing, 87
- with tricycle landing gear, *89*

Fly Back Ejection Seat, 409–410
Flying Printed Circuit EW Aircraft, 229–230
Flying Radar Reflector aircraft, 225–226
Flying Radar Target Advanced Technology Demonstrator (FLYRT ATD)
- deployment concept, 42–43
- fiber-optic laser rate gyro, *45*
- final demonstration launch, *53*
- flight tests, 48, **50–52**
- fuselage, 44, *46*
- FX 63-137 airfoil, 43
- helicopter captive carry test, 47, *48*
- in anechoic chamber, 38
- low wing components, *42*
- low wing configuration, 40, *41*, 46
- multi wing configuration, *41*
- NDM utility aircraft, 47
- on launch sequence, *49*
- propulsion system, 44
- pylon-mounted high wing configuration, 39, 40
- Saft silver-zinc primary battery, 44
- solid fuel rocket motor, 39
- technical characteristics, **54**
- three-view drawing, *53*
- vehicle-based captive carry rig test, *48*
- wind tunnel models, 39
- wing construction, *45*

Flying Wing Research Testbed (FLYRT)
- airfoil, 133
- flight test program, 133
- technical characteristics, **134**
- three-view drawing, *134*
- wing tip modification, 134

Foch, Richard J. (Foch, Rick), 29, 227–228, 350–351
Fortran-based flight simulator, 27
Fuel cell aircraft
- Experimental Fuel Cell (XFC), 384–394
- Ion Tiger, 395–402
- Jake Spirit, 402–404
- Spider Lion, 381–384

G

Geometric dihedral, 456
Ghost
- airframe, 187
- internal systems, *188*
- payload, 187
- servos, 188
- technical characteristics, **190**
- three-view drawing, *189*
- wings, 188

Golden Eagle, 210
Ground-launched fixed wing mission aircraft
- Counter-UAS UAS, 215–216
- Crawler Delivery Aircraft, 177
- Dragon Eye, 201–206
- Flying Radar Reflector aircraft, 225–226
- Ghost, 186–190
- Golden Eagle, 210
- Moth, 193–195
- Nexus, 196–200
- Para Eyes, 191–193
- Sea Airborne Lead Line (SEA ALL), 207–209
- Sea Quell, 210–213
- Sea Wind, 216–220
- Searcher, 185–186
- SENDER (*See* Self-Navigating Drone Expendable/Recoverable (SENDER)
- SENDER 2, 174–176
- Spotlight, 213–215
- Swallow, 178–184
- Trade Wind, 220–224

H

Helicopter captive carry test, 47, 48
High Aspect Ratio (HAR) CG Shift, 148–149
High Point
- coaxial counter-rotating rotors, 329
- Dragon Warrior/Vantage, 330–336
- electric tether system, 328–329
- technical characteristics, **330**
- three-view drawing, *329*

Hill, Maynard, 29

I

Insertion of Cargo Expendable (ICE)
- airframe configuration, 98–99
- three-view drawing, *99*
- top and bottom views, 99

Interim Active Electronic Decoy (IAED)
- design, 30
- fuselage, 31
- technical characteristics, **33**
- thermoplastic airframe, 31
- three-view drawing, *32*

Interim Small Unit Remote Scouting System (ISURSS), 200
Ion Dasch
as green aircraft, 419
flight control system, 422
hydrogen storage, 423
landing gear, 422
military uses, 419
motors and propellers, 421
technical characteristics, **424**
three-view drawing, *424*
Ion Tiger
Cloud Cap autopilot, 398
flight tests, 398, 401
fuel cell system, *396*
initial displaced tail, *397*
LH_2 fuel systems, 400
propulsion, 399
technical characteristics, **401**
three-view drawing, *401*

J
Jake Spirit, 402–404

K
Kitty Hawk
deployment, 352–353, *352*
design range, 353
features of, 354
flight tests, 353, 354
plank flying wing, 352
propulsion system, 353
technical characteristics, **355**
three-view drawing, *355*
Kitty Hawk 2, 358–360
Klemm, Frank, 2

L
Laser tethered decoy system, 428
LAURA Joined Wing
captive carry rig tests, 117
flight testing, 118–119
technical characteristics, **120**
three-view drawing, *119*
wind tunnel tests, 117
wing design, 117
LAURA Tandem Wing
flight control arrangement, 121
flight testing, 122–123
technical characteristics, **124**
three-view drawing, *124*
wing design, 121
LAURA Three Surface
captive carry rig tests, 115
technical characteristics, **116**
three-surface configuration, 114
three-view drawing, *115*
twin-boom configuration, 114
wind tunnel tests, 113
LAURA Variable Span
airfoils, 125
airframe, 127
flight test, 127
fuel consumption, 128
technical characteristics, **129**
three-view drawing, *129*
wind tunnel testing, 126
wings retracting and deployment system, 125–126
wings-folded configuration, 127
Lawrence Livermore National Laboratory (LLNL), 72
Libelle
coaxial rotor, 304, 308
deployment concept, *305*
elevator, 308
flight testing, 307
operational concept, *305*
packaging concept, 306
rotorcraft, 307
technical characteristics, **308**
three-view drawing, *308*
Liebeck, Robert, 73
Lift-to-drag ratio (L/D), 456, 458
Lockheed P-3 Orion aircraft, 79
Long Duration Expendable Decoy (LODED)
actuators, 24–25
design endurance, 21
electromagnetic and radio frequency interference, 25–26
flight test program, 26, *26*, 28
Fortran-based flight simulator, 27
launcher concept, *22*
nose payload, 27
packaging concept, *22*
propulsion, 25
radio controlled (RC) tests, 28
rail launcher system, 26–27
technical characteristics, **29–30**
three-view drawing, *29*
unconventional control system, 23–24
wing configuration, 22, 27–28
winglet orientation, *23*
Low Altitude/Airspeed Unmanned Research Aircraft (LAURA)
configurations and designers, **109**
demonstrators, 106
flight test program, 112

fuselage internal arrangement, *108*
low-speed wind tunnel, 110
LRN airfoils, 107
multiple aircraft configurations, 113
on captive carry rig, *112*
optimal aircraft configuration, 107
three-view drawing, *109*
wind tunnel models, *110*, 111
Low Cost Micro Autopilot (LCMA), 446–447
Low Radar Cross Section (RCS), 136
Lunenburg lenses, 262

M

M-wing drop model, 319
Mach number, 455
Machine vision portion, 440
Magnetic anomaly detector (MAD) payload, 67
Magpie, 297
Man-Operated Light Expendable (MOLE), 413–415
Manned aircraft
Fly Back Ejection Seat, 409–410
Ion Dasch, 418–425
Man-Operated Light Expendable (MOLE), 413–415
Personal Extraction Aircraft (PEA), 418
Rapid Descent Vehicle, 411–413
Submersible Manned Air-Launched Expendable (SMAX), 416–417
Veeness, 410–411
Marine Corps Warfighting Laboratory (MCWL), 201, 330–331
Mars Airborne Geophysical Explorer (MAGE), 351
Mars Astronaut Companion (MAC)-1, 360–362
Mars Helicopter, 378–379
Matador
external features, 365
hydrazine engine, 362
in vertical wind tunnel testing, *366*
internal features, 365
packaging into entry vehicle, 364
propeller-driven flyer, 363
rocket-powered flyer, 363
technical characteristics, **367**
three-view drawing and planform details, *367*
with rocket propulsion, *362*
Micro Air Vehicle Demonstrators, 240–241
Micro air vehicles (MAVs)
Biplane Insectoid Travel Engine (BITE)-Wing, 246–249
Close-In Covert Autonomous Disposable Aircraft (CICADA), 251–259
Cube of Sky, 241–242
Dragonfly, 239–240
Flying Printed Circuit EW Aircraft, 229–230
Micro Air Vehicle Demonstrators, 240–241
Micro Tactical Expendable (MITE), 230–233
Nexus Surrogate MAV, 242–243, *242*
Pink Eye, 234–235
Semi-Autonomous Mobile Platform Expendable (SAMPLE), 244–250
Silicon Bug, 236–237
Winged Autonomous Silent Platform (WASP), 237–239
Micro Tactical Expendable (MITE)
control systems, 232
designs, *231*
payload, 231
propulsion, 230
radar-jamming, 232
technical characteristics, **233**
three-view drawing, *233*
Montgomery, John A., 1–3
Moth
camera, 194
flight tests, *193*, 194
Kosovo Kluge, 193
technical characteristics, **195**
three-view drawing, *195*

N

NASA 723, 356–357
Naval Air Systems Command (NAVAIR), 79
Naval Research Laboratory (NRL)
history and development, 1–5
technology development
aerodynamic design, 7
autonomous operation, 5
configuration design, 6
payloads, 7
propulsion systems, 6–7
structures, 6
NDM aircraft
NDM I, 270–272
NDM II, 271–273
NDM III, 273–274
Nexus
concept of, 196
Insitu Integrator, 200

modular design, 197
payload bay, 198
technical characteristics, **199**
three-view drawing, *199*
Nexus Surrogate MAV, 242–243
NRL Raptor
cantilever wing, 74
configuration and deployment concept, 73–74, *74*
early design concept model, *73*
Scaled Composites LLC, 72
solar-powered and liquid-fueled Raptors, 72
three-view drawing, *75*

O
Offboard Countermeasures (OCM), 1–2
Omni-Directional Inflatable Radar Reflector (ODIRR), 226
Optionally Piloted Aircraft (OPA), 290

P
Para Eyes
launching, 191, 192
technical characteristics, **193**
three-view drawing, *192*
Payloads, 7
Pectenopter, 249–250
Pendulus, *66*, *67*, 70–71
radio controlled demonstrators, 68–69
Spratt Controlwing, 68
three-view drawing, *70*
Personal Extraction Aircraft (PEA), 418
Pink Eye, 234–235
Planetary flight
Kitty Hawk, 350–355
Kitty Hawk 2, 358–360
Mars Astronaut Companion (MAC)-1, 360–362
Mars Helicopter, 378–379
Matador, 362–367
NASA 723, 356–357
Titan Flyer (AVIATR), 368–378
Venus Flyers, 378
PLICATE Wing ASW UAS, 130–131
Power line recharging, energy harvesting by, 429–432
Power requirements, 459–460
Project names selection, 451–452
Propulsion systems, 6–7
Pylon-mounted high wing configuration, 39, 40

R
Radar chaff, 1
Radar Expendable Active Decoy (READY)
design model, 36
fuselage, 37
tail section, 37
technical characteristics, **38**
three-view drawing, *37*
Radio Frequency (RF) decoys, 2
Ramwing
airfoil, 155
deployment, *156*
high Reynolds number device, 155
technical characteristics, **157**
three-view drawing, *157*
Rapid Descent Vehicle, 411–413
Research vehicles
CG Shift Bird aircraft, 151–154
CG Shift Demonstrators, 143–145
CG Shift DT, 150–151
First Skew Wing CG Shift, 145–146
Flying Wing Research Testbed (FLYRT), 132–134
High Aspect Ratio (HAR) CG Shift, 148–149
LAURA (*See* Low Altitude/Airspeed Unmanned Research Aircraft (LAURA))
LAURA Joined Wing, 116–120
LAURA Tandem Wing, 120–124
LAURA Three Surface, 113–116
LAURA Variable Span, 125–129
PLICATE Wing ASW UAS, 130–131
Ramwing, 155–157
Sail-A-Plane, 160–165
Second Skew Wing CG Shift, 146–147
Semi-Autonomous Research Aircraft (SARA), 140–143
Semi-Rigid Delta Wing, 131–132
Senior Telemaster, 278
Shadow Displaced Tail (DT), 135–139
Shearwing, 104–105
Venture program, 158–160
Reynolds number, 455
RnR Products SB-XC, *442*, 443, **445**
Rotary Wing Decoy (ROWDY) 1, 322–323
Rotary Wing Decoy (ROWDY) 2, 341
Ruddervator, 457

S
Saft silver-zinc primary battery, 44
Sail-A-Plane
in flight and sailing mode, 160
in flying configuration, 163
in sailing configuration, *164*
operation concept, 162

subscale models, 162
technical characteristics, **165**
three-view drawing, *165*
water testing, *164*
wing flooding flight configuration, *161*
Sailwing AED
flight performance, 13
in gliding flight, 12–13
technical characteristics, **14**
three-view drawing, *14*
wing twist distribution, *13*
Samara, 244–246
Samara 2, 245
SB-XC cross-country model sailplane, 443, 445
Science Instrumentation Environmental Remote Research Aircraft (SIERRA)
airframe, 292
technical characteristics, **293**
three-view drawing, *292*
Scientific Payload Insertion Device Electric Rotor (SPIDER)
acoustic altimeter, *338*
forward fuselage, *337*
propulsion system, 337
technical characteristics, **339**
three-view drawing, *338*
Sea Airborne Lead Line (SEA ALL), 207–209
Sea Quell
motor controller, 212
propulsion system, 212
prototype, 211
technical characteristics, **213**
three-view drawing, *212*
wing, 211
Sea Wind
air deployment, *218*
airframe, 217
control surfaces, 219
fuselage, 217
technical characteristics, **220**
three-view drawing, *219*
Searcher, 185–186
Second Skew Wing CG Shift, 146–147
Self-Navigating Drone Expendable/Recoverable (SENDER)
air launch, 170
airframe, 168, 169
electric propulsion, 170
fuselage, 168
ground dolly, 170–171
launch rail, 171
launching methods, 170–171
payload, 169–170
ruddervators, 169
technical characteristics, **173**
three-view drawing, *172*
Self-Navigating Drone Expendable/Recoverable (SENDER) 2, 174–176
Semi-Autonomous Mobile Platform Expendable (SAMPLE)
Biplane Insectoid Travel Engine (BITE)-Wing, 246–249
Delphinopter, 250
Pectenopter, 249–250
Samara, 244–246
Semi-Autonomous Research Aircraft (SARA)
electric propulsion, 140
electromagnetic interference (EMI), 141
flight test program, 141
fuselage, 140
technical characteristics, **142**
three-view drawing, *142*
Semi-Rigid Delta Wing, 131–132
Shadow Displaced Tail (DT)
displaced tail, 135
features, 136
low-RCS, 135
tail control surfaces, 138
technical characteristics, **139**
three-view drawing, *139*
wind tunnel testing, 136–137
Shear Wing RF Decoy
design, 33
technical characteristics, **34**
three-view drawing, *34*
Shearwing
fuselage, 104
small demonstrator, 104–105
technical characteristics, **105**
three-view drawing, *104*
wind tunnel model, 105
Ship-launched decoys and targets
Active Electronic Decoy
designs, 9–10
DSI AED, 15–16
E-Systems AED, 19–20
Interim Active Electronic Decoy (IAED), 30–33
Sailwing AED, 12–14
Spinner AED, 10–12
Teledyne Brown AED, 17–18
FLYRT ATD (*See* Flying Radar Target Advanced Technology Demonstrator (FLYRT ATD))

LODED (*See* Long Duration Expendable Decoy (LODED))
Radar Expendable Active Decoy (READY), 36–38
Shear Wing RF Decoy, 33–34
Ship-Launched Electronic Decoy (SLED), 34–36
Submarine-launched Tactical Airborne Reconnaissance Expendable (STARE), 57–59
Transonic Missile Engagement Target (TMET), 55–56
Vertically Launched Imaging Infrared Decoy Technology (VLIIRDT), 60–62
Ship-Launched Electronic Decoy (SLED)
booster and sustainer motor, 35
engineering mock-up, 34
longer-endurance flight, 35
technical characteristics, **36**
three-view drawing, *35*
Ship's Tethered Autogyro Platform (STAPL)
actuators, 300
airframe, 300
flight tests, 301
instability problems, 302
propulsion system, 300
side and top views, 303
technical characteristics, **303**
winch and control hut, 301, 302
Silicon Bug, 236–237
Slo-Comet, 261
Small Ships Compatible Decoy (SSCD) Autogyro
ballistic launch, 314
deployment, *316*
drop model, 317, 318
initial autogyro concept, 315
M-wing drop model, *313*, *319*
payload, 313
rotor spin-up wind tunnel test, 316, *317*
single-blade rotorcraft, 315
three-view drawing, *321*
tip propeller concept, *314*
vertical body and tails, 318, *319*
Solid fuel rocket motor, 39
Spider Lion
battery technology, 382
fuel cell balance, 382–383
technical characteristics, **384**
Spinner AED
technical characteristics, **12**
two-view drawing, *11*
Spotlight, 213–215
Spratt Controlwing, AAED-FF, 68
Stop Rotor
fixed wing aircraft mode, 343
motor controller, 343
propeller, 344
swashplateless rotor system, 342
technical characteristics, **347**
three-view drawing, *346*
wind tunnel testing, 344
Submarine-launched Tactical Airborne Reconnaissance Expendable (STARE), 57–59
Submersible Manned Air-Launched Expendable (SMAX), 416–417
Subsystem technology
advanced turbine-driven generator, 432–438
Autonomous Locator of Updrafts (ALOFT), 442
energy harvesting by power line recharging, 429–432
laser tethered decoy system, 428
Low Cost Micro Autopilot (LCMA), 446–447
vision-based recovery system, 439–442
Swallow
ABCD concept, 178–179
bio agent detector, 180
flight testing, 180–181
on rail launcher, *182*
payload, 178–179
propeller arrangement, 181
propulsion system, 182
tail, 180
technical characteristics, **184**
three-view drawing, *184*
with sniper-detection payload, *183*
Swashplateless rotor system, 342
System cost, 462

T

Tactical Electronic Warfare (TEW) Division, 1
Teledyne Brown AED
deployment concept, *17*
technical characteristics, **18**
three-view drawing, *18*
wind tunnel testing, *18*
wing/tail design, *17*
Telemaster
8-foot Senior Telemaster, 278
night flights, 278–279
technical characteristics, **280**
twelve-foot, *277*

Test Radar Calibration Experiment (TRACE)
  technical characteristics, **277**
  test radars, 276
  three-view drawing, *276*
  wing, 275
Thermoplastic airframe, 31
Thorpe P10B, 296
Threat Radar Active Countermeasure Expendable (TRACE), 339–340
Time management, 450–451
Titan Flyer (AVIATR)
  antenna, 370, 371
  ASRGs, 377
  automatic flight control system, 375–376
  climb-collect-glide profile, 371
  communications antenna, 377
  conceptual design, *370*
  deployment, 372, 373
  design features, 369
  displaced tail, 370
  fuselage, 374
  inertial measurement unit, 376
  initial design, 369
  internal layout, *375*
  mission profile, 377
  multiple-view drawing, *372*
  packaging challenges, *370*
  propeller, 374
  structural arrangement, *376*
  technical characteristics, **378**
  wing unfolding, 374
Titan Tornado program (*See* Unmanned Titan Tornado program)
Torqueless Stop Rotor, 345, 346
Trade Wind
  high-speed variant, *223*
  long-endurance configuration, *222*
  optimization program, 224
  payload, 221
  propulsion system, 223
  wing modules, 221
Transonic Missile Engagement Target (TMET), 55–56
Traveling wave tube (TWT) amplifiers, 9

U
UASUSA Tempest, 298
Unmanned Titan Tornado program
  airframes, 289
  FAA requirement, 289
  technical characteristics, **291**
  three-view drawing, *290*
  wing, 289
Unscented Kalman filter (UKF) method, 444
Utility aircraft
  Arcturus T-15, T-16, and T-20, 293–294
  Arsea, 261–264
  BAI Tern, 295
  Baseline Aircraft Radio Controlled (BARC), 267–270
  Dakota, 280–284
  Dakota 2, 285–288
  Drone Aircraft Radio Controlled (DARC), 264–267
  Magpie, 297
  NDM aircraft, 270–274
  Science Instrumentation Environmental Remote Research Aircraft (SIERRA), 291–293
  Telemaster, 277–280
  Test Radar Calibration Experiment (TRACE), 275–277
  Thorpe P10B, 296
  UASUSA Tempest, 298
  Unmanned Titan Tornado program, 288–291

V
Veeness, 410–411
Vehicle Research Section (VRS), technology development, 2–3, *3*
Vehicle-based captive carry rig test, 48
Venture program
  flight tests, 159
  fuel consumption, 159
  technical characteristics, **160**
  three-view drawing, *159*
  unfold during flight, 158
Venus Flyers, 378
Vertical Takeoff and Landing (VTOL) and Rotorcraft
  Commander's Overhead Vehicle for Elevated Reconnaissance (COVER), 309–313
  Dragon Warrior/Vantage, 330–336
  Eager, 323–327
  Expendable Radar Active Countermeasure Electric Rotorcraft (ERACER), 340–341
  High Point, 328–330
  Libelle, 304–308
  Rotary Wing Decoy (ROWDY) 1, 322–323
  ROWDY 2, 341
  Scientific Payload Insertion Device Electric Rotor (SPIDER), 336–339

Ship's Tethered Autogyro Platform (STAPL), 299–303
Small Ships Compatible Decoy (SSCD) Autogyro, 313–321
Stop Rotor, 342–347
Threat Radar Active Countermeasure Expendable (TRACE), 339–340
Vertically Launched Imaging Infrared Decoy Technology (VLIIRDT), 60–62
Vision-based recovery system, 439–442

W
Weight estimation, 457
Wing sweep, 461
Winged Autonomous Silent Platform (WASP)
flying wing airfoils, 237
technical characteristics, **238**
three-view drawing, *238*

X
XFC (*See* Experimental Fuel Cell (XFC))

Aircraft Index

For details about individual aircraft, see the Subject Index.

A

Active Electronic Decoy (AED), 9–10
Air Deployed Long Endurance Reconnaissance (ADLER), 96
Air Launched Integrated Countermeasure, Expendable (ALICE), 91
Airborne Active Expendable Decoy-Forward Fired (AAED-FF), 63
Arcturus T-15, 294
Arcturus T-16, 293
Arcturus T-20, 294
Arsea, 261

B

BAI Tern, 295
Baseline Aircraft Radio Controlled (BARC), 267
Biplane Insectoid Travel Engine (BITE)-Wing, 246

C

CG Shift Bird aircraft, 151, 154
CG Shift Demonstrators, 143
CG Shift DT, 150
CICADA II, 255
CICADA III, 257
Close-In Covert Autonomous Disposable Aircraft (CICADA) I, 251, 252
Commander's Overhead Vehicle for Elevated Reconnaissance (COVER), 309
Counter-UAS UAS, 215
Crawler Delivery Aircraft, 177
Cube of Sky, 241

D

Dakota, 280
Dakota 2, 285, 287
Delphinopter, 250
Disposable Air Vehicle Experiment (DAVE), 100
Dragon Eye, 201, 204
Dragon Warrior/Vantage, 330
Dragonfly, 239
Drone Aircraft Radio Controlled (DARC), 264
DSI AED, 15
Duster, 94

E

E-Systems AED, 19
Eager, 323
Expendable Multi-mission Marine Aircraft (EMMA), 76
Expendable Radar Active Countermeasure Electric Rotorcraft (ERACER), 340
Experimental Fuel Cell (XFC), 384
Extender, 79

F

Fairchild Sailwing AED, 12, 13
Fairchild Spinner AED, 10
First Skew Wing CG Shift, 145
Flight-Insertable Detection Expendable for Reconnaissance (FINDER), 85, 87
Fly Back Ejection Seat, 409
Flying Printed Circuit EW Aircraft, 229
Flying Radar Reflector aircraft, 225
Flying Radar Target Advanced Technology Demonstrator (FLYRT ATD), 38, 48
Flying Wing Research Testbed (FLYRT), 132, 134

G

Ghost, 186, 189
Golden Eagle, 210

H
High Aspect Ratio (HAR) CG Shift, 148
High Point, 328

I
IAED-2, 32
Insertion of Cargo Expendable (ICE), 98
Interim Active Electronic Decoy (IAED), 30
Ion Dasch, 418
Ion Tiger, 395

J
Jake Spirit, 402

K
Kitty Hawk, 350, 352, 354
Kitty Hawk 2, 358

L
LAURA Joined Wing, 116
LAURA Tandem Wing, 120, 123
LAURA Three Surface, 113
LAURA Variable Span, 125
Libelle, 304
Lockheed P-3 Orion aircraft, 79
Long Duration Expendable Decoy (LODED), 21, 23
Low Altitude/Airspeed Unmanned Research Aircraft (LAURA), 106, 108, 112

M
Magpie, 297
Man-Operated Light Expendable (MOLE), 413
Mars Astronaut Companion (MAC)-1, 360
Mars Helicopter, 378
Matador, 362, 364
Micro Air Vehicle Demonstrators, 240
Micro Tactical Expendable (MITE), 230
Moth, 193

N
NASA 723, 356
NDM I, 270, 271
NDM II, 271
NDM III, 273
Nexus, 196
Nexus Surrogate MAV, 242
NRL Raptor, 71, 73, 74

P
Para Eyes, 190, 192
Pectenopter, 249
Pendulus, 66
Personal Extraction Aircraft (PEA), 418
Pink Eye, 234
PLICATE Wing ASW UAS, 130

R
Radar Expendable Active Decoy (READY), 36
Ramwing, 155
Rapid Descent Vehicle, 411
Rotary Wing Decoy (ROWDY) 1, 322
Rotary Wing Decoy (ROWDY) 2, 341
RQ-14A Dragon Eye, 201

S
Sail-A-Plane, 160, 162, 164
Sailwing AED, 12, 13
Samara, 244
Samara 2, 245
Science Instrumentation Environmental Remote Research Aircraft (SIERRA), 291
Scientific Payload Insertion Device Electric Rotor (SPIDER), 336
Sea Airborne Lead Line (SEA ALL), 207
Sea Quell, 211
Sea Wind, 216
Searcher, 185
Second Skew Wing CG Shift, 146
Self-Navigating Drone Expendable/ Recoverable (SENDER), 167, 172
Semi-Autonomous Mobile Platform Expendable (SAMPLE), 244
Semi-Autonomous Research Aircraft (SARA), 140
Semi-Rigid Delta Wing, 131
SENDER 2, 174, 175
Shadow Displaced Tail (DT), 135, 137
Shear Wing RF Decoy, 33
Shearwing, 104
Ship-Launched Electronic Decoy (SLED), 34
Ship's Tethered Autogyro Platform (STAPL), 299
Silicon Bug, 236
Small Ships Compatible Decoy (SSCD) Autogyro, 313
Spider Lion, 381, 383
Spinner AED, 10
Spotlight, 213
Stop Rotor, 342
Submarine-launched Tactical Airborne Reconnaissance Expendable (STARE), 57

Submersible Manned Air-Launched Expendable (SMAX), 416
Swallow, 178, 181

T
Teledyne Brown AED, 17
Telemaster, 277
Test Radar Calibration Experiment (TRACE), 275
Thorpe P10B, 296
Threat Radar Active Countermeasure Expendable (TRACE), 339
Titan Flyer (AVIATR), 368
Titan Tornado, 288
Trade Wind, 220
Transonic Missile Engagement Target (TMET), 55

U
UASUSA Tempest, 298
Unmanned Titan Tornado program, 288

V
Veeness, 410
Venture program, 158
Venus Flyers, 378
Vertically Launched Imaging Infrared Decoy Technology (VLIIRDT), 60

W
Winged Autonomous Silent Platform (WASP), 237

Supplemental Materials

A complete listing of titles in the Library of Flight series is available from AIAA's electronic library, Aerospace Research Central (ARC) at arc.aiaa.org. Visit ARC frequently to stay abreast of product changes, corrections, special offers, and new publications.

AIAA is committed to devoting resources to the education of both practicing and future aerospace professionals. In 1996, the AIAA Foundation was founded. Its programs enhance scientific literacy and advance the arts and sciences of aerospace. For more information, please visit www.aiaafoundation.org.